Handbook of Essential Pharmacokinetics, Pharmacodynamics and Drug Metabolism for Industrial Scientists

Handbook of Essential Pharmacokinetics, Pharmacodynamics and Drug Metabolism for Industrial Scientists

Younggil Kwon, Ph.D.

Bioneer Life Science
San Diego, California

Kluwer Academic / Plenum Publishers
New York, Boston, Dordrecht, London, Moscow

ISBN 0-306-46234-6

233 Spring Street, New York, N.Y. 10013

http://www.wkap.nl/

10 9 8 7 6 5 4 3 2 1

A C.I.P. record for this book is available from the Library of Congress

Printed in the United States of America

To my wife, Heekyung, and
two daughters, Jessica and Jennifer

Preface

In the pharmaceutical industry, the incorporation of the disciplines of pharmacokinetics, pharmacodynamics, and drug metabolism (PK/PD/DM) into various drug development processes has been recognized to be extremely important for appropriate compound selection and optimization. During discovery phases, the identification of the critical PK/PD/DM issues of new compounds plays an essential role in understanding their pharmacological profiles and structure–activity relationships. Owing to recent progress in analytical chemistry, a large number of compounds can be screened for their PK/PD/DM properties within a relatively short period of time. During development phases as well, the toxicology and clinical study designs and trials of a compound should be based on a thorough understanding of its PK/PD/DM properties.

During my time as an industrial scientist, I realized that a reference work designed for practical industrial applications of PK/PD/DM could be a very valuable tool for researchers not only in the pharmacokinetics and drug metabolism departments, but also for other discovery and development groups in pharmaceutical companies. This book is designed specifically for industrial scientists, laboratory assistants, and managers who are involved in PK/PD/DM-related areas. It consists of thirteen chapters, each of which deals with a particular PK/PD/DM issue and its industrial applications. Chapters 3 and 12 in particular address recent topics on higher throughput *in vivo* exposure screening and the prediction of pharmacokinetics in humans, respectively. Chapter 8 covers essential information on drug metabolism for industrial scientists. The important equations are highlighted and the commonly used terms in PK/PD/DM are summarized in a glossary at the end of the book. I hope that all those who consult this book find it useful as an easy-to-understand reference for identifying, analyzing, and addressing PK/PD/DM related issues in their respective fields of research.

Younggil Kwon,
Bioneer Life Science Co.
San Diego, California

Acknowledgments

I would like to express my gratitude to Drs. Bonnie Mangold and Francis Tse and Ms. Elina Dunn for their valuable comments and suggestions, and to Mr. Michael F. Hennelly for his encouragement and support.

Contents

List of Figures

CHAPTER 1

CHAPTER 2

CHAPTER 4

CHAPTER 5

CHAPTER 6

CHAPTER 7

CHAPTER 8

CHAPTER 9

CHAPTER 10

CHAPTER 11

CHAPTER 12

CHAPTER 13

1

Introduction

Pharmacokinetics and pharmacodynamics are the important fields of pharmaceutical sciences for investigating disposition profiles and the pharmacological efficacy of drugs in the body under various experimental and clinical conditions (Caldwell *et al.*, 1995 and Cocchetto and Wargin, 1980).

Pharmacokinetics (PK) is the study of the way drug molecules behave in the body after administration. Four distinctive yet somewhat interrelated processes occur between the administration and the elimination of a drug from the body: after oral administration, drug molecules are absorbed into the portal vein via the enterocytes from the gastrointestinal lumen, pass through the liver and the lungs, reach the systemic circulation, and then further distribute into various tissues and organs via blood vessels, some of which may have metabolic or excretory activity for eliminating the drug. These sequential events are called the ADME processes of the drug after administration, i.e., *a*bsorption, *d*istribution, *m*etabolism, and *e*xcretion, as illustrated in Fig. 1.1.

The purpose of pharmacokinetics is to study ADME processes of drugs in the body by examining the time course of drug concentration profiles in readily accessible body fluids such as blood, plasma, urine, and/or bile. Basically, all of a drug's pharmacokinetic parameters, such as clearance, volume of distribution, mean residence time, and half-life, can be estimated from its concentration-*vs.*-time profiles in plasma (or blood). It is important to realize that pharmacokinetic interpretations of drug exposure profiles are simply descriptions of the phenomenology of the ADME processes, and, thus, there might possibly be many different interpretations of the pharmacokinetic properties of a drug based on the same plasma drug concentration profiles.

Pharmacodynamics (PD) is the study of the relationships between the concentration of a drug at the effect site(s), where target enzymes or receptors are located, and the magnitude of its pharmacological efficacy. Let us consider an anticoagulant drug as an example. As the drug's effect site is the systemic circulation, its pharmacodynamics elucidates the relationship between its concentration in blood (the effect site) and the extent of its anticoagulant effect (pharmacological effect).

When the effect site is not in plasma and the drug concentration in the plasma (or blood) is different from that in the effect site, the kinetic relationship between pharmacokinetics and pharmacodynamics becomes an important component in

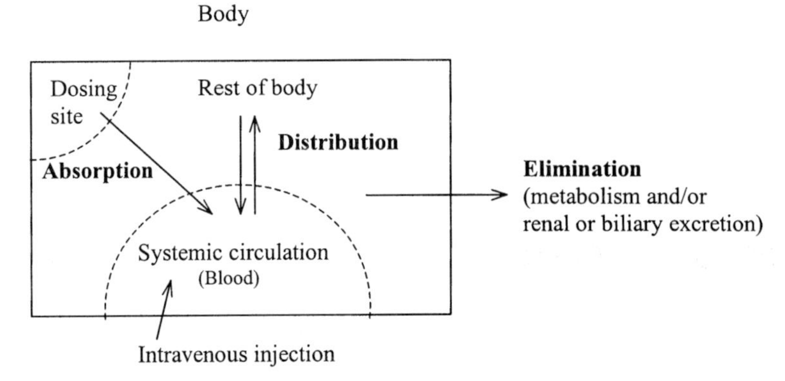

Figure 1.1. A schematic description of the pharmacokinetic behavior of a drug. Distribution and elimination processes are often referred as to disposition processes.

correlating the drug's concentration in plasma and the pharmacological endpoints observed. The kinetic relationship between drug concentrations in plasma and in the effect site can be arrived at by exploring various pharmacokinetic/pharmacodynamic (PK/PD) models (Fig. 1.2). For reasonable PK and PD studies, one must have a

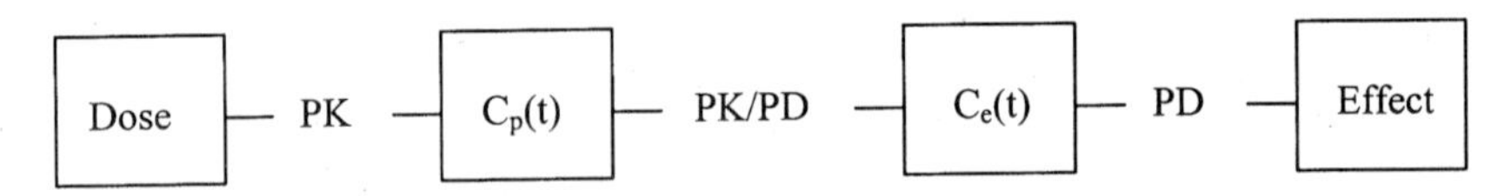

Figure 1.2. A schematic description of the pharmacokinetics (PK), the pharmacodynamics (PD), and the pharmacokinetic/pharmacodynamic (PK/PD) relationships of a drug. $C_p(t)$ and $C_e(t)$ are the concentrations of the drug in the plasma (sampling site) and the effect site, respectively.

thorough understanding of the conditions and assumptions under which the experiments are carried out as well as of PK and PD models employed, as the validity of virtually all the PK and PD data interpretations depends on the scientific soundness and physiological relevance of those assumptions and conditions.

REFERENCES

Caldwell J. *et al.*, An introduction to drug disposition: the basic principles of absorption, distribution, metabolism and excretion, *Toxicol. Pathol.* **23**: 102–114, 1995.

Cocchetto D. M. and Wargin W. A., A bibliography for selected pharmacokinetic topics, *Drug Intel. Clin. Pharmacol.* **14**: 769–776, 1980.

2

Pharmacokinetic Study Design and Data Interpretation

Pharmacokinetic data interpretation can be viewed primarily as an effort to deduce what has happened to a drug in the body after administration based on the time course of its exposure in biological fluids such as plasma or blood. Reliability of data obtained from *in vivo* pharmacokinetic studies depends on the validity of study design and execution and on sample collection, handling, and assay. Selection of proper data analysis methods is equally important in understanding pharmacokinetic characteristics of a drug. In this chapter, useful information and guidelines for intravenous and oral administration studies in animals as well as data interpretation are discussed.

2.1. INTRAVENOUS ADMINISTRATION OF DRUGS

2.1.1. Utility of Intravenous Administration Studies

The plasma exposure profiles of a drug after intravenous dosing provide critical information on its pharmacokinetic properties including:

(i) Systemic clearance and volume of distribution at steady state. An estimate of the systemic clearance of a drug can be obtained from plasma (or blood) concentration–time profiles after *intravenous* injection. It can also be estimated after dosing the drug by a route other than intravenous injection, as long as its bioavailability is complete. However, an estimate of the volume of distribution of the drug at steady state cannot be obtained from exposure data after administration by any route other than intravenous injection.

(ii) Terminal half-life of a drug. The terminal half-life of a drug following intravenous injection is governed by disposition (distribution and elimination) processes of the drug in the body. The terminal half-life estimated after administration by the route other than intravenous injection can be affected not only by disposition but also by absorption (or input) processes from the site of administration.

(iii) Reference exposure levels for estimates of bioavailability. The area under the plasma concentration *vs.* time curve (AUC) after intravenous injection is commonly used as a reference for estimating the bioavailability of a drug by a route other than intravenous injection.

2.1.2. General Considerations for Intravenous Administration Studies

Important considerations and suggestions for intravenous dosing studies are summarized below.

(i) Bolus injection vs. short infusion. In general, intravenous (or intraarterial) injection of a drug is assumed to be bolus administration completed within a few seconds, unless otherwise indicated. If injection takes more than 1 min, it should be considered a short infusion.

(ii) Dosing solution. In general, isotonic sterile water at pH 6.8 is the most desirable dosing vehicle for intravenous injection. Although an aqueous vehicle is generally preferred, because of the limited water solubility of some compounds or their chemical instability in water the use of various organic cosolvents is not uncommon. Nonaqueous vehicles such as, e.g., dimethyl sulfoxide (DMSO), ethanol, polyethylene glycol (PEG) 400, and vegetable oil or solubilizing agents such as β-cyclodextrin are often used with sterile water to enhance compound solubility in a dosing vehicle, especially during drug discovery. In this case, the effects of organic vehicles or solubilizing agents on pharmacokinetic profiles of a compound (such as inhibition of metabolism and hemolysis of blood) and on its pharmacological and toxicological responses should be considered. In general, the amount of organic cosolvent should not exceed 20% of the total injection volume. The pH of a dosing vehicle can be slightly acidic or basic to optimize aqueous solubility. However, caution should be taken to adjust pH to enhance aqueous solubility of a compound, because the alteration of pH may result in chemical instability. The viscosity of a dosing vehicle should be maintained such that it allows ease of injection (syringe-ability) and optimal fluidity.

(iii) Dosing volume. In case of bolus injection, it is important to have a suitable dosing volume. If the dosing volume is too large, it may take more time to inject, and if it is too small, there can be difficulties in preparation and administration of the dosing solution. The maximum volume for single bolus injection is approximately 1 ml/kg body weight for laboratory animals such as rabbits, monkeys, and dogs. In small animals such as mice and rats, larger volumes of up to 0.3 and 0.5 ml, respectively, per animal can be used. In small laboratory animals, continuous 24-hr intravenous infusion should not exceed 4 ml/kg body weight/hr (see Chapter 13).

2.1.3. Sample Collection after Intravenous Administration

(ii) Blood—Sampling time points (Fig. 2.1).

• The entire concentration-*vs.*-time curve. Seven (at least five) time points are recommended.

• The early distribution phase. At least two time points within a short period of time after injection, typically less than 15 min after administration, are recommended for reliable estimation of imaginary plasma concentration at time zero [$C_p(0)$].

• The terminal phase. At least three time points during the terminal phase have to be obtained for reliable estimation of the terminal half-life of a drug. As a rule of thumb, three or four time points during the terminal phase are selected such that the interval between the first and the last is more than twice the estimated terminal half-life based on them.

(ii) Blood— Volume. In general, no more than 10% of the total blood volume in the body can be drawn as an acceptable weekly maximum in small laboratory animals; 20% is the maximum that can be taken acutely without serious hemorrhagic shock and tissue anoxia. In the latter case 3–4 weeks should be allowed for recovery and/or a proper quantity of blood should be infused.

NOTE: IMAGINARY PLASMA CONCENTRATION AT TIME ZERO AFTER INTRAVENOUS INJECTION [$C_p(0)$]. The $C_p(0)$ of a drug can be estimated by connecting the first two data points after intravenous injection and extrapolating back to the y-axis on a semilog scale (Fig. 2.1). There is, of course, no drug in the plasma at the sampling site at time zero because at the moment of injection, the drug has not yet been delivered to the sampling site. However, an estimate of $C_p(0)$ is a useful value for calculating the area under the concentration–time curve from time zero to the first sampling time point and the apparent volume of distribution of the drug in the central compartment (V_c; see Multicompartment model). V_c, an imaginary space in the body where drug molecules in plasma reach rapid equilibrium upon injection,

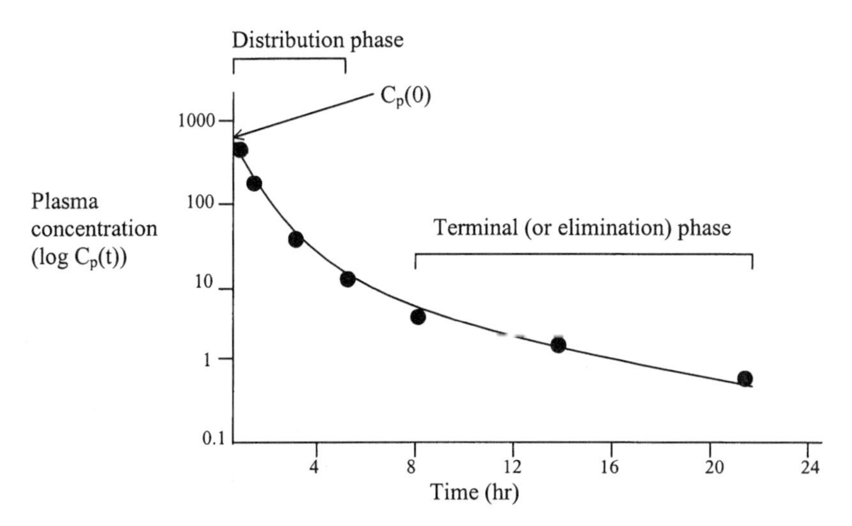

Figure 2.1. Example of a plasma drug concentration-*vs.*-time profile after intravenous injection of a drug. In general, the first two time points shortly after injection are used to estimate $C_p(0)$, and at least three time points during the terminal phase are needed to calculate of the terminal half-life of a drug.

can be estimated by dividing an intravenous dose by $C_p(0)$. In small laboratory animals it may take only a few seconds before the distributional equilibrium of the drug between the plasma and the central compartment is achieved.

(ii) Urine. Collection of urine from laboratory animals over an extended period of time (usually up to 24 hr in small animals) can also provide useful pharmacokinetic information, such as renal clearance and metabolic profiles of a drug. In general, it is easier to identify metabolites in urine than in blood owing to their higher concentrations in the former.

Renal clearance (Cl_r) can be calculated by dividing the amount of the parent drug excreted in the urine by AUC, regardless of the route of administration. The difference between the systemic clearance (Cl_s; see Chapter 6) and Cl_r is nonrenal clearance (Cl_{nr}):

$$Cl_{nr} = Cl_s - Cl_r \tag{2.1}$$

where Cl_{nr} represents clearances of a drug in the body other than by the kidney, such as elimination by, e.g., the liver, lung, intestine, blood, or brain. In general, Cl_{nr} is assumed to be similar or equal to hepatic clearance because the liver is the major eliminating organ for most drugs.

2.2. ORAL ADMINISTRATION OF DRUGS

2.2.1. Utility of Oral Administration Studies

Oral administration is the most popular and acceptable route for drug administration. Important pharmacokinetic parameters estimated from plasma exposure profiles after oral administration of drug are given below.

(i) C_{max} *and* t_{max}. C_{max} is the highest drug concentration observed after oral administration, and t_{max} is the time at which C_{max} is observed.

(ii) Terminal half-life. The terminal half-life of a drug after oral administration can be affected by both its absorption and disposition rates, and it is usually similar to or longer than that following intravenous injection.

(iii) Bioavailability. Bioavailability of a drug after oral administration is determined by dose-normalized AUC from time zero to infinity ($AUC_{0-\infty}$) after oral administration compared to that after administration of the drug via a reference route, usually intravenous injection.

2.2.2. General Considerations for Oral Administration Studies

(i) Dosing volume. Drug solution or suspension can be administered by oral gavage. In small laboratory animals such as rats, up to 10 ml/kg body weight can be dosed in a fasted condition. Approximately 5 ml/kg is considered acceptable for

oral administration in small animals under a fed condition. A solution formulation of a compound is most desirable for oral administration, but suspension can also be given when necessary (see Chapter 13).

(ii) Food intake. Concomitant food intake can alter the rate and extent of absorption of orally dosed drugs. In addition, when the drug is subject to enterohepatic circulation, its exposure profiles in animals with restricted food intake can be significantly different from those in animals with free access to food.

(iii) Water intake. Restrictions on water intake are sometimes required to reduce variability in exposure levels, especially when nonaqueous dosing vehicles such as polyethylene glycol (PEG) 400 are used to increase solubility of water-insoluble drugs in a dosing vehicle. In such cases, water intake may cause precipitation of the drug and subsequently reduce the extent of its absorption.

(iv) Coprophagy. In rodents, coprophagy (feeding on their own feces) can have significant effects on drug absorption profiles. Coprophagy can be avoided either by using tail caps or by conducting the experiments in metabolism cages, where the feces can be separated from the animals.

(ii) Dose levels. At least three different dose levels have to be examined over the intended therapeutic range to test for the presence of potential nonlinear pharmacokinetics. In most cases, however, one or two dose levels may be sufficient to determine preliminary pharmacokinetic profiles of a compound during drug discovery.

2.2.3. Sample Collection after Oral Administration—Blood

(i) Sampling time points (Fig. 2.2).

• The entire concentration *vs.* time curve. Seven (at least five) time points are recommended.

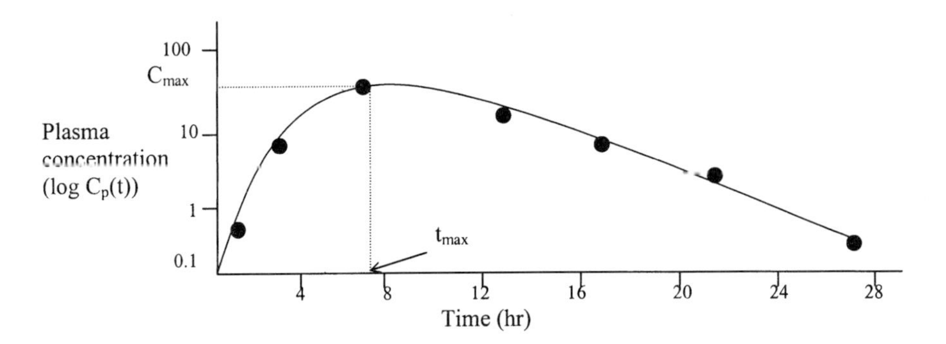

Figure 2.2. Example of a plasma drug concentration *vs.* time profile after oral administration of a drug. In general, at least one time point before and three time points after t_{max} are desirable for reliable characterization of oral exposure profiles of a drug.

• Before and after t_{max}. At least one time point before and three time points after t_{max}.

• The terminal phase. At least three time points during the terminal phase for half-life estimation, of which the interval is greater than twice the estimated terminal half-life.

• Estimate of $AUC_{0-\infty}$. Preferably, select time points over three half-lives beyond t_{max} for reliable $AUC_{0-\infty}$ estimation.

(ii) Volume. Less than 10% of the total blood volume from small laboratory animals within a week. Refer to the suggestions for intravenous administration.

2.3. DATA INTERPRETATION

There are basically two different approaches—compartmental and noncompartmental—to analyzing plasma drug concentration-*vs.*-time profiles for estimating pharmacokinetic parameters (Balant and Gex-Fabry, 1990; Gerlowke and Jain, 1983; Gillespie, 1991; Zierler, 1981). The noncompartmental approach is more commonly used for simple pharmacokinetic data interpretation in the pharmaceutical industry.

2.3.1. Compartmental Approach

The compartmental approach (or compartmental model) views the body as being composed of a number of pharmacokinetically distinct compartments. Each compartment can be thought of as an imaginary space in the body representing a combination of various tissues and organs, among which concentrations of a drug are in rapid equilibrium. Anatomical composition of the compartment is unknown and in most cases its analysis is of little value. The number of compartments in a model is empirically determined depending on plasma drug concentration time profiles. The compartmental model is designed to:

1. Provide a conceptual understanding of distributional behaviors of a drug between the plasma (or blood) and other tissues or organs in the body.
2. Empirically assess the changes in physiological processes such as membrane transport or metabolism without thorough mechanistic investigations.
3. Estimate various pharmacokinetic parameters such as rate constants, clearance, and apparent volumes of distribution.

The compartmental approach requires mathematical data analysis, usually nonlinear regression methods, to estimate the parameters used in models by fitting the model to the plasma concentration–time profile. Several computer programs for nonlinear regression are commercially available (e.g., PCNONLIN). The first step in the compartmental approach to data analysis is to determine the number of compartments required for the model.

2.3.1.1. One- vs. Multicompartment Models

When drug molecules are administered, the drug is initially localized at the administration site before further distribution into different regions of the body. For instance, if upon intravenous injection the distribution of drug molecules from the injection site, i.e., venous blood, throughout the body occurs instantaneously, the body may behave as if it is one pharmacokinetically homogeneous compartment for the drug. In this case, the plasma drug concentration–time profile exhibits a monophasic decline on a semilogarithmic scale (plasma drug concentrations on a log scale and time on a linear scale), and can be readily described with a one-compartment model.

When the distribution of a drug from the plasma into certain organs or tissues is substantially slower than to the rest of the body, multicompartment models, i.e., a central compartment and one or more peripheral (or tissue) compartments, should be considered. In general, it is expected that the distribution of a drug from the plasma into the highly perfused organs or tissues such as the liver, kidneys, or spleen is much faster than to those organs with a limited blood supply such as fat, muscle, skin, or bone. The central compartment represents the systemic circulation and those highly perfused organs and tissues, whereas the peripheral compartment(s) represents the poorly perfused organs and tissues. In a multicompartment system, the plasma drug concentration-*vs.*-time profile exhibits a multiphasic decline on a semilogarithmic scale. The intercompartmental distribution of a drug can be conceptually viewed as a pharmacokinetic expression of drug transport actually occurring between tissues and organs via blood vessels and/or membranes, and is generally assumed to follow first-order kinetics.

NOTE: FIRST-ORDER PHARMACOKINETICS. A first-order pharmacokinetic process is one in which the rate of change of concentration of a drug in biological fluids is directly proportional to its concentration. For instance, under a first-order kinetic condition, the rate of change in plasma drug concentration can be described as a function of the concentration [$C_p(t)$], i.e., $dC_p(t)/dt = k \cdot C_p(t)$, where k is a first-order rate constant. First-order pharmacokinetics is often called linear pharmacokinetics (see Chapter 10).

2.3.1.2. One-Compartment Model Analysis

The simplest compartment model is a one-compartment model, in which the entire body is viewed as a single kinetically homogeneous compartment. A schematic description of the one-compartment model with first-order elimination of a drug after intravenous dosing is shown in Fig. 2.3.

The amount of drug present in the body at any given time t [A(t)] in a one-compartment model is described in Eq. (2.2):

$$A(t) = C_p(t) \cdot V \tag{2.2}$$

where $C_p(t)$ and V are the drug concentration in the plasma and the apparent volume of distribution (see Chapter 5), respectively. The equation describing the

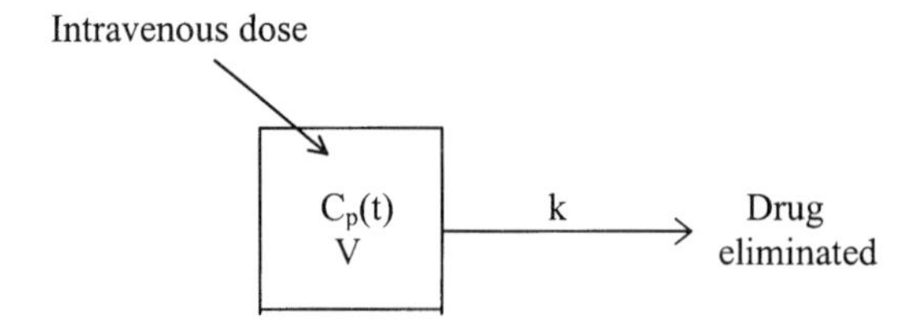

Figure 2.3. One-compartment model with first-order elimination after intravenous administration of a drug. $C_p(t)$ is the drug concentration in plasma at time t, k is a first-order elimination rate constant, and V is the apparent volume of distribution of the compartment.

plasma drug concentration $[C_p(t)]$ at time t in a one-compartment model after intravenous bolus injection is

$$C_p(t) = \underbrace{\frac{D_{iv}}{V}}_{C_p(0)} \cdot e^{-k \cdot t} \quad (2.3)$$

where $C_p(0)$ is an imaginary drug concentration at time zero (Fig. 2.1) and k is a first-order elimination rate constant. Equation (2.3) can be fitted to the plasma drug concentration–time data for estimates of V and k. The systemic clearance (Cl_s, see Chapter 6) and half-life ($t_{1/2}$) of a drug can be estimated from these parameters through the following equations:

$$Cl_s = k \cdot V \quad (2.4)$$

$$t_{1/2} = \frac{0.693}{k} \quad (2.5)$$

(a) Drug Concentration in Plasma and Tissues. It is important to note that one-compartment behavior of plasma drug concentrations does not necessarily imply that the drug is at the same concentration in all the tissues and organs in the body. It means rather that the drug concentrations in different tissues or organs are in instantaneous equilibrium with those in the plasma upon drug administration into the systemic circulation, establishing the constant concentration ratios between the plasma and the various tissues. When this occurs, the rate of change of drug concentration in the plasma can directly reflect the change in drug concentration in tissues with differences in concentrations corresponding to the magnitude of the accumulation between plasma and tissues (Fig. 2.4).

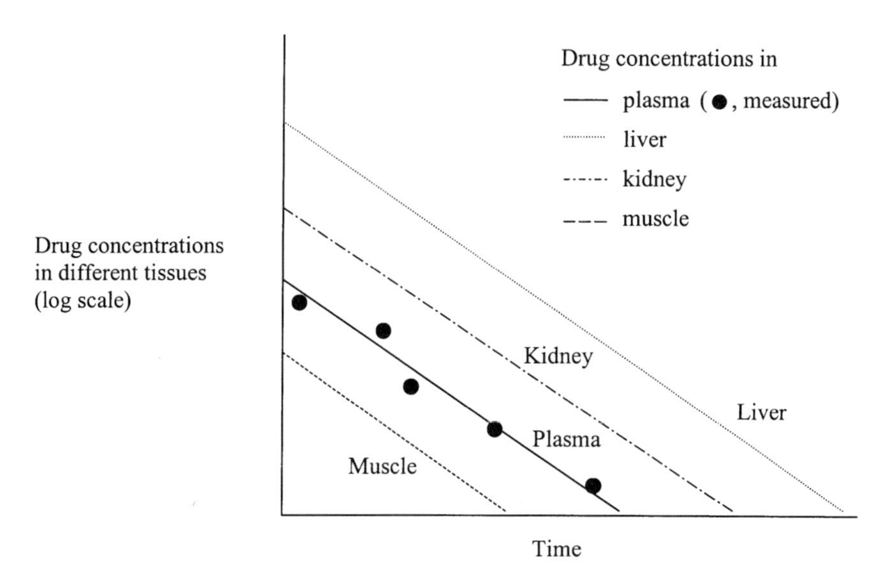

Figure 2.4. Hypothetical concentration–time profiles of a drug in plasma and various tissues when its plasma drug concentration–time profile shows a monophasic decline on a semilog scale and thus can be readily described by a one-compartment model. Drug concentrations in plasma are experimentally determined, whereas those in tissues are assumed.

(b) Relationships among Monophasic Decline, the Monoexponential Equation, and the One-compartment Model. In a one-compartment system, a monoexponential equation for a plasma drug concentration–time profile and a monophasic decline on a semilog scale after intravenous injection are necessary and sufficient conditions for each other (Fig. 2.5).

(c) Plasma Concentration–Time Plot on a Linear or a Semilogarithmic Scale. When the disposition of a drug after intravenous injection follows linear kinetics with a one-compartment system, a concentration–time profile [$C_p(t)$ *vs.* t] will be curvilinear on a linear scale. If the same data are plotted on a semilog scale, the plot of log $C_p(t)$ *vs.* t becomes a straight line and shows a monophasic decline (Fig. 2.6). When two- or three-compartment models are required for drug disposition after intravenous injection, the concentration–time profile on a semilog scale shows a bi- or triphasic decline, respectively, with a straight line during the terminal phase. On a linear scale, however, the plasma drug concentration–time plots will be curvilinear with little distinction between two- and three-compartment models. Conversion of the linear scale of plasma concentration–time data to a semilogarithmic scale thus makes it possible to determine the number of compartments needed for data analysis based on a visual inspection of the plots.

NOTE: NATURAL LOG *VS.* COMMON LOG. The base of a natural logarithm is *e* (=2.718), whereas the base of the common logarithm is 10. The relationship between the natural log and the common log is

$$\ln C_p(t) = 2.303 \log C_p(t) \tag{2.6}$$

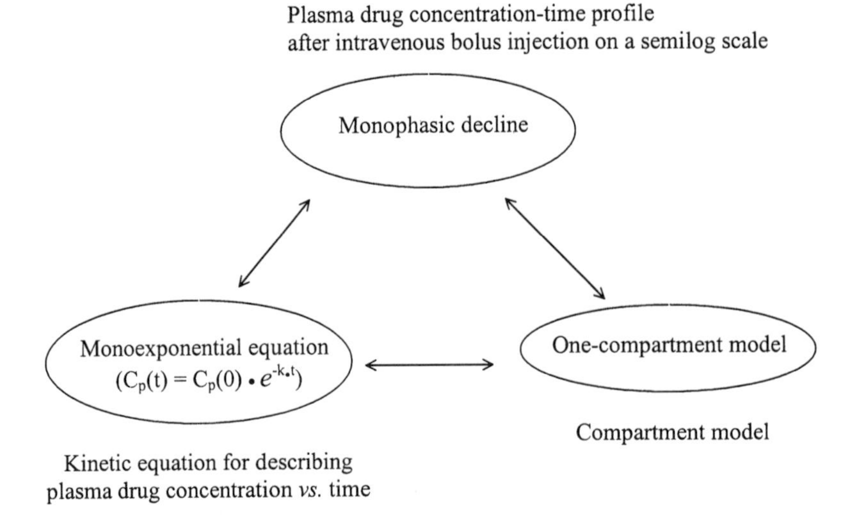

Figure 2.5. Relationships among a monophasic decline of a plasma drug concentration–time profile after intravenous injection on a semilog scale, a monoexponential equation, and a one-compartment model.

Taking the natural or common logarithms of both sides of Eq. (2.3), gives

$$\ln C_p(t) = \ln C_p(0) - k \cdot t \tag{2.7}$$

or

$$\boxed{\log C_p(t) = \log C_p(0) - \frac{k}{2.303} \cdot t} \tag{2.8}$$

2.3.1.3. Multicompartment Model Analysis

The number of compartments required to describe drug disposition profiles can vary depending on how often samples are collected and how fast after administration the drug is distributed throughout the body. Let us consider a drug for which the initial distribution into the blood pool and highly perfused organs takes place within 5 min after intravenous injection, followed by slower distribution into the rest of the body, and the first plasma sample is collected more than 5 min after injection. In this case, the drug exposure profile will show a monophasic decline so that a one-compartment model may be considered suitable for model-fitting. However, if several additional blood samples are obtained within the first 5 min, the entire plasma drug concentration–time profile may exhibit a biphasic decline on a semilog scale, and a two-compartment rather than a one-compartment model would be more suitable.

There are three different types of two-compartment models and seven three-compartment models, depending on the compartment(s) responsible for drug elim-

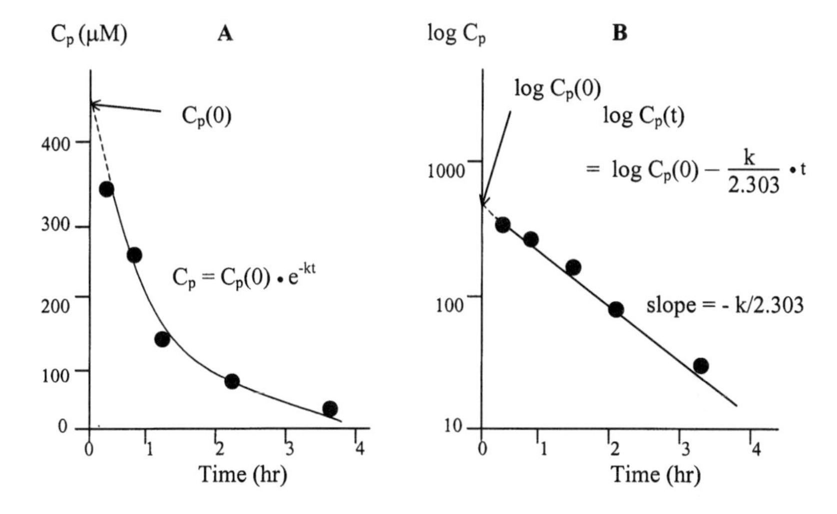

Figure 2.6. Plasma drug concentration *vs.* time profiles after intravenous injection of a hypothetical drug on linear (A) or semilogarithm (B) scales, with plasma drug concentration–time data being describable with a monoexponential equation.

ination. In the absence of any experimental evidence, it is usually assumed that drug elimination takes place exclusively from the central compartment. This is because in most drugs, the major sites of elimination are the liver (metabolism and biliary excretion) and the kidney (urinary excretion), both of which are well perfused with blood and thus readily accessible to a drug in plasma. The most commonly used two-compartment model for drug disposition after intravenous administration is shown in Fig. 2.7.

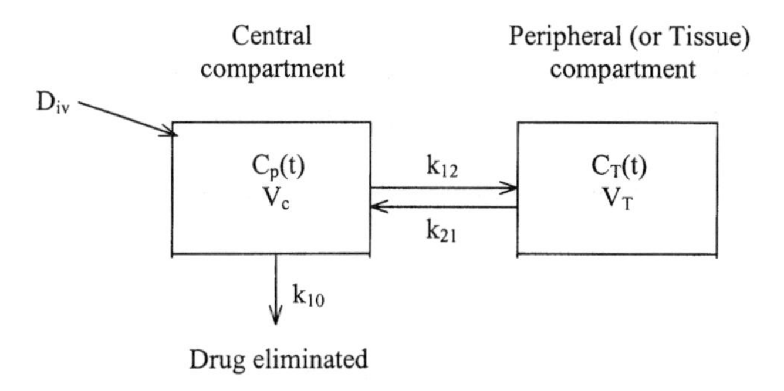

Figure 2.7. Two-compartment model with first-order elimination of a drug from the central compartment after intravenous administration. $C_p(t)$ and $C_T(t)$ are drug concentrations in the plasma and the peripheral compartment at time t, respectively; D_{iv} is an intravenous drug dose; k_{12} and k_{21} are the first-order rate constants for distribution of the drug from the central to the peripheral compartments, and vice versa, respectively; k_{10} is the first-order elimination rate constant from the central compartment; and V_c and V_T are the volumes of distribution of the central and the peripheral compartments, respectively.

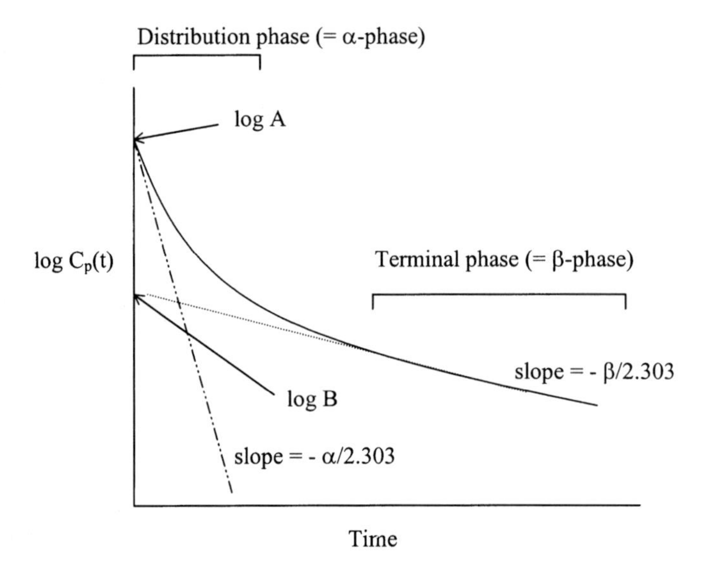

Figure 2.8. Biexponential decline of a $\log C_p(t)$ *vs.* t plot after intravenous bolus injection when drug disposition can be described using a two-compartment model.

In a two-compartment model under linear conditions, plasma drug concentration–time data after intravenous injection exhibiting a biphasic curve on a semilog scale (Jusko and Gibaldi, 1972) can be described by the following biexponential equation:

(2.9)
$$\boxed{C_p(t) = A \cdot e^{-\alpha \cdot t} + B \cdot e^{-\beta \cdot t}}$$

Estimates of A, B, α, and β can be obtained from the intercepts and slopes of a plasma concentration–time plot after intravenous administration of a drug by curve-fitting with the method of residuals or nonlinear regression using a computer program (Fig. 2.8). Those parameters can be used to estimate $C_p(0)$, V_c, and microconstants such as k_{12}, k_{21}, and k_{10}.

At time zero,

(2.10)
$$C_p(0) = A + B$$

Therefore,

(2.11)
$$V_c = \frac{D_{iv}}{C_p(0)} = \frac{D_{iv}}{A + B}$$

The relationships between the microconstants, i.e., k_{12}, k_{21}, and k_{10}, and A, B, α, and β are as follows:

$$k_{21} = \frac{A \cdot \beta + B \cdot \alpha}{A + B} \tag{2.12}$$

$$k_{10} = \frac{\alpha \cdot \beta}{k_{21}} \tag{2.13}$$

$$k_{12} = \alpha + \beta - k_{21} - k_{10} \tag{2.14}$$

From k_{12}, k_{21}, k_{10}, and V_c, the systemic clearance (Cl_s, see Chapter 6) and the volume of distribution at steady state (V_{ss}, see Chapter 5) can be also calculated:.

$$Cl_s = k_{10} \cdot V_c \tag{2.15}$$

and

$$V_{ss} = V_c \cdot (1 + k_{12}/k_{21}) \tag{2.16}$$

(a) Distribution and Terminal Phases. When the disposition of a drug can be described using a two-compartment model under linear conditions, the plasma drug concentration–time profile after intravenous injection will show a biphasic decline on a semilog scale (Fig. 2.8). The initial sharply declining phase of the exposure profile is often called the "distribution or α-phase" and the later phase, shown as a shallower straight line, is called "terminal or β-phase" (also known as the postdistribution, pseudodistribution equilibrium, or elimination phase) (Riegelman *et al.*, 1968). During the distribution phase, the decrease in the plasma drug concentration is due mainly to the initial rapid distribution of the drug from the plasma into well-perfused organs and tissues. The pseudodistribution equilibrium is achieved at some time after drug administration when the ratios of the amounts of drug between the plasma pool and all other body tissues become constant. During this phase, the decrease in the plasma drug concentration is due primarily to the elimination of the drug from the body, and exhibits a straight line on a semilog scale (Fig. 2.8).

(b) Drug Levels in a Peripheral Compartment. Concentrations of a drug in a peripheral compartment increase rapidly during the distributional phase following intravenous injection and decrease gradually in parallel with drug concentrations in the plasma during the terminal phase (Fig. 2.9). The shape of the curve can vary depending on drug distribution and elimination rates (Gibaldi *et al.*, 1969).

(c) Relationships among Biphasic Decline, the Biexponential Equation, and the Two-Compartment Model. If the fall in the plasma drug concentrations on a semilog scale after intravenous injection of a drug is biphasic (an initial rapid decline followed by a slower decrease) or triphasic, two- or three-compartment models respectively, may be suitable (Fig. 2.10). However, a multiphasic decline of the plasma drug concentration profile does not necessarily mean that the body behaves

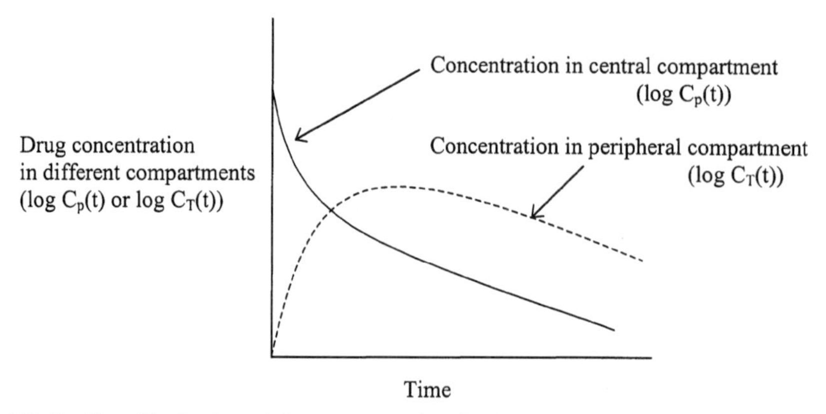

Figure 2.9. Semilogarithmic plots of drug concentrations in the central [——, measured concentrations in plasma, $C_p(t)$] and peripheral [······, projected concentrations in tissues, $C_T(t)$] compartments. The extent of drug concentrations in the peripheral compartment can vary, depending on the rate of drug distribution between the plasma and the tissues.

in a multicompartment fashion in relation to the drug. For instance, exposure profiles of drugs with one of the following disposition characteristics can also exhibit a biexponential decline, even if the body behaves as a single compartment for drug distribution.

• Nonlinear protein binding. At high drug concentrations during the early time points after intravenous injection, the fraction of the drug not bound to plasma protein can be higher owing to binding saturation than that during the later time points. Unless intrinsic clearance becomes saturated, drug clearance is generally faster when there is less protein binding than otherwise. This more rapid clearance can cause a steeper decline in drug concentrations during the initial phase as compared to the later phase. As concentrations decrease, protein binding of the drug becomes more extensive, causing slower clearance, which is reflected in a shallower slope of the concentration profile with time.

• Product inhibition. Metabolite(s) of a drug can inhibit clearance mechanisms of the parent drug. The effects of metabolite(s) on drug clearance shortly after drug administration may be negligible because there is not much metabolite formation. However, once a sufficient quantity of metabolite(s) is accumulated, drug clearance can be significantly impaired, resulting in a slower decline of plasma drug concentrations during the later phase.

• Cosubstrate depletion. Depletion of cosubstrate required for elimination (e.g., metabolism) of drug after a certain period of time can result in an apparent biphasic decline in the drug concentration profile.

• Pharmacokinetic differences of enantiomers. When a drug is administered as a racemic mixture and pharmacokinetic behaviors of the enantiomers of the drug are different, it is possible to have apparent biphasic profiles of drug concentrations in plasma when determined as a racemate.

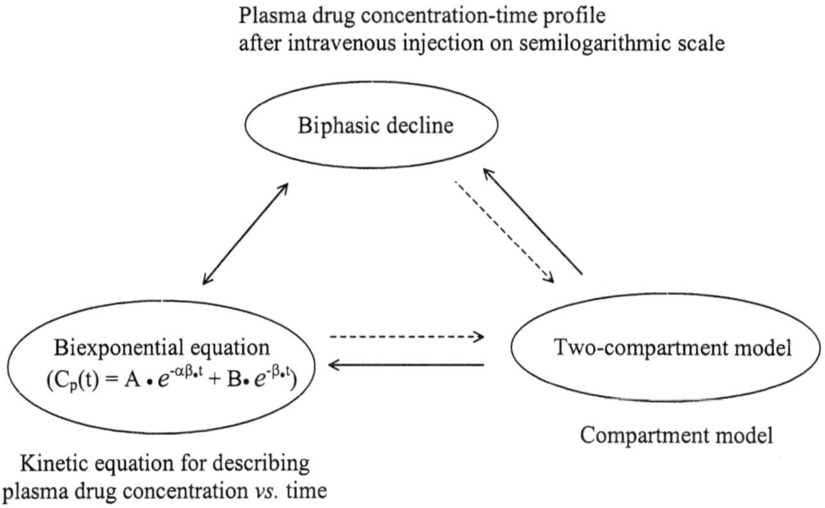

Figure 2.10. Relationships among a biphasic decline of a plasma drug concentration–time profile after intravenous injection on a semilogarithmic scale, a biexponential equation, and a two-compartment model. A solid arrow from the two-compartment model to the biphasically declining plasma concentration–time plot implies that if the body behaves as two compartments, a plasma drug concentration profile will be biphasic. However, a biphasic exposure profile does not necessarily mean that the body behaves as two compartments, as indicated with a dotted arrow.

2.3.1.4. Model Selection

The most important factor in selecting a pharmacokinetic model to fit the experimental data is its physiological relevance to kinetic behaviors of the drug. Especially when there is experimental evidence suggesting particular drug distribution patterns or elimination routes, pharmacokinetic models that can accommodate those findings should be considered. For instance, if the data suggest that the elimination of a drug occurs mainly via hepatic metabolism with a biphasically declining plasma concentration profile after intravenous injection, a two-compartment model with elimination of the drug from the central compartment rather than from the peripheral compartment would be more reasonable.

Many different compartmental models can be used for the same data. The most complicated model with numerous compartments and parameters for the data is not necessarily the best model for the characterization of drug pharmacokinetic profiles. A rule of thumb for model selection is "the principle of parsimony." That is, the simpler the model, the better it can be. There are several statistical approaches to identifying the most appropriate pharmacokinetic model among those available for the same data.

(a) Akaike Information Criterion (AIC). The most well known method for model selection is the so-called Akaike information criterion (AIC) value estimation (Akaike, 1974). An AIC value for a particular model can be obtained as follows:

$$\text{AIC value} = n \cdot \ln(\text{WSS}) + 2 \cdot m \quad (2.17)$$

where n and m are, respectively, the number of data points and parameters used in the model, and WSS is the weighted sum of squares estimated as

$$\text{WSS} = \sum_{i=1}^{i=n} (Y_{\text{obs},i} - Y_{\text{calc},i})^2 \cdot W_i \tag{2.18}$$

where W_i is a weighting factor for fitting the model to the experimental data (drug concentrations) and can be $1/Y$ or $1/Y^2$, $Y_{\text{obs},i}$ is the observed y-value (measured drug concentration), and $Y_{\text{calc},i}$ is the calculated y-values (estimated drug concentration according to the model). Among different models, the model yielding the lowest AIC value (highest negative in the case of negative values) is the most appropriate model for describing the data.

(b) Schwarz Criterion. The Schwarz criterion (SC) is similar to the ACI criterion (Schwarz, 1978), and its value is calculated as follows:

$$\text{SC value} = n \cdot \ln(\text{WSS}) + m \cdot \ln(n) \tag{2.19}$$

Similarly to the AIC criterion, the model yielding the lowest SC value is the most appropriate model.

2.3.2. Noncompartmental Approach

The noncompartmental approach for data analysis does not require any specific compartmental model for the system (body) and can be applied to virtually any pharmacokinetic data. There are various noncompartmental approaches, including statistical moment analysis, system analysis, or the noncompartmental recirculatory model. The main purpose of the noncompartmental approach is to estimate various pharmacokinetic parameters, such as systemic clearance, volume of distribution at steady state, mean residence time, and bioavailability without assuming or understanding any structural or mechanistic properties of the pharmacokinetic behavior of a drug in the body. In addition, many noncompartmental methods allow the estimation of those pharmacokinetic parameters from drug concentration profiles without the complicated, and often subjective, nonlinear regression processes required for the compartmental models. Owing to this versatility and ruggedness, the noncompartmental approach is a primary pharmacokinetic data analysis method for the pharmaceutical industry. Moment analysis, the most commonly used noncompartmental method, is discussed below.

2.3.2.1. Moment Analysis

Statistical moment analysis has been used extensively in chemical engineering to elucidate diffusion characteristics of chemicals in liquid within tubes. Similar concepts were applied to pharmacokinetics to analyze drug disposition and to estimate pharmacokinetic parameters (Yamaoka *et al.*, 1978). The plasma concentration–time profile of a drug can be thought of as a statistical distribution curve, for which the first two moments (zero and first) are defined as the area under the plasma concentration–time curve (AUC) and as the mean residence time (MRT), a mean time interval during which a drug molecule resides in the body before being excreted. According to moment analysis, the AUC and MRT of a drug can be calculated from plasma drug concentration–time profiles, regardless of the route of administration, as follows:

$$AUC_{0-\infty} = \int_0^\infty C_p(t)\,dt \tag{2.20}$$

$$MRT = \frac{AUMC_{0-\infty}}{AUC_{0-\infty}} = \frac{\int_0^\infty t \cdot C_p(t)\,dt}{\int_0^\infty C_p(t)\,dt} \tag{2.21}$$

where AUMC is the area under the first-moment curve of the plasma drug concentration–time curve from time zero to infinity.

(a) Units of AUC, AUMC, and MRT.
AUC: concentration · time, e.g., $\mu g \cdot hr/ml$ or $\mu M \cdot hr$.
AUMC: (concentration · time) · time, e.g., $\mu g \cdot hr^2/ml$ or $\mu M \cdot hr^2$.
MRT: time, e.g., hr

(b) Pharmacokinetic Implications of AUC and AUMC. AUC is an important pharmacokinetic parameter in quantifying the extent of exposure of a drug and of its clearance from the body. AUC is considered a more reliable parameter for assessing the extent of overall exposure of a drug than individual drug concentrations. AUMC is used for assessing the extent of distribution, i.e., the volume of distribution at steady state and the persistence of a drug in the body.

(c) Estimating AUC and AUMC.

(i) Linear trapezoidal method. The linear trapezoidal method is the one most well-known for estimating AUC and AUMC. For instance, AUC over two adjacent time points, t_1 and t_2, (AUC_{t1-t2}, Fig. 2.11) can be approximated as the area of a trapezoid formed by connecting the adjacent points with a straight line [Eq. (2.22)]. An estimate of AUC over an extended period of time can be obtained by adding the areas of a series of individual trapezoids. Estimating AUC by the linear trapezoidal method should be done on a linear scale.

$$AUC_{t1-t2} = \text{Area of a trapezoid between } t_1 \text{ and } t_2$$
$$= (t_2 - t_1) \cdot \frac{C_2 + C_1}{2} \tag{2.22}$$

↑ *Adjacent time points (time interval)*

↖ *Concentrations of drug corresponding to the time points (mean concentration)*

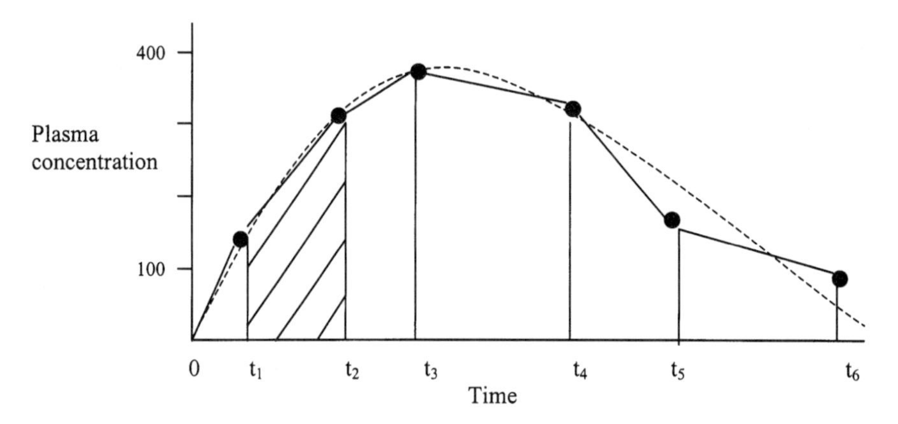

Figure 2.11. Estimate of AUC by the linear trapezoidal method on a linear scale. AUC between t_1 and t_2 is shown as a hatched area. A discrepancy can be seen between a plot from interpolation of data points (solid line) and a nonlinear regression plot (dotted line) fitted to the individual data.

The advantages and disadvantages of the linear trapezoidal method are as follows:

1. Advantages: (a) Easy to use. (b) Reliable for slow declining or ascending curves.
2. Disadvantages: (a) Owing to the linear interpolation between data points, it tends to over- or underestimate the true AUC, depending, respectively, on the concave or convex shape of the curve (Fig. 2.11). (b) Error-prone whenever there is a sharp bending in concentration values between time points. (c) Error-prone for data points with a wide interval.

AUMC can be also estimated with the linear trapezoidal approximation from the area under the curve of *the product of concentration and time* $[C_p(t) \cdot (t)]$ *vs.* time on a linear scale. An example for AUC and AUMC calculation with the linear trapezoidal method after oral administration of a hypothetical drug is shown in Table 2.1. When the concentration of a drug in plasma at the last sampling time point (t_{last}) is not zero, $AUC_{0-\infty}$ can be estimated by combining AUC from time zero to t_{last} ($AUC_{0-t_{last}}$) using the trapezoidal method and AUC from t_{last} to infinity ($AUC_{t_{last}-\infty}$) estimated using the following equation:

$$AUC_{t_{last}-\infty} = \frac{C^*}{\lambda_z} \tag{2.23}$$

where C^* is the estimated drug concentration at t_{last}, and λ_z is the slope of the terminal phase of the plasma drug concentration–time profile on a semilog scale. C^* and λ_z can be obtained using an appropriate linear regression method with the last few (usually three) data points during the terminal phase. An estimate of $AUMC_{t_{last}-\infty}$ can be obtained from

$$AUMC_{t_{last}-\infty} = \frac{C^* \cdot t}{\lambda_z} + \frac{C^*}{\lambda_z^2} \tag{2.24}$$

Table 2.1. Estimates of AUC and AUMC from 0 to 7 hr Based on the Linear Trapezoidal Method with Plasma Drug Concentrations after Oral Administration of a Hypothetical Drug

Sampling time (hr)	Plasma drug concentration (ng/ml)	Plasma drug concentration × time (ng · hr/ml)	AUC[a] (ng · hr/ml)	AUMC[a] (ng · hr²/ml)
0	0	0	0	0
1	100	100	50	50
2	200	400	150	250
3	300	900	250	650
4	200	800	250	850
6	100	600	300	1400
7	0	0	50	300
			AUC_{0-7}: 1050	$AUMC_{0-7}$: 3500

[a]AUC or AUMC between adjacent time points.

Similarly, $AUMC_{0-\infty}$ can be obtained by adding $AUMC_{0-t_{last}}$ calculated using the linear trapezoidal method and $AUMC_{t_{last}-\infty}$ estimated.

(ii) Log trapezoidal method. The so-called log trapezoidal method assumes that the concentration values vary linearly within each sampling interval. AUC_{t1-t2} can be estimated as follows:

$$AUC_{t1-t2} = (t_2 - t_1) \cdot \frac{C_2 - C_1}{\ln(C_2/C_1)} \quad (2.25)$$

Equation (2.25) is most appropriate for an exponentially declining concentration–time profile. The method is error-prone in an ascending curve, near a peak, or in a steeply descending multiexponential curve, and it cannot be used if the concentration is zero or if the two values are equal. There are several other methods for estimating AUC. For instance, the Lagrange method uses a cubic polynomial equation $[C_p(t) = a + b \cdot t + c \cdot t^2 + d \cdot t^3]$ instead of the linear function, and the Spline method uses piecewise polynominals for curve-fitting (Yeh and Kwan, 1978).

NOTE: ESTIMATED CONCENTRATION AT THE LAST TIME POINT (C*). Drug concentration ($C_{t_{last}}$) measured at the last time point (t_{last}) is analytically most error-prone, because $C_{t_{last}}$ is generally closest to the limit of quantification of an assay. It is thus considered to be more reliable to use a concentration C* at t_{last} estimated using a proper linear regression method with the last few (usually three) data points for calculation of $AUC_{t_{last}-\infty}$ (Fig. 2.12).

2.3.2.2. Estimating Pharmacokinetic Parameters with Moment Analysis

(a) Clearance. The systemic clearance (Cl_s) of a drug (see Chapter 6) can be estimated as the reciprocal of the zero moment of a plasma concentration–time

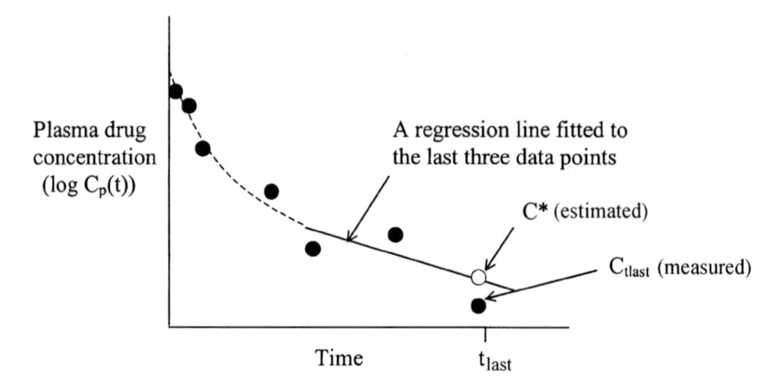

Figure 2.12. Estimated concentration (C*, ○) at the last time point (t_{last}) according to linear regression with the last three data points measured. $C_{t_{last}}$ is the actual concentration measured at t_{last}.

curve after intravenous administration (AUC_{iv}) normalized by an intravenous dose (D_{iv}) as shown below:

$$Cl_s = \frac{D_{iv}}{AUC_{iv}} \tag{2.26}$$

(b) Volume of Distribution at Steady State. The volume of drug distribution at steady state (Vd_{ss}) (see Chapter 5) can be estimated as the product of MRT after intravenous bolus injection (MRT_{iv}) and Cl_s:

$$Vd_{ss} = MRT_{iv} \cdot Cl_s = \frac{AUMC_{iv}}{AUC_{iv}} \cdot \frac{D_{iv}}{AUC_{iv}} \tag{2.27}$$

NOTE: RELATIONSHIP AMONG AUC, Cl_s AND V_{ss}. The $AUC_{0-\infty}$ of a drug inversely reflects the extent of Cl_s, but does not have a direct correlation with the size of V_{ss}. This is because Cl_s affects only the $AUC_{0-\infty}$, whereas V_{ss} is governed by both $AUC_{0-\infty}$ and $AUMC_{0-\infty}$ [Eq. (2.26) and (2.27)]. Therefore, it is true that a drug with a smaller $AUC_{0-\infty}$ after intravenous injection has a faster Cl_s than one with a larger $AUC_{0-\infty}$ at the same dose. However, the drug with the smaller $AUC_{0-\infty}$ does not necessarily have a greater V_{ss}. Let us assume that there are two drugs (A and B) and that both $AUC_{0-\infty}$ and $AUMC_{0-\infty}$ of drug A are smaller than those of drug B (Table 2.2) after intravenous injection at 3 mg/kg in rats (Fig. 2.13). Cl_s and V_{ss} estimated based on $AUC_{0-\infty}$ and $AUMC_{0-\infty}$ (Table 2.2) indicate that Cl_s of drug A is greater than that of drug B, reflected by its lower $AUC_{0-\infty}$, whereas V_{ss} of drug B is greater than that of drug A, despite the fact that $AUC_{0-\infty}$ of drug B is greater than that of drug A.

(c) Bioavailability. Bioavailability (F) of a drug generally refers to the fraction of a dose administered via a route other than intravenous injection that reaches the

Table 2.2. Summary of Pharmacokinetic Parameters for Drugs A and B

Parameters	A	B
D_{iv} (mg/kg)	3	3
$AUC_{0-\infty}$ (μg·hr/ml)	2	3
$AUMC_{0-\infty}$ (μg·hr^2/ml)	1	4.5
Cl_s (ml/min/kg)	25	16.7
V_{ss} (liter/kg)	0.75	1.5

systemic circulation. For instance, F after oral administration can be estimated as the ratio of the dose-normalized zero moments ($AUC_{0-\infty}$) after oral and intravenous administration (see Chapter 4):

$$F = \frac{D_{iv} \cdot AUC_{po}}{D_{po} \cdot AUC_{iv}} \quad (2.28)$$

where D_{iv} and D_{po} are intravenous and oral doses, and AUC_{iv} and AUC_{po} are $AUC_{0-\infty}$ after intravenous and oral administration of the drug, respectively.

(d) Mean Residence Time. The mean residence time (MRT) is the average time spent by a single drug molecule in the body before being excreted via elimination processes, regardless of the route of administration. When a drug disappearance curve exhibits a monophasic decline after intravenous injection on a semilog scale, its MRT_{iv} is the time required for 63.2% of the dose to be eliminated from the body. The MRT values after administration by routes other than intravenous bolus injection are always greater than MRT_{iv}. Differences in MRT values following administration via these other routes and MRT_{iv} can be viewed as the average time required for drug molecules to reach the systemic circulation from the site of administration. For instance, a difference between MRT after oral administration (MRT_{po}) and MRT_{iv} is the mean absorption time (MAT; see Chapter 4), represen-

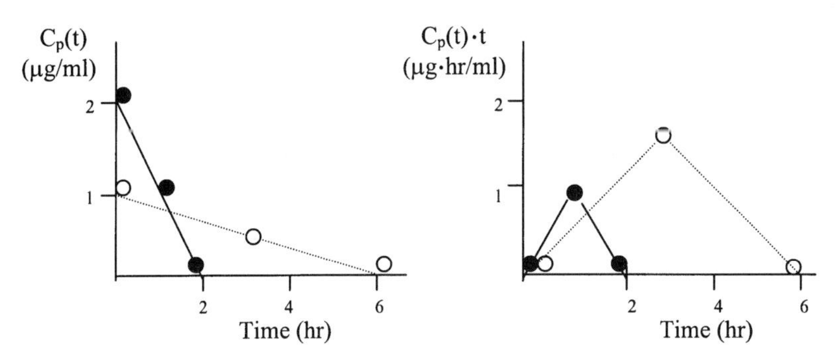

Figure 2.13. Plasma drug concentration–time profiles (left) and drug first-moment curves (right) for drugs A(●) and B (○) on a linear scale.

ting the average time required for the drug to reach the systemic circulation from the gastrointestinal tract after oral administration:

$$\boxed{\text{MAT} = \text{MRT}_{\text{po}} - \text{MRT}_{\text{iv}}} \tag{2.29}$$

$$\text{MRT}_{\text{po}} = \frac{\text{AUMC}_{\text{po}}}{\text{AUC}_{\text{po}}}$$

If drug absorption can be assumed to be a single first-order kinetic process, the absorption rate constant (k_a) after oral dosing can be estimated as the reciprocal of MAT:

$$\boxed{k_a = 1/\text{MAT}} \tag{2.30}$$

(e) Half-Life. The half-life ($t_{1/2}$) of a drug generally implies the *terminal* half-life during the terminal phase where a plot of log $C_p(t)$ *vs.* time exhibits a straight line (Gibaldi and Weintraub, 1971). The half-life of a drug is the period of time over which its concentration in plasma decreases by half from a reference concentration at any given time point. When drug disappearance shows a monophasic decline after intravenous injection, $t_{1/2}$ is proportional to MRT_{iv}:

$$t_{1/2} = 0.693 \cdot \text{MRT}_{\text{iv}} \tag{2.31}$$

When a plasma drug concentration–time plot exhibits a bi- or a triphasic decline on a semilog scale, $t_{1/2}$ is longer than $0.693 \cdot \text{MRT}_{\text{iv}}$ (Kwon, 1996).

(i) Estimating half-life. There are several ways to estimate $t_{1/2}$ of a drug from its plasma concentration–time profile.

- Visual inspection of the plasma concentration–time profile. A rough estimate of $t_{1/2}$ can be obtained from a plasma concentration–time profile simply by eyeballing a time interval over which the concentration decreases by half from any reference time point.

- Curve fitting. In general, three data points during the terminal phase are used, over which the time interval is greater than at least twice the estimated $t_{1/2}$ based on those points. The slope (λ_z) during the terminal phase of a plasma drug concentration–time plot on a semilog scale is inversely related to $t_{1/2}$:

$$\boxed{t_{1/2} = 0.693/\lambda_z} \tag{2.32}$$

where λ_z is equal to k [Eq. (2.5)] or β [Eq. (2.9)] when a plasma drug concentration–time plot exhibits a monophasic or a biphasic decline, respectively.

• Estimation between two data points. The following equation can be used to estimate $t_{1/2}$ between two drug concentrations (C_1 and C_2) at two different time points (t_1 and t_2). In this case, the estimated $t_{1/2}$ indicates how much time it would take for a drug concentration to decrease by half, if C_1 decreases to C_2 from t_1 to t_2:

$$t_{1/2} = \frac{(0.693)\cdot(t_2 - t_1)}{\ln(C_1/C_2)} \tag{2.33}$$

(ii) Pharmacokinetic implications of half-life. The terminal half-life of a drug is probably the most important parameter in assessing the *duration* of drug exposure.

• Relationship between terminal half-life and efficacy of drug. If there is a direct correlation between plasma exposure levels of a drug and its pharmacological response, absolute exposure levels during the terminal phase and $t_{1/2}$ can be important in assessing the duration of its efficacy. Let us assume that drugs A and B have the same *in vitro* potency, but that $AUC_{0-\infty}$ of drug A is greater than that of drug B after intravenous injection, while the exposure levels of B during the terminal phase are higher with a longer $t_{1/2}$ than those of A (Fig. 2.14). If there are direct relationships among EC_{50}, *in vivo* efficacy, and plasma drug levels, drug B may be more desirable for a longer duration of efficacy than drug A, despite A's greater AUC.

• Significance of half-life after multiple dosing. Regardless of the route of administration, $t_{1/2}$ of a drug after multiple dosing becomes close to that during the true terminal phase after single dosing; i.e., $t_{1/2}$ after multiple dosing is dictated by the true terminal $t_{1/2}$ after single dosing. It is not uncommon to see the apparent $t_{1/2}$ of a drug after multiple doses being longer than that after a single dose. This can be simply because the true terminal $t_{1/2}$ after a single dose cannot be readily measured owing to assay limitations and/or inadequate sampling time points (Fig. 2.15).

• Time to reach steady state after multiple dosing. The time required to reach steady state exposure levels of a drug after multiple dosing is directly related to its

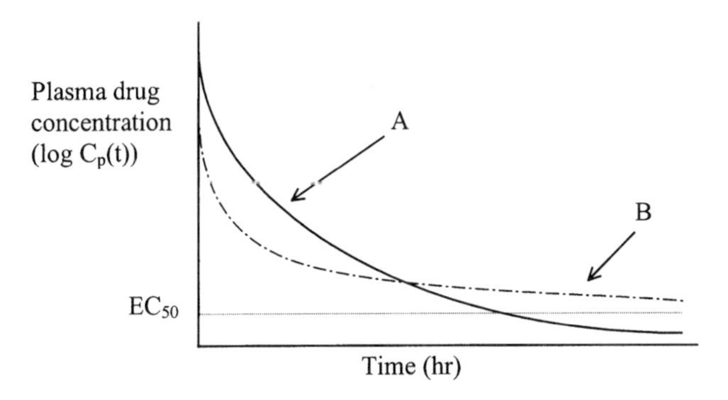

Figure 2.14. Plasma drug concentration profiles of drugs A and B with the same EC_{50}. $AUC_{0-\infty}$ of drug A is greater than that of drug B, whereas exposure levels of drug B during the terminal phase are higher with a longer terminal half-life as compared to those of drug A.

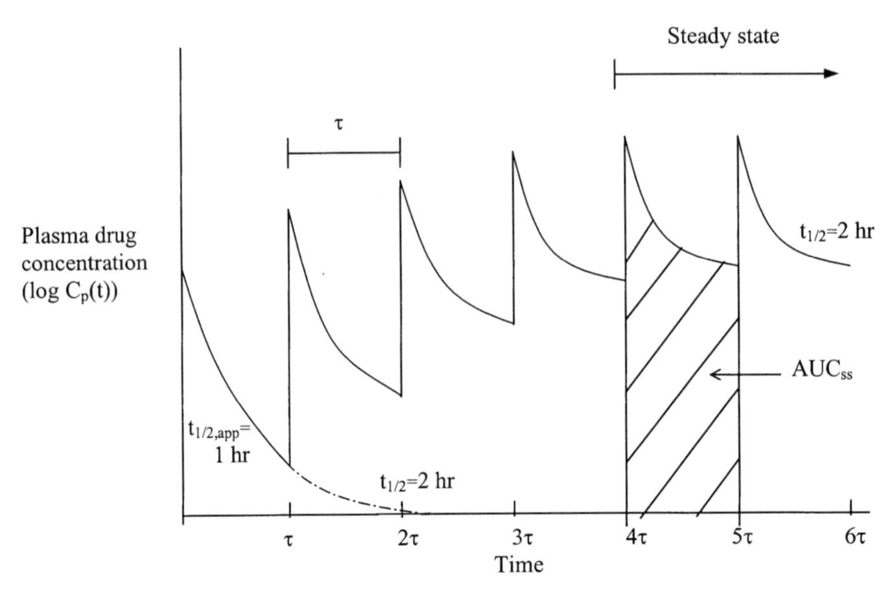

Figure 2.15. A schematic description of changes in plasma drug exposure profiles and apparent half-lives after multiple dosing. The plots show that owing to insufficient data collection and/or detection limitations of assay the apparent terminal half-life ($t_{1/2,app} = 1$ hr) estimated after a single dose (up to time τ) can be shorter than that ($t_{1/2} = 2$ hr) after multiple dosing. τ is the dosing interval, and AUC_{ss} is the AUC between dosing intervals after reaching steady state.

$t_{1/2}$ after a single dose. It usually takes about five half-lives to reach steady state drug concentrations after multiple dosing for a drug exhibiting one-compartment kinetic characteristics, regardless of the dose or the dosing interval, under linear conditions. For instance, if $t_{1/2}$ of a drug after a single dose is 10 hr, steady state concentrations upon multiple dosing will be achieved after approximately 50 hr, regardless of how much or how often the drug has been dosed during that period. The size of the dose and the dosing interval determine the extent of the steady state drug concentrations after multiple dosing, depending on clearance of the drug but not on the time to reach the steady state.

NOTE: ACCUMULATION FACTOR AFTER MULTIPLE DOSING. The accumulation factor (R) reflects how much drug is accumulated in the body at steady state after multiple dosing as compared to that after single dosing. The value of R can be estimated by dividing AUC over the dosing interval (τ) at steady state after multiple dosing (AUC_{ss}) divided by AUC from 0 to τ ($AUC_{0-\tau}$) after the first dose. Instead of AUC_{ss}, $AUC_{0-\infty}$ after a single dose can be used, since $AUC_{0-\infty}$ is equal to AUC_{ss}. (Fig. 2.15):

$$R = \frac{AUC_{ss}}{AUC_{0-\tau}} = \frac{AUC_{0-\infty}}{AUC_{0-\tau}} \tag{2.34}$$

The average plasma drug concentration at steady state ($C_{avg,ss}$) after multiple dosing, which is $AUCs_{ss}$ divided by τ, can be also estimated by dividing $AUC_{0-\infty}$ by τ:

$$C_{avg,ss} = \frac{AUC_{ss}}{\tau} = \frac{AUC_{0-\infty}}{\tau} \tag{2.35}$$

These equations enable the estimation of R and $C_{avg,ss}$ after multiple dosing, based on drug exposure levels after *a single dose.*

(iii) Assay limitation and half-life. An arbitrary study protocol for blood sampling over a fixed period without consideration of appropriate time points can lead to a significant underestimate of the true $t_{1/2}$ of a drug. Another factor that makes an accurate estimate of $t_{1/2}$ difficult is assay sensitivity for drug concentrations during the terminal phase. It is not uncommon to observe apparently longer terminal $t_{1/2}$ of a drug at a higher dose level, compared to that at a lower dose level. This may be due to nonlinear pharmacokinetics of a drug at a higher concentration; however, it can be due simply to an inability to accurately measure drug concentrations at a later time point because of limited assay sensitivity at a lower dose level (Fig. 2.16). It is, therefore, important to measure and compare $t_{1/2}$ of a drug over an extended period of time during the terminal phase at different dose levels, in order to obtain a reliable estimate of $t_{1/2}$. Another approach is to measure $t_{1/2}$ at elevated exposure levels after multiple dosing, which is the true $t_{1/2}$ of a drug, assuming that there are no dose- or time-dependent pharmacokinetic variations.

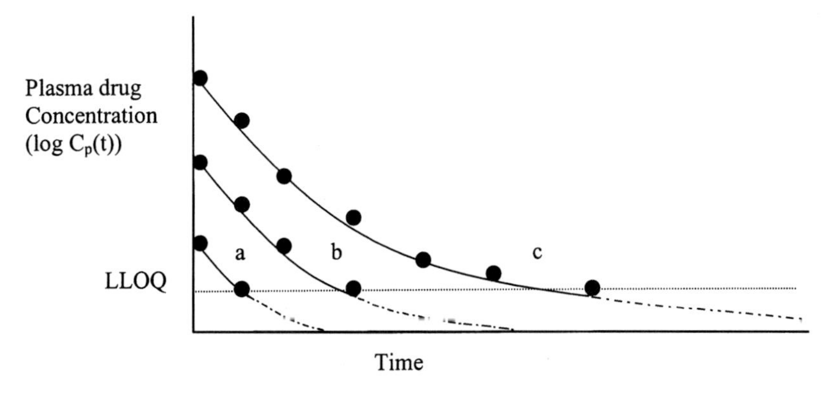

Figure 2.16. Plasma drug concentration–time profiles of a hypothetical drug at three different dose levels (dose levels: a < b < c) under linear conditions. Estimates of terminal half-lives based on plasma drug concentrations appear to be shorter at the low and medium dose levels (a and b) compared to the high dose level (c), albeit there are no dose-dependent changes in drug disposition. The apparently shorter half-life of a drug at lower dose levels is due to limited assay sensitivity. The dotted lines below LLOQ at a and b indicate actual drug concentrations, which decrease in parallel to those at c. LLOQ is the lower limit of quantitation of the assay.

REFERENCES

Akaike H., A new look at the statistical model identification, *IEEE Trans. Automat. Control* **19**: 716–723, 1974.

Balant L. P. and Gex-Fabry M., Physiological pharmacokinetic modelling, *Xenobiotica* **20**: 1241–1257, 1990.

Gerlowke L. E. and Jain R. K., Physiologically based pharmacokinetic modeling: principles and applications, *J. Pharm. Sci.* **72**: 1103–1127, 1983.

Gibaldi M. *et al.*, Relationship between drug concentration in plasma or serum and amount of drug in the body, *J. Pharm. Sci.* **58**: 193–197, 1969.

Gibaldi M. and Weintraub H., Some considerations as to the determination and significance of biological half-life, *J. Pharm. Sci.* **60**: 624–626, 1971.

Gillespie W. R., Noncompartmental versus compartmental modelling in clinical pharmacokinetics, *Clin. Pharmacokinet.* **20**: 253–262, 1991.

Jusko W. J. and Gibaldi M., Effects of change in elimination on various parameters of the two-compartment open model, *J. Pharm. Sci.* **61**: 1270–1273, 1972.

Kwon Y., Volume of distribution at pseudo-distribution equilibrium: relationship between physiologically based pharmacokinetic parameters and terminal half-life of drug, *Pharm. Sci.* **2**: 387–388, 1996.

Riegelman S. *et al.*, Shortcomings in pharmacokinetic analysis by conceiving the body to exhibit properties of a single compartment, *J. Pharm. Sci.* **57**: 117–123, 1968.

Schwarz G., Estimating the dimension of model, *Ann. Stat.* **6**: 461–464, 1978.

Yamaoka K. *et al.*, Statistical moments in pharmacokinetics, *J. Pharmacokinet. Biopharm.* **6**: 547–557, 1978.

Yeh K. C. and Kwan K. C., A comparison of numerical integrating algorithms by trapezoidal, Lagrange, and Spline approximation, *J. Pharmacokinet. Biopharm.* **6**: 79–97, 1978.

Zierler L., A critique of compartmental analysis, *Ann. Rev. Biophys. Bioeng.* **10**: 531–562, 1981.

3

New Approaches for High Throughput *In Vivo* Exposure Screening

Rapid turn-around of high-quality pharmacokinetic data in animals has long been recognized as a critical, yet potentially rate-limiting step during drug discovery in the pharmaceutical industry. In particular, the recent advent of combinatorial chemistry, which has dramatically increased the number of compounds synthesized during discovery, has triggered a reevaluation of the conventional one-at-a-time approach as primary *in vivo* exposure screening (Kubinyi, 1995 and Tarbit and Berman, 1998). Recent advances in liquid chromatography/mass spectrometry/mass spectrometry (LC/MS/MS)-based quantitative analytical techniques have made it possible to increase the throughput of *in vivo* exposure screening to a significant extent, and several innovative experimental approaches have been introduced to facilitate it. Those include:

1. N-in-1 (cassette or cocktail) dosing.
2. Postdose pooling (or cocktail analysis).
3. AUC estimation from one pooled sample.
4. Continuous sampling.

Depending on study needs and assay capability, a researcher can use only one of the above approaches or combine them for exposure screening in animals. Brief backgrounds and study design strategies for each method are discussed below.

3.1. N-IN-1 (CASSETTE OR COCKTAIL) DOSING

The N-in-1 approach implies administration of a mixture of several compounds in one dosing vehicle to animals as opposed to dosing individual compounds in one vehicle to individual animals at a time. Modern analytical methods such as LC/MS/MS allow the simultaneous measurement of concentrations of multiple compounds coexisting in biological samples with little method development time

(Beaudry *et al.*, 1998 and Olah *et al.*, 1997). There are several important factors to be considered for the N-in-1 dosing approach with LC/MS/MS (Frick *et al.*, 1998).

1. Number of compounds

• Mass spectrometry sensitivity. The number of compounds dosed should be determined based on assay sensitivity and selectivity as the more compounds that are included, the less sensitive the assay will be. The problem of sensitivity is further conpounded because as more drugs are combined in one dosing vehicle, a smaller amount of any individual drug is usually given.

• Drug–drug interaction. Any carrier or enzyme-mediated pharmacokinetic process such as cytochrome P450-mediated metabolism is potentially subject to drug–drug interaction, depending on compounds and their concentrations, and the possibility of drug–drug interaction increases as the number of compounds dosed increases.

2. Grouping of compounds

• Compounds with similar physicochemical properties. Compounds with similar aqueous solubility and ionizability (acid or base) are preferable for grouping in the same vehicle, primarily in order to maintain homogeneity and reproducibility of a dosing formulation and sample preparation for assay.

• Compounds with minimal interference in mass spectrometry assay. Preferably, compounds producing the same product (or daughter) ion should be avoided in the same dosing regimen, when the multiple reaction monitoring (MRM) mode is used for the mass spectrometry assay. Compounds in which molecular weights differ by 16 (potential oxidative metabolite) from other compounds should be also avoided because similar fragmentation patterns of potential oxidative metabolites of the other compounds can interfere the MS assay. Compounds best ionized in a different ion mode (positive *vs.* negative ion mode) of MS should not be combined.

3. Dose

• Constant and low dose level. The total dose of compounds should be held constant and as low as possible to minimize potential drug–drug interaction during the various stages of drug absorption and disposition.

4. Reference compound

• Reduce study variability. The inclusion in each study of a reference compound with known exposure profiles as a potential indicator of drug–drug interaction and a biological internal standard minimizes both intra-/interanimal and experimental variabilities among studies. Changes in exposure profiles of this compound might indicate potential drug–drug interactions among compounds examined and/or dosing errors. A similar approach has been used to assess the activities of metab-

Table 3.1. Advantages and Disadvantages of N-in-1 Dosing as Compared to Dosing with Individual Compounds for *In Vivo* Exposure Screening

Advantages	Disadvantages
Rapid screening with more compounds	Needs LC/MS/MS for analysis
Fewer samples for assay	Potential drug–drug interaction
Fewer animals required for studies	More problematic data analysis

olizing enzymes *in vivo* (Frye *et al.*, 1997) and to estimate the extent of membrane permeability of compounds in Caco-2 cells (Taylor *et al.*, 1997).

3.2. POSTDOSE POOLING (OR COCKTAIL ANALYSIS)

As an alternative to N-in-1 dosing, plasma samples collected from different animals after dosing individual drugs can be combined for assay. This method may be useful if significant drug–drug interactions are expected among compounds when they are dosed in one vehicle, despite the fact that its use requires more animals and resources for the study (Kuo *et al.*, 1998).

3.3. AUC ESTIMATION FROM ONE POOLED SAMPLE

For the extent of drug exposure after oral administration, $AUC_{0-t_{last}}$ is often considered more relevant than C_{max}. The conventional technique for estimating $AUC_{0-t_{last}}$ is to measure plasma concentrations at each time point and calculate AUC with those individual concentrations at different time points. Another interesting approach to the higher throughput of oral exposure is to prepare one *pooled* sample by combining different aliquots of the individual samples at all time points and to calculate $AUC_{0-t_{last}}$ by multiplying its concentration (C_{pool}) with a sampling time interval between time zero and t_{last} (Hop *et al.*, 1998).

$$AUC_{0-t_{last}} = C_{pool} \cdot t_{last} \tag{3.1}$$

In other words, plasma samples from the same animal at different time points are pooled in a weighted ratio that reflects the size of their respective time interval. The advantages of this method is that far fewer samples have to be analyzed and pooled AUC values can still be calculated for each animal. The disadvantage is that the entire concentration *vs.* time profile for the compound cannot be obtained. Mathematical manipulation to calculate proper fractions (f_0, f_1, $f_2, \ldots, f_{last}$) of the total volume of the pooled sample needed for aliquoting individual samples at each time point (0, t_1, $t_2, \ldots, t_{last}$) is based on the linear trapezoidal rule for estimating AUC, as seen in Table 3.2 (Hamilton *et al.*, 1981).

For comparison of AUC values between the sample pooling method and the

Table 3.2. Fractions of Total Volume of a Pooled Sample Required for Aliquoting Individual Samples at Each Time Point for Estimation of $AUC_{0-t_{last}}$

	Fraction of total volume of a pooled sample required for each sampling time point							
Time[a]	0	t_1	t_2	t_3	t_4	t_5	t_6	t_7
t_1	$\frac{1}{2}$	$\frac{1}{2}$	—	—	—	—	—	—
t_2	$t_1/2t_2$	$t_2/2t_2$	$(t_2\text{-}t_1)/2t_2$	—	—	—	—	—
t_3	$t_1/2t_3$	$t_2/2t_3$	$(t_3\text{-}t_1)/2t_3$	$(t_3\text{-}t_2)/2t_3$	—	—	—	—
t_4	$t_1/2t_4$	$t_2/2t_4$	$(t_3\text{-}t_1)/2t_4$	$(t_4\text{-}t_2)/2t_4$	$(t_4\text{-}t_3)/2t_4$	—	—	—
t_5	$t_1/2t_5$	$t_2/2t_5$	$(t_3\text{-}t_1)/2t_5$	$(t_4\text{-}t_2)/2t_5$	$(t_5\text{-}t_3)/2t_5$	$(t_5\text{-}t_4)/2t_5$	—	—
t_6	$t_1/2t_6$	$t_2/2t_6$	$(t_3\text{-}t_1)/2t_6$	$(t_4\text{-}t_2)/2t_6$	$(t_5\text{-}t_3)/2t_6$	$(t_6\text{-}t_4)/2t_6$	$(t_6\text{-}t_5)/2t_6$	—
t_7	$t_1/2t_7$	$t_2/2t_7$	$(t_3\text{-}t_1)/2t_7$	$(t_4\text{-}t_2)/2t_7$	$(t_5\text{-}t_3)/2t_7$	$(t_6\text{-}t_4)/2t_7$	$(t_7\text{-}t_5)/2t_7$	$(t_7\text{-}t_6)/2t_7$

[a] t_1, t_2, t_3, t_4, t_5, t_6, t_7 are sample time points.

conventional method with individual concentrations, let us assume that plasma samples at 0, 0.25, 0.5, 1, 2, 4, 7, and 24 hr postdose after oral administration of a drug have concentrations of 0, 1, 3, 10, 7, 5, 4, and 1 μg/ml, respectively. If the volume of a single pooled sample is set for 480 μl, the volumes of plasma samples at different time points to prepare the pooled sample are 2.5, 5, 7.5, 15, 30, 50, 200, and 170 μl at 0, 0.25, 0.5, 1, 2, 4, 7, and 24 hr, respectively (Table 3.3). Thus, in theory, C_{pool} is equal to 3.338 μg/ml, as shown below, based on the fractions (f_t) of the total volume of a pooled sample:

$$\begin{aligned} C_{pool} &= C_0 \cdot f_0 + C_1 \cdot f_1 + C_2 \cdot f_2 + C_3 \cdot f_3 + C_4 \cdot f_4 + C_5 \cdot f_5 + C_6 \cdot f_6 + C_7 \cdot f_7 \\ &= 0 \cdot 0.005 + 1 \cdot 0.01 + 3 \cdot 0.015 + 10 \cdot 0.03 + 7 \cdot 0.063 + 5 \cdot 0.104 \\ &\quad + 4 \cdot 0.417 + 1 \cdot 0.354 \\ &= 3.338\ \mu g/ml \end{aligned}$$

where $C_0, C_1, \ldots, C_7$ are the concentrations of the drug at 0, 0.25, ..., 24 hr postdose, respectively. In other words, the measured C_{pool} value should be close to 3.338 (g/ml, and thus $AUC_{0-t_{last}}$ based on C_{pool} is approximately 80.112 $\mu g \cdot hr/ml$:

$$AUC_{0-t_{last}} = C_{pool} \cdot t_{last} \approx 3.338 \cdot (24\text{–}0) = 80.112\ \mu g \cdot hr/ml$$

AUC estimated using the conventional method with concentrations of individual

Table 3.3. Fractions of the Volume of a Pooled Sample Required from Individual Samples at Various Times Postdose

	Fractions of the volume of a pooled sample for each sampling time point (hr)							
	0	0.25	0.5	1	2	4	7	24
24	0.25/48	0.5/48	0.75/48	1.5/48	3/48	5/48	20/48	17/48
Fraction (f_t)	f_0	f_1	f_2	f_3	f_4	f_5	f_6	f_7

samples at different time points based on the linear trapezoidal rule is 80.375 $\mu g \cdot hr/ml$, which is in good agreement with the value using C_{pool}. This so-called "pooling method" can be combined with N-in-1 dosing to further reduce the number of samples required for assay.

3.4. CONTINUOUS SAMPLING METHOD

Instead of intermittent sampling followed by subsequent sample pipetting for a pooled sample, continuous blood withdrawal from animals at a suitable flow rate has been explored to obtain a single sample for each animal (Humphreys *et al.*, 1998). The major advantage of this method over the sample pooling method is a reduction in the time required for sample collection and processing (pipetting). $AUC_{0-t_{last}}$ can be calculated by multiplying the concentration of a single sample obtained from continuous withdrawal and the sample withdrawal period:

$$AUC_{0-t_{last}} = C_{ss_{cw}} \times P_w \tag{3.2}$$

where $C_{ss_{cw}}$ is a drug concentration of a single sample obtained from the continuous withdrawal and P_w is a withdrawal period.

REFERENCES

Beaudry F. *et al.*, *In vivo* pharmacokinetic screening in cassette dosing experiments: the use of on-line Prospekt® liquid chromatography/atmospheric pressure chemical ionization tandem mass spectrometry technology in drug discovery, *Rapid Commun. Mass Spectrom.* **12**: 1216–1222, 1998.

Frick L. W. *et al.*, Cassette dosing: rapid *in vivo* assessment of pharmacokinetics, *Pharm. Technol. Today* **1**: 12–18, 1998.

Frye R. F. *et al.*, Validation of the five-drug "Pittsburgh cocktail" approach for assessment of selective regulation of drug-metabolizing enzymes, *Clin. Pharmacol. Ther.* **62**: 365–376, 1997.

Hamilton R. A. *et al.*, Determination of mean valproic acid serum level by assay of a single pooled sample, *Clin. Pharmacol. Ther.* **29**: 408–413, 1981.

Hop C. E. C. A. *et al.*, Plasma-pooling methods to increase throughput for *in vivo* pharmacokinetic screening, *J. Pharm. Sci.* **87**: 901–903, 1998.

Humphreys W. G. *et al.*, Continuous blood withdrawal as a rapid screening method for determining clearance and oral bioavailability in rats, *Pharm. Res.* **15**: 1257–1261, 1998.

Kubinyi H., Strategies and recent technologies in drug discovery, *Pharmazie* **50**: 647-662, 1995.

Kuo B-S. *et al.*, Sample pooling to expedite bioanalysis and pharmacokinetic research, *J. Pharm. Biomed. Anal.* **16**: 837–846, 1998.

Olah T. V. *et al.*, The simultaneous determination of mixtures of drug candidates by liquid chromatography/atmospheric pressure chemical ionization mass spectrometry as an *in vivo* drug screening procedure, *Rapid Commun. Mass Spectrom.* **11**: 17–23, 1997.

Tarbit M. H. and Berman J., High-throughput approaches for evaluating absorption, distribution, metabolism and excretion properties of lead compounds, *Curr. Opin. Chem. Biol.* **2**: 411–416, 1998.

Taylor E. W. *et al.*, Intestinal absorption screening of mixtures from combinatorial libraries in the Caco-2 model, *Pharm. Res.*, **14**: 572–577, 1997.

4

Absorption

Oral dosing is the most common route for the administration of drugs and most of the drugs given orally are generally designed to show systemic pharmacological efficacy rather than local effects in the gastrointestinal (GI) tract. To achieve desirable systemic exposure levels, i.e., plasma or blood drug concentrations, many pharmacokinetic studies are concerned with the bioavailability of drugs after oral administration. As the drug passes down the GI tract, part of the dose may not be available for absorption owing to poor aqueous solubility, limited membrane permeability, and/or chemical or biological degradation, i.e., limited absorption. Drug molecules absorbed into the intestinal membranes can then be further subject to intestinal and/or hepatic elimination before reaching the systemic circulation, i.e., first-pass elimination. A thorough understanding of the quantitative contributions of these two processes during absorption is important for enhancing the oral bioavailability of drugs. In this chapter, various physiological factors and physicochemical properties of drug molecules that are critical for oral absorption, factors affecting the first-pass elimination, and various experimental approaches for assessing oral bioavailability are discussed.

4.1. RATE-LIMITING STEPS IN ORAL DRUG ABSORPTION

For drugs orally dosed in solid dosage forms such as tablets or capsules, there are two distinctive processes during absorption: *dissolution* of solid drug particles to drug molecules in the GI fluid and *permeation* of the drug molecules across intestinal membranes (Fig. 4.1). Depending on the relative magnitude of the rates of these two processes, one of them can be rate-limiting in overall drug absorption.

4.1.1. Dissolution Rate-Limited Absorption

As a prerequisite for oral absorption, the drug must be present in aqueous solution except in the case of pinocytosis or for the lymphatic absorption pathways. When a drug is administered in solid dosage formulations such as tablets, *disintegration* of the dosage form into small solid particles in a suspension should occur prior to the dissolution of the particles. In general, disintegration occurs much faster than

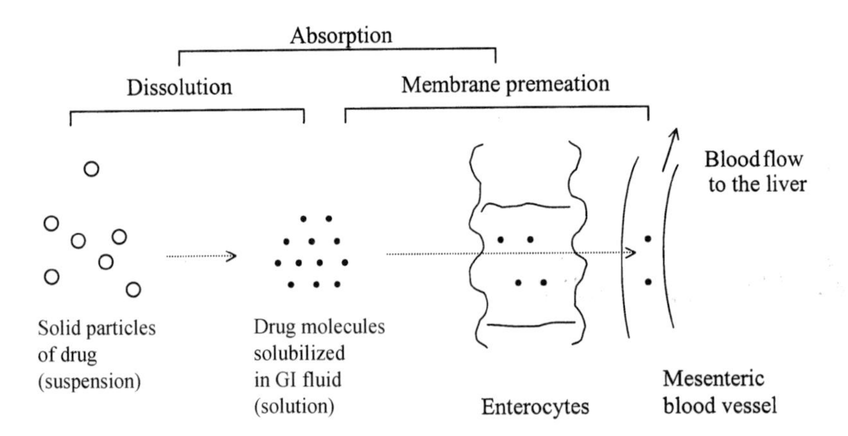

Figure 4.1. Potential rate-limiting steps in drug absorption processes after oral administration of solid dosage forms such as tablets or capsules.

dissolution. For most drugs with high lipophilicity, the rate of absorption can be governed primarily by the dissolution rate of the drug particles. The surface area of the particles, the aqueous solubility of drug, the pH of the GI fluid, and the extent of mixing in the GI tract are the important factors affecting the dissolution rate of solid drug particles.

• Dissolution. The dissolution rate of drug molecules from solid drug particles into a surrounding aqueous medium is a function of the aqueous solubility of the drug, the surface area of the particles, and the dissolution rate constant, and is expressed by the Noyes–Whitney equation:

$$dC/dt = k \cdot S \cdot (C_s - C_t) \tag{4.1}$$

where dC/dt is the rate of dissolution of a solid drug particle; k is the dissolution rate constant; S is the surface area of the dissolving solid drug particle; C_s is the saturation concentration of the drug in the diffusion layer, which can be close to the maximum solubility of drug, as the diffusion layer is considered saturated with drug; and C_t is the concentration of the drug in the surrounding dissolution medium at time t (Fig. 4.2). As a solid drug particle undergoes dissolution, the drug molecules on its surface are the first to diffuse into the solution adjacent to the particle, generating a saturated layer of drug solution that envelops the particle surface. From this drug-saturated solution layer, which is called a diffusion layer, drug molecules dissolve into the surrounding medium, and the layer is continuously replenished with newly diffused drug molecules from the surface of the particle.

The dissolution rate of a drug can be increased by:

1. Increasing the surface area of the particles by reducing particle size (grinding, jet-milling, etc.).
2. Increasing the aqueous solubility of the drug (elevating temperature, changing pH in the case of ionizable drugs, etc.).

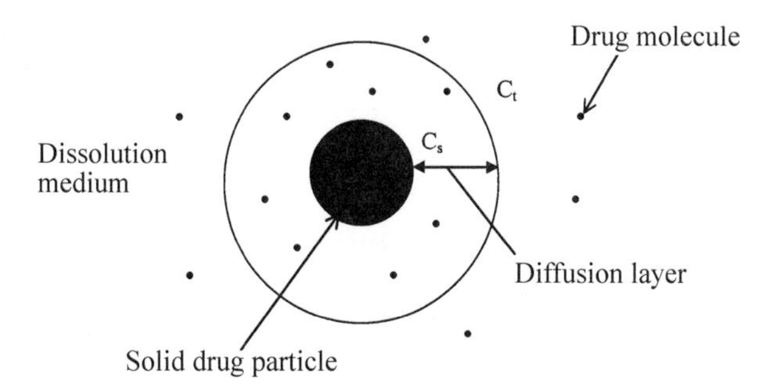

Figure 4.2. Schematic description of drug dissolution from solid drug particles in suspension into surrounding dissolution medium. C_s is the saturation concentration of a drug in the diffusion layer and C_t is the concentration in the surrounding dissolution medium at time t.

3. Increasing the dissolution rate constant (agitating the medium, increasing temperature, decreasing viscosity of the medium, etc.).

4.1.2. Membrane Permeation Rate-Limited Absorption

If the dissolution process is very rapid, the absorption rate of a drug could be dependent primarily on its ability to transport across the intestinal membrane. For highly water-soluble compounds, the membrane permeation can become critical in overall absorption owing to their limited ability to partition into the lipid bilayers of the enterocyte membranes (Fig. 4.3).

4.1.2.1. Permeation

The permeation rate of a drug via the intestinal membrane after oral administration can be expressed as a function of the intestinal membrane permeability (P_{int}), the effective surface area (S_{int}) of intestinal membrane available for permeation of drug molecules, and the concentration of the drug (C_{int}) in the GI fluid:

$$\text{Permeation rate} = P_{int} \cdot S_{int} \cdot C_{int} \tag{4.2}$$

A fraction of oral dose (F_{abs}) absorbed into the portal vein after dosing in an aqueous solution, i.e., no dissolution process, can be estimated using Eq. (4.3) assuming the absence of gut microflora and intestinal metabolism. The utility of Eq. (4.3) to *in vivo* oral drug absorption is rather limited, however, because in most cases, estimates of P_{int} and S_{int} obtained from *in vitro* or *in situ* experiments might not be relevant for *in vivo* conditions:

$$F_{abs} = 1 - e^{-P_{int} S_{int}/Q_{int}} \tag{4.3}$$

4.1.2.2. Permeability

The permeability of a drug reflects how readily the drug molecules pass through membranes. Three major factors governing the permeability of compounds are

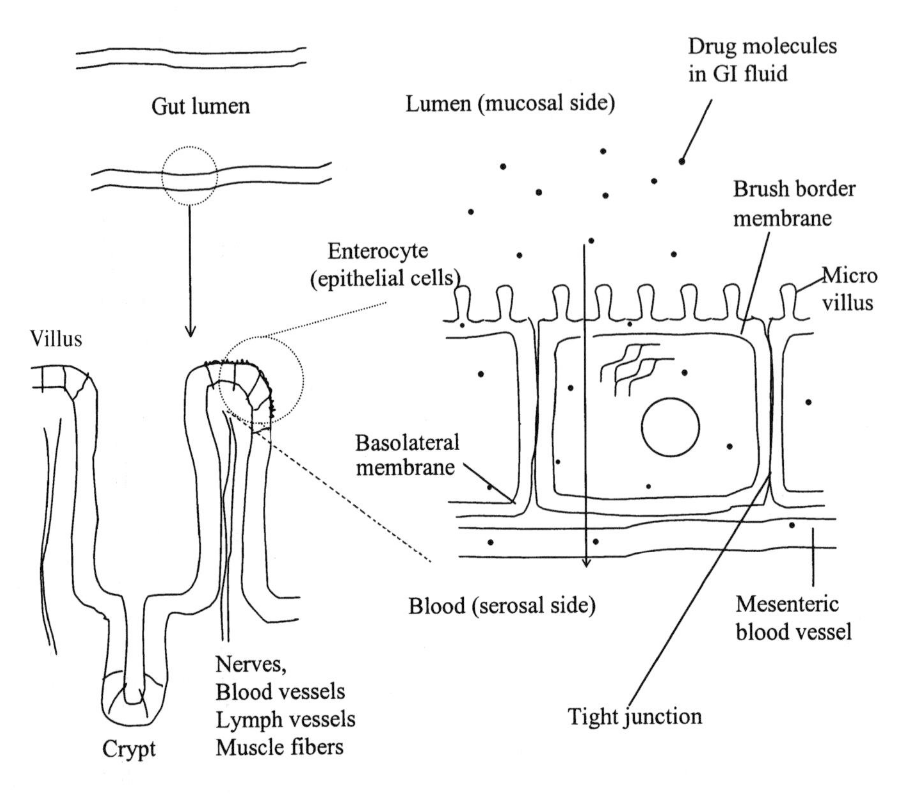

Figure 4.3. Intestinal villus and epithelial cells. The apical surface of the small-intestinal epithelium is covered with villi (0.5–1.0 mm long), which have a single layer of columnar epithelial cells along the surface connected by tight junctions forming barriers between the lumen and serosal capillaries. The villous cells have microvilli (100 nm in diameter), which form brush border membranes, increasing the cell surface area. The average life span of mucosal cells is 2–5 days. The arrow indicates an absorptive direction of drug from the luminal side of the gut to the basolateral side.

lipophilicity, molecular size, and charge. Permeability has a unit of velocity, i.e., distance/time. The product of permeability and surface area available to permeation of a drug can be viewed as distributional clearance with the units of flow rate, i.e., volume/time.

4.2. FACTORS AFFECTING ORAL ABSORPTION

The following is a summary of physiological and physicochemical factors of drugs that affect the rate and extent of oral absorption.

4.2.1. Physiological Factors

(i) Gastric motility and residence time. A small surface area of the stomach may be less favorable for drug absorption compared to the small intestine, the major

absorption site in the GI tract. However, the gastric residence time of a drug can be critical for immediate-release solid dosage forms such as regular tablets and capsules. An important physiological factor dictating the gastric residence time of a drug is gastric motility (Dressman, 1986; Kararli, 1995; Walter-Sack, 1992), which has two distinct modes depending on the presence of food (fasted *vs.* fed states). Gastric motility during the fasted state has three different phases, phases I, II, and III, which repeat periodically (e.g., every 2 hr in humans). Phase I, a quiescent phase, accounts for about half of the fasting cycle period, during which there is little contractile activity. In phase II, irregular contractions start to occur and gradually increase in amplitude and frequency. These progress into a maximal amplitude and frequency of contraction, which is referred to as phase III. During phase III, the strong contraction can expel the entire gastric content into the small intestine (this process is the so-called *housekeeper wave*). Phase III activity in the stomach is usually associated with the initiation of a *migrating motility complex* (MMC) in the duodenum, which then proceeds to migrate through the small intestine toward the ileum. At the end of phase III activity, stomach motility reverts back to the quiescent phase. When food enters the stomach, i.e., in the fed state, the contractions of the stomach return to a level lower than that of phase III. These regular tonic contractions of the stomach propel food toward the antrum while grinding and mixing it with gastric secretions. Implications of gastric residence time on drug absorption should be considered in conjunction with dosage forms. For instance, the gastric residence time of drugs given in liquid dosage forms will depend on the liquid emptying time and the total volume of liquid administered. Since large objects empty only during phase III of the fasted state, the gastric residence time of nondisintegrating solid dosage forms will depend on the frequency of phase III activity, if drug is given in the fasted state, and the time-restoring phase III activity if given in the fed state.

(ii) pH in the gastrointestinal tract. The pH ranges in the human stomach and intestines from the duodenum to the colon are about 1 to 3 and 5 to 8, respectively. For acidic or basic drugs, the un-ionized forms of drugs, if sufficiently lipophilic, are better absorbed than their ionized counterparts. How large a fraction of a drug exists in the un-ionized form in the GI fluid depends on both the drug dissociation constant (K_a) and the pH of the GI fluid. For instance, acidic drugs with pK_a between 4 and 8 exist predominantly as un-ionized forms at the low pH of gastric fluid, may be absorbed in part from the stomach, and can be partially un-ionized in the intestine. Very weak acids ($pK_a > 8$) are essentially un-ionized throughout the entire GI tract. On the other hand, most bases are poorly absorbed in the stomach, as they are largely ionized there at low pH. Weak bases ($pK_a < 5$) are essentially un-ionized throughout the intestine. In some cases, the drug itself can induce changes in the pH of the GI fluid (Dressman *et al.*, 1993).

(iii) Intestinal surface area and transit time. The entire GI tract is lined with a continuous sheet of epithelial cells. The stomach lacks the villus structure of other areas of the GI tract. Owing to the large surface area it offers for absorption by numerous microvilli and relatively long transit time for drug molecules to pass through, the small intestine is considered to be the major site of oral absorption for

most drugs. The colon has a longer transit time than the small intestine; however, the more viscous contents of the colon (low permeability) and the lack of villi (smaller surface area) tend to offset the effects of the longer transit time for drug absorption. In addition, cellular morphology and functions of epithelial cells are quite variable in different segments of the GI tract. Some sites in the GI tract are primarily involved in selective absorption of various nutrients, but not others. Some areas are better equipped for secretion than absorption, while others perform both functions (Hunt and Groff, 1990).

(iv) Food. Food intake stimulates GI secretions including hormones and bile salts, which lower gastric pH, delay stomach emptying, and increase GI transit time. Fluid volume and the quantity of dietary fat in the meal appear to be the primary food-related factors affecting drug absorption. For instance, fluids ingested with the meal can increase the available gastric volume up to as much as 1.5 l in humans. The increased secretion of bile salts induced by a fatty meal can enhance the stability of the emulsion phase within the gut lumen, which will increase absorption of lipophilic drugs. The absorption profiles of hydrophilic drugs, however, appear not to be significantly altered when these drugs are given with a fatty diet. Limited studies have suggested that dietary protein could induce an increase in splanchnic and hepatic blood flow, whereas dietary fat does not alter hepatic blood flow (Baijal and Fitzpatrick, 1996; Winne, 1980; Zhi *et al.*, 1995).

(v) Intestinal microflora. There are almost 400 different microorganisms in the GI tract. Some of the microflora residing in the GI tract can metabolize a variety of drugs, which can reduce the amount available for absorption. Hydrolysis of esters and amides, reduction of double bonds, and nitro and diazo groups, dehydroxylation, dealkylation, deamination, acetylation, and esterification are some of metabolic reactions mediated by gut microflora.

(vi) Other factors.

1. Wetting of drug particles by the gastric juice or the intestinal fluid.
2. Blood circulation to the site of absorption.
3. Active transporters.
4. Disease state.

4.2.2. Physicochemical Factors of Drugs

(i) Hydrophilicity and lipophilicity. A balance between hydrophilicity and lipophilicity of a drug is important in oral absorption. In general, log $D_{7.4}$ values of compounds between -0.5 and 2 are considered to be optimal for oral absorption.

(ii) Ionizability and charge. The un-ionized form of a drug is better absorbed than its ionized counterpart, and a drug's ionizability is influenced by both the pK_a of the drug and the pH of the GI fluids.

(iii) Chemical stability. Hydrolysis can occur for ester or amide moieties of a drug at acidic or alkaline pH in the GI tract.

(iv) Particle size in suspension. Reduction of particle sizes of solid drug particles usually enhances the dissolution rate of the particles in suspension by increasing their surface area.

(v) Polymorphism of crystalline forms. Many different crystalline (polymorphic) forms of a drug can exist, each of which has a different energy level and different physicochemical properties, including, e.g., melting point, solubility, density, and refractive index. Owing to different kinetic solubilities, dissolution rates can also vary among polymorphic particles. When oral absorption of a drug is limited by dissolution rate, the polymorphic form of the solid drug particles can be important in determining oral bioavailability. In general, an amorphous (metastable) form of a drug has a higher kinetic solubility as compared with crystalline forms. An amorphous form is unstable and converts to the more stable crystalline form with low kinetic solubility during manufacturing and storage, which rather limits its commercial potential.

(vi) Molecular size. Absorption pathways can be affected by the size of drug molecules. Paracellular transport via tight junctions between enterocytes can be an important absorption pathway for small highly water-soluble molecules with molecular weights below 200. As the molecular weight increases, transcellular transport (passive diffusion or active transport) becomes more important.

(vii) Complexation. The rate and extent of oral absorption of many drugs such as tetracycline derivatives and cefdinir, an oral cephalosporin, can be significantly impaired if they form water-insoluble complexes with polyvalent metal ions such as Ca^{+2}, Mg^{+2}, Fe^{+3}, or Al^{+3}, which are often present in food (Hörter and Dressman, 1997).

4.2.3. Effects of the pH and pK_a of a Drug on Absorption (pH-Partition Theory)

Owing to the lipoidal properties of the membrane, passively diffused drugs must undergo partitioning from the aqueous GI fluids into the membrane and eventually into the blood. This concept of absorption by partitioning processes of a drug between water and lipid at different pH is known as the "pH-partition theory" of drug absorption, and it addresses relationships among three different factors affecting partition processes of oral absorption of a compound: the dissociation constant and lipophilicity of the compound and the pH at the absorption site. For an ionizable compound, its un-ionized form is considered to be better partitioning into lipophilic membranes than the ionized counterpart. The ionizability of a compound in aqueous solution is a function of both the dissociation constant (K_a) of the compound and the pH of the surrounding solution. The dissociation constant is often expressed as a pK_a (the negative logarithm of the acidic dissociation constant) for both acidic and basic compounds. The relationship between the pH and pK_a of a compound is described by the Henderson–Hasselbalch equation as shown in Eqs. (4.4) and (4.5).

For an acidic compound [HA]:

Concentration of un-ionized compound (acid)

(4.4) $$pK_a = pH + \log([HA]/[A^-])$$

Concentration of ionized compound (salt)

For a basic compound [B]:

Ionized (salt)

(4.5) $$pK_a = pH + \log([BH^+]/[B])$$

Un-ionized (base)

A ratio of the concentrations between the un-ionized ([HA] and [B]) and ionized ($[A^-]$ and $[BH^+]$) forms of acidic and basic compounds in aqueous solution at different pH can be obtained based on these equations. For instance, a ratio between [HA] and $[A^-]$ of an acidic compound with pK_a of 4 in aqueous solution at pH of 7 would be 0.001, which means that the concentration of A^- is 1000-fold greater than that of HA in aqueous solution at pH 7:

pKa (↓) *pH* (↓) *Ratio between un-ionized and ionized forms* (↓)

$$4 = 7 + \log([HA]/[A^-])$$

Let us assume that there are a carboxylic acid (R–COOH) with pK_a of 4 and a primary amine ($R–CH_2–NH_2$) with pK_a of 9. The ratios between the ionized and the un-ionized concentrations of these compounds in aqueous solution at different pH are summarized in Table 4.1, according to the Henderson–Hasselbalch equation. In the stomach (pH $\approx$ 2), a ratio of the ionized ($R–COO^-$) to the un-ionized (R–COOH) forms of the carboxylic acid would be 1:99, i.e., most of the acid is un-ionized, whereas the primary amine would exist mainly in ionized form ($R–CH_2–NH_3^+$) with a negligible amount of the un-ionized ($R–CH_2$-NH_2) form. On the other hand, in the intestine (pH $\approx$ 6), carboxylic acid exists primarily as the ionized form (ionized:un-ionized = 99:1) and so does the amine (ionized:un-ionized = 99.9:0.1). Therefore, owing to its lower pH, absorption of acidic compounds in the stomach is favored over that in the intestine, although intestinal absorption of compounds, including acids, is quantitatively more important than stomach absorption because of the longer transit time and larger surface area. The pK_a values of several common structural moieties of organic compounds are shown in Table 4.2.

Table 4.1. Ratios of Ionized and Un-Ionized Forms of an Acid (e.g., R–COOH) with pK_a of 4 or a Base (e.g., $R-CH_2-NH_2$) with pK_a of 9 at Different pH

	Ionized: Un-ionized ≈	
pH	Acid ($pK_a = 4$)	Base ($pK_a = 9$)
13	↑	
12	Decreasing	0.1:99.9
11	un-ionized form	1:99
10		10:90
9[a]		50:50
8		90:10
7	99.9:0.1	99:1
6	99:1	99.9:0.1
5	90:10	
4[a]	50:50	
3	10:90	Decreasing
2	1:99	un-ionized form
1	0.1:99.9	↓

[a]Note that at the pH value equal to the pK_a of the compounds, the ratios between un-ionized and ionized forms of the compounds become unity. At pH higher than pK_a, there are more ionized molecules than un-ionized molecules for the acid and vice versa for the base.

Table 4.2. The pK_a Values of Common Structural Moieties in Organic Compounds

Chemical moiety	Structure	Acid or base	pK_a[a]
Carboxylic acid	R–COOH	Acid	
	(R: aliphatic)	4–5	
	(R: aromatic)	9–10	
Phenol	$R-C_6H_4OH$	Acid	10
	(R: aliphatic)		
Sulfonic acid	$R-SO_3H$	Acid	<1
Sulfonamide	$C_6H_5-SO_2-NH-R$	Acid	
	R = H	10	
	R = aromatic or heterocycles		5–7
Hydroxamic acid	R–CO–NHOH	Acid	9
Amine	$R-NH_2$, R_2NH, R_3N	Base	
	(R: aliphatic)		9–10
	(R: aromatic)		4–5
Pyridine	C_5H_5N	Base	5.2
N-oxide	R_3N-O	Base	
	(R: aliphatic)		4.6
	C_5H_5N-O (pyridine N-oxide)		0.8
Quaternary Ammonium salt	R_4-N^+	Polar cation	Fully ionized Over pH 1 to 13
Alcohol	RCH_2-OH	Neutral	
Ether	R–O–R′	Neutral	
Ketone	R–CO–R′	Neutral	
Ester	R–COO–R′	Neutral	
Amide	R–CONH–R′	Neutral	

[a]The lower the pK_a of an acid, the more acidic the compound, whereas the higher the pK_a of a base, the more basic the compound.

4.2.4. Partition and Distribution Coefficients

Important pharmacokinetic properties of a compound such as metabolism, membrane transport (distribution), and passive absorption can be influenced by several of its physicochemical properties, including lipophilicity [partition (P) and distribution coefficients (D)], molecular weight and surface area (Krarup, 1998; Palm *et al.*, 1996), the ionization state, and the hydrogen-binding capacity (Lipinski *et al.*, 1997). In particular the lipophilic characteristic of a drug has been recognized as one of the important factors governing the extent of protein binding, metabolism, and absorption (Lee *et al.*, 1997; Testa *et al.*, 1997).

The partition coefficient (P, or log P as generally described) of a compound is defined as the ratio of the concentrations of the un-ionized compound in organic and aqueous phases at equilibrium. The partition coefficient can be viewed as an indicator of intrinsic lipophilicity in the absence of ionization or dissociation of the compound. Octanol is the most widely used organic phase for log P measurement of organic compounds.

Distribution coefficient (D, or log D as generally described) is defined as the overall ratio of organic and aqueous phases of a compound, ionized and un-ionized, at equilibrium. When the compound is partially ionized in the aqueous phase, not only the partition equilibrium of un-ionized compound between the aqueous and the organic phases but also the dissociation equilibrium between un-ionized and ionized compound within the aqueous phase will be established. Only the un-ionized form is considered to distribute between the aqueous and the organic phases. These processes are elucidated in the following scheme:

$$[\text{Ionized drug}]_{\text{aq phase}} \underset{}{\overset{\text{Dissociation equilibrium}}{\rightleftharpoons}} [\text{Un-ionized}]_{\text{aq phase}} \overset{\text{Partition equilibrium}}{\rightleftharpoons} [\text{Un-ionized}]_{\text{org phase}}$$

Let us consider the partitioning of an organic acid (HA) between organic and aqueous phases at a certain pH. The equilibrium processes of ionized (A^-) and un-ionized (HA) forms between aqueous and organic phases can be described as follows:

$$[A^-]_{\text{aq phase}} \rightleftharpoons [HA]_{\text{aq phase}} \rightleftharpoons [HA]_{\text{org phase}}$$

The partition coefficient (P) and the distribution coefficient (D) of HA can be expressed as

$$P = \frac{[HA]_{\text{org phase}}}{[HA]_{\text{aq phase}}} \tag{4.6}$$

$$D = \frac{[HA]_{\text{org phase}}}{[HA]_{\text{aq phase}} + [A^-]_{\text{aq phase}}} \tag{4.7}$$

Since the partition coefficient refers only to equilibrium of un-ionized compound between the phases, it is *pH-independent*, whereas the distribution coefficient is

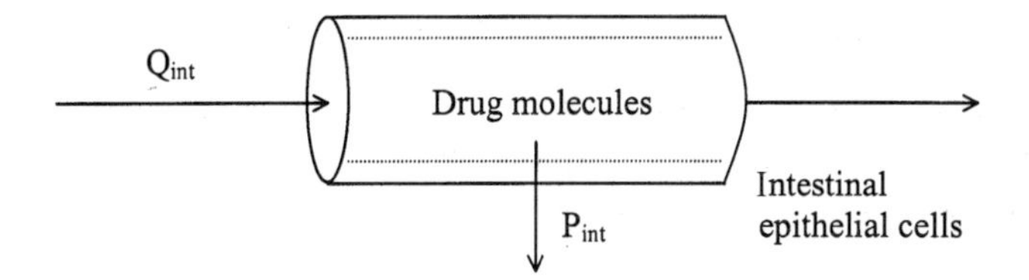

Figure 4.4. Schematic description of permeation of drug molecules via intestinal epithelial cells. Q_{int} and P_{int} represent, respectively, intestinal fluid flow rate (i.e., intestinal fluid volume divided by intestinal transit time) and apparent permeability of drug across intestinal epithelial cells.

pH-dependent because the degree of ionization in the aqueous phase is affected by both the pH and the pK_a of the compound. A rough estimate of log D of a compound at any given pH can be obtained by subtracting one unit from its log P for every unit of pH above or below its pK_a as acid or base, respectively [Eq. (4.8)]. Note that log P is always greater than log D. An estimate of log P of a compound can be obtained through a number of mathematical methods, such as the fragmental constant method developed by Hansch and Leo (1979):

$$\log D \approx \log P - \Delta|pK_a - pH| \tag{4.8}$$

(i) clog P and Mlog P. The clog P and Mlog P of a compound are the log P estimates of the compound calculated with the methods developed by the Medicinal Chemistry Department, Pomona College, CA, and by Moriguchi *et al.* (1992), respectively.

(ii) log D and oral absorption. In general, log $D_{7.4}$ values between −0.5 and 2 are considered to be optimal for oral absorption of compounds. Compounds with log $D_{7.4}$ values smaller than −0.5 or greater than 2 tend to have poor oral absorption owing to limited membrane permeation (low lipophilicity) or poor aqueous solubility (low hydrophilicity), respectively (Smith *et al.*, 1990). The log $D_{7.4}$ is the octanol/aqueous buffer distribution coefficient of compound at pH 7.4 uncorrected for the degree of ionization.

4.3. BIOAVAILABILITY

Bioavailability is considered to be one of the most important pharmacokinetic parameters of any drug developed for extravascular administration.

4.3.1. Definition

Oral bioavailability is a relative term used to describe the rate and extent of absorption after oral administration of a drug compared to that after its administration via a reference route, usually intravenous bolus injection.

UNIT: Bioavailability has no unit. Often, it is expressed as a percent.

4.3.2. Factors Affecting Bioavailability and the First-Pass Effect

Oral bioavailability of a drug is affected by both the extent of its *absorption* into enterocytes from the gut lumen and the extent of its *presystemic elimination* by the intestine and the liver before it reaches the systemic circulation.

4.3.2.1. Incomplete Absorption

Bioavailability can be less than unity, owing to incomplete absorption of a drug from the GI tract. Various physiological and physiochemical factors of a drug that affect intestinal absorption after oral administration have already been discussed.

4.3.2.2. Presystemic Elimination (= First-Pass Effect)

After being absorbed into enterocytes from the gut lumen, drug molecules pass into the portal circulation, and then through the liver and the lung prior to entering the systemic circulation, where blood samples are normally taken. During these absorption processes, a significant portion of a drug can be eliminated by metabolism within the enterocytes, metabolism and/or biliary excretion in the liver, and metabolism in the lung, for the first time, before reaching the systemic circulation. This process is known as the "first-pass or presystemic effect (elimination)." Since elimination by the lung is generally thought to be minimal, the pulmonary first-pass effect after oral administration of a drug is considered to be negligible. Drugs given intravenously also have to first pass through the lung before reaching the systemic circulation. Thus, the first-pass effect in the lung after oral administration is not taken into account for an estimate of bioavailability when drug concentrations after *intravenous* injection are used for exposure comparisons. A schematic illustration of blood circulation and the anatomic arrangement of various organs is shown in Fig. 4.5.

The extent of the presystemic intestinal or hepatic elimination of drug can be affected by:

1. Site of absorption: If the site of absorption of a drug is different from the site of metabolism in the intestine, first-pass intestinal metabolism may not be significant.
2. Intracellular residence time of drug molecules in enterocytes: The longer the drug molecules stay in the enterocytes prior to entering the mesenteric vein, the more extensive the metabolism of the drug in enterocytes will be.
3. Diffusional barrier between the splanchnic bed and the enterocytes: The lower the diffusibility of the drug from the enterocytes to the mesenteric vein, the longer the residence time of the drug within the enterocytes.
4. Mucosal and portal blood flow: Blood in the splanchnic bed can act as a sink to carry drug molecules away from the enterocytes once they are absorbed, which reduces their intracellular residence time in the enterocytes. Factors causing changes in portal blood flow rate can also affect the extent of presystemic hepatic elimination (see Chapter 6).

istration at the same sampling site, usually, venous blood (Cassidy and Houston, 1980; Kwan, 1998; Pang, 1986).

When oral bioavailability of a drug is determined compared to exposure after *intravenous* injection, F_l cancels out because just like the oral dose, the entire intravenous dose also has to first pass through the lung before reaching the systemic circulation. Oral bioavailability of a drug referred to exposure levels after *intravenous* administration (F) is thus only a function of F_a, F_g, and F_h:

$$F = F_a \cdot F_g \cdot F_h \tag{4.12}$$

The methods for estimating F_a, F_g, and F_h separately are discussed below.

(a) Fraction of Dose Absorbed into the Enterocytes from the Intestinal Lumen after Oral Administration (F_a). When oral bioavailability of a drug is poor, information on how much drug is actually absorbed into the enterocytes or into the portal vein after oral administration becomes critical in order to distinguish between the extent of drug absorption and first-pass effects. Study with a radiolabeled drug makes it possible to estimate the actual amount of drug absorbed. There are two different approaches to discern these processes.

(i) Mass balance. Urine, bile, and feces from animals dosed with a radiolabeled drug can be collected over an extended period of time. Total radioactivity of the drug and its metabolites in the urine and bile samples reflects the actual amount of the drug absorbed into the enterocytes from the gut. When the metabolites produced within the enterocytes are released back into the gut lumen and/or the drug is subject to enterohepatic circulation, the total radioactivity found in urine and bile from bile-duct-cannulated animals may be different from the actual amount of drug absorbed in normal animals. Bile-duct-cannulation surgery can also alter animal physiology (liver function, blood protein contents, etc) and the absorption profile of a drug.

(ii) AUC or radioactivity comparison in urine. A ratio of dose-normalized AUC values of total radioactivity, or total radioactivity found in urine between oral and intravenous administration of a radiolabeled compound, approximates to the fraction of the dose absorbed after oral dosing. This method requires intravenous data.

(b) Fraction of the Amount of a Drug Absorbed into the Enterocytes that Escapes Presystemic Intestinal Elimination (F_g). It is difficult to estimate F_g experimentally; however, a product of F_a and F_g ($F_a \cdot F_g$, the fraction of a dose absorbed into the portal blood after oral administration) can be estimated. If F_a is measured experimentally, e.g., from a mass balance study with a radiolabeled compound, F_g can be calculated by dividing $F_a \cdot F_g$ by F_a.

(c) Fraction of the Dose Absorbed into the Portal Blood after Oral Administration ($F_a \cdot F_g$). The following four different approaches can be used to estimate $F_a \cdot F_g$. Advantages and disadvantages of each method are summarized in Table 4.3.

Table 4.3. Advantages and Disadvantages of Different Methods for Estimating $F_a \cdot F_g$

Methods	Advantages	Disadvantages
AUC comparison between P.O. and I.P. administration	More accurate estimate of $F_a \cdot F_g$ in the presence of systemic intestinal metabolism of drug	Needs P.O. and I.P. data Experimental difficulties in I.P. dosing Difficult to validate the assumptions (see the text)
AUC comparison between P.O. and I.V. administration	Extra information available when F is known	Needs P.O. and I.V. data Difficult to validate the assumptions (see the text)
Mass balance (Fick's principle)	Needs P.O. data only	Surgical difficulties and complications associated with the serial bleeding from the portal vein Difficult to measure portal blood flow rate Underestimation of $F_a \cdot F_g$ in the presence of systemic intestinal metabolism
Clearance method	More accurate estimation of $F_a \cdot F_g$ in the presence of systemic intestinal metabolism of drug	Needs P.O. and I.V. data Surgical difficulties and complications in serial bleeding from the portal vein Difficult to validate the assumptions (see the text)

(i) AUC comparison between oral and intraportal vein administration. $F_a \cdot F_g$ can be estimated by comparing dose-normalized AUC values after oral and intraportal vein [or intraperitoneal (I.P.)] administration of a drug (Cassidy and Houston, 1980):

$$F_a \cdot F_g = \frac{AUC_{po,0-\infty} \cdot D_{ip}}{AUC_{ip,0-\infty} \cdot D_{po}} \tag{4.13}$$

where D_{ip} and D_{po} are doses after intraportal vein and oral administration, and $AUC_{ip,0-\infty}$ and $AUC_{po,0-\infty}$ are $AUC_{0-\infty}$ in the systemic plasma after intraportal vein and oral administration, respectively. This approach is valid only when the extent of hepatic clearance is the same for intraportal vein and oral dosing.

(ii) AUC comparison between oral and intravenous administration. In the absence of systemic intestinal metabolism of a drug, $F_a \cdot F_g$ can be estimated from the plasma drug concentration–time profiles after intravenous and oral administration without portal blood sampling after oral administration, based on the following assumptions: (1) linear kinetics (administration route-independent clearance); (2)

hepatic clearance is the only elimination pathway after intravenous administration, although there can be intestinal first-pass elimination after oral administration; and (3) blood drug concentrations are equal to plasma drug concentrations:

(4.14)
$$F_a \cdot F_g = \frac{Q_h \cdot AUC_{po} \cdot D_{iv}}{(Q_h\text{-}Cl_s) \cdot AUC_{iv} \cdot D_{po}}$$

where Q_h is hepatic blood flow rate.

(iii) Mass balance. The amount of drug in the portal vein after oral administration is the sum of the amount of the newly absorbed drug into the portal vein from the gut and the amount of drug coming from the mesenteric artery supplying blood to the intestine:

Amount of drug in the portal vein over a short period of time (dt) ↓ — *Amount of newly absorbed drug over* dt ↓ — *Amount of drug coming from the mesenteric artery over* dt ↓

(4.15)
$$Q_{pv} \cdot C_{po,pv}(t) \cdot dt = A(t) + Q_{pv} \cdot C_{po,sys}(t) \cdot dt$$

where Q_{pv} is portal vein blood flow rate, and hence $Q_{pv} \cdot dt$ represents the total volume of blood flowing through the portal vein over a short period of time, dt, from time t. A(t) is the amount of drug newly absorbed from the intestine showing up in the portal vein over dt. $C_{po,pv}(t)$ and $C_{po,sys}(t)$ are the drug concentrations in the portal blood and systemic circulation (usually venous blood) at time t after oral administration, respectively. Drug concentrations in the systemic circulation can be used instead of those in the mesenteric artery for estimating the amount of a drug in the mesenteric artery, because in most cases they can be assumed to be the same in both regions. Figure 4.6 illustrates the relationships among different drug concentrations at various anatomical locations after oral administration.

Integrating Eq. (4.15) from 0 to ∞ gives an estimate of the total amount of drug absorbed into the portal vein (A_a) after oral administration:

(4.16)
$$A_a = Q_{pv} \cdot (AUC_{po,pv} - AUC_{po,sys})$$

where $AUC_{po,pv}$ and $AUC_{po,sys}$ are, respectively, $AUC_{0-\infty}$ of a drug in the portal vein blood and systemic blood (or plasma when both plasma and blood concentrations are the same) after oral administration (Fujieda *et al.*, 1996). Therefore, $F_a \cdot F_g$ can be expressed as

(4.17)
$$F_a \cdot F_g = \frac{Q_{pv} \cdot (AUC_{po,pv} - AUC_{po,sys})}{D_{po}}$$

• Portal blood sampling. Portal blood samples can be collected from portal-vein-cannulated animals for serial bleeding or individual animals at different time

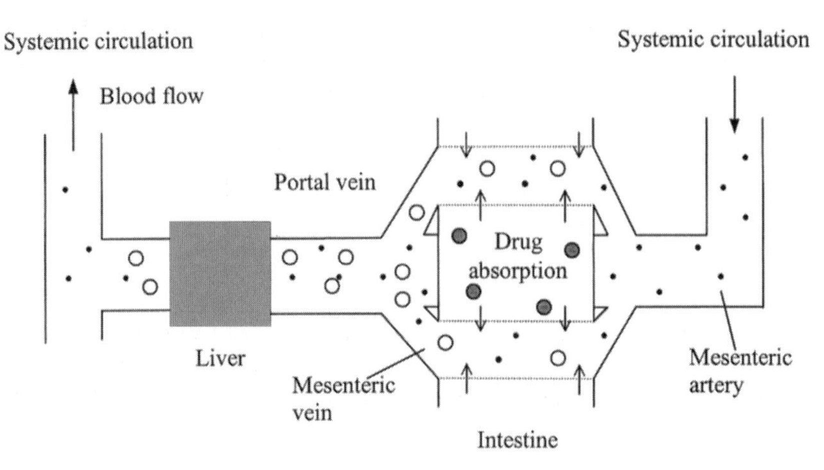

Figure 4.6. Schematic description of relationships among newly absorbed drug molecules (○) from drug particles or molecules (●) in the intestine and previously absorbed drug molecules (•) coming from the systemic circulation via the mesenteric artery. A difference in drug concentrations between the portal vein and the mesenteric artery is due to the newly absorbed drug.

points by terminal bleeding to avoid portal-vein cannulation (see Appendix C). Portal-vein-cannulation surgery may cause some physiological changes in, e.g., portal blood flow rate or the amount of albumin in the blood, which may affect drug disposition profiles, whereas sampling from individual animals at different time points by terminal bleeding may result in a large interanimal variability in exposure.

• Limitations of the mass balance method. (1) The estimate of A_a can vary depending on the portal blood flow rate used (the values published in the literature are often used). (2) A_a can be underestimated when a drug is subject to systemic intestinal metabolism. Since the previously absorbed drug returning to the mesenteric artery from the systemic circulation is subject to intestinal metabolism during vascular intestinal transit prior to reaching the portal circulation, the difference between the portal vein and systemic blood concentrations underestimates the true concentration of newly absorbed drug into the portal vein. The clearance method is more suitable for estimating $F_a \cdot F_g$, when significant systemic intestinal elimination of the drug is suspected.

(iv) Clearance method. The clearance method was derived based on a three-compartment model (systemic blood, intestine, and liver compartments). Important assumptions for the model include: (1) linear and route-independent kinetics, (2) intestinal and/or hepatic clearance only, and (3) instantaneous and homogeneous distribution of the drug within the compartments. Estimates for A_a and $F_a \cdot F_g$ can be obtained using the following equations (Kwon, 1996):

$$\boxed{A_a = Cl_b \cdot AUC_{po,pv}} \tag{4.18}$$

(4.19)
$$F_a \cdot F_g = \frac{Cl_b \cdot AUC_{po,pv}}{D_{po}}$$

where Cl_b is the systemic blood clearance (or systemic plasma clearance when blood and plasma concentrations are the same).

NOTE: HOW TO DETECT SYSTEMIC INTESTINAL METABOLISM OF DRUG. For some drugs, intestinal metabolism can play an important role in the elimination of a drug from the systemic circulation. One of the following findings can be indicative of the presence of systemic intestinal metabolism of a drug: (1) $F_a \cdot F_g$ estimated using Eq. (4.17) is smaller than F estimated using Eq. (4.9), which is an impossible outcome in the absence of systemic intestinal elimination. (2) $F_a \cdot F_g$ estimated using Eq. (4.17) is significantly smaller than the estimate using Eq. (4.13) or Eq. (4.19). When a drug is subject to substantial systemic intestinal metabolism, $F_a \cdot F_g$ based on the AUC comparison or the clearance methods becomes more accurate than the mass balance method.

(d) The Fraction of the Amount of a Drug Entering the Liver after Oral Administration that Escapes Presystemic Hepatic Elimination (F_h).

(i) AUC comparison between intraportal and intravenous administration. F_h can be estimated by comparing dose-normalized AUC values of systemic plasma (or blood) drug concentration profiles after intraportal and intravenous administration (Cassidy and Houston, 1980):

(4.20)
$$\boxed{F_h = \frac{AUC_{ip} \cdot D_{iv}}{AUC_{iv} \cdot D_{ip}}}$$

(ii) Clearance method. F_h can be also calculated according to the clearance method (Kwon, 1996):

(4.21)
$$F_h = \frac{AUC_{po,pv} - AUC_{po,sys}}{AUC_{po,pv}}$$

(e) The fraction of the Amount of Drug Entering the Lung after Oral Administration that Escapes the Presystemic Pulmonary Elimination (F_l). F_l can be estimated by comparing dose-normalized AUC in systemic blood after intravenous and intraarterial administration of drug:

(4.22)
$$\boxed{F_l = \frac{AUC_{iv} \cdot D_{ia}}{AUC_{ia} \cdot D_{iv}}}$$

AUC_{ia} is $AUC_{0-\infty}$ in systemic blood after intraarterial administration, and D_{ia} is the intraarterial dose. In the absence of pulmonary metabolism, AUC_{ia} and AUC_{iv} should be the same.

(f) Relationships between F_s *and* F_a, F_g, F_h *and* F_l. F_s can be expressed as a function of $F_a \cdot F_g$, F_h, and F_l with corresponding AUC comparisons [Eqs. (4.13), (4.20), and (4.22), respectively], assuming the same dose, as follows:

$$F_s = \underbrace{\frac{AUC_{po}}{\cancel{AUC_{pv}}}}_{F_a \cdot F_g} \cdot \underbrace{\frac{\cancel{AUC_{pv}}}{\cancel{AUC_{iv}}}}_{F_h} \cdot \underbrace{\frac{\cancel{AUC_{iv}}}{AUC_{ia}}}_{F_l} = \frac{AUC_{po}}{AUC_{ia}} \tag{4.23}$$

When AUC_{ia} and AUC_{iv} are the same, i.e., no pulmonary elimination, F_s becomes equal to F, oral bioavailability (AUC_{po} divided by AUC_{iv}).

(g) Oral Absorption Pathways Avoiding Intestinal or Hepatic First-Pass Effects

(i) Lymphatic delivery in the GI tract. The lymph from the GI tract is collected in the thoracic lymph duct without passing through the liver, before entering the bloodstream. Therefore, the drug absorbed via lymphatic vessels in the GI tract can avoid hepatic first-pass effects, although presystemic elimination of the drug by enterocytes may still occur (Muranishi, 1991). The actual amount of drug delivered via lymphatic pathways can be rather limited owing to the slow flow rate of lymph (see Chapter 13).

(ii) Rectal administration. Blood vessels from the lower part of the rectum connect with the inferior vena cava instead of merging into the portal vein, so that a drug administered in a suppository via a rectal route can avoid hepatic first-pass effects.

4.3.4. Estimating the Rate of Absorption

The rate of drug absorption after oral administration can be assessed from plasma drug concentration–time profiles with curve-fitting or moment analysis of *in vivo* data. *In vitro* or *in situ* experiments such as Caco-2 cell permeation or intestinal perfusion studies can also provide information regarding the rate of intestinal absorption of an orally dosed drug.

4.3.4.1. In Vivo Experiments

The absorption rate constant (k_a), which reflects how fast drug molecules transport across intestinal epithelial cells and reach the systemic circulation after oral dosing, can be estimated by curve-fitting or moment analysis.

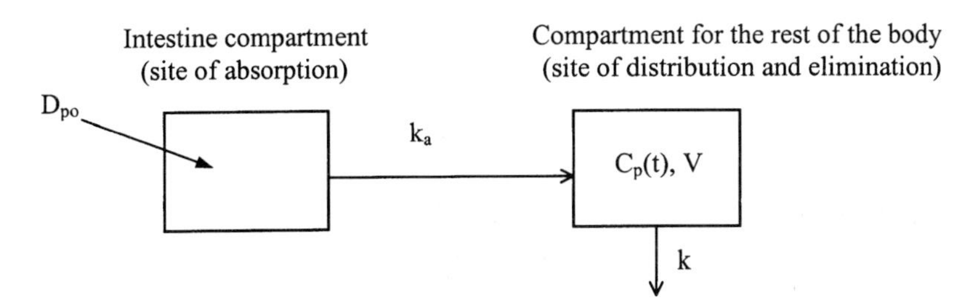

Figure 4.7. Two-compartment model for drug absorption and disposition. $C_p(t)$: plasma concentration at time t after oral administration, D_{po}: oral dose, k_a: absorption rate constant, k: elimination rate constant, V: apparent volume of distribution.

(a) Curve Fitting. The value of k_a can be estimated by fitting a proper compartmental model to the plasma drug concentration–time profile after oral administration of a drug, with the method of residuals or a nonlinear least-square regression program such as WinNonlin® (Pharsight, NC).

(i) Bateman equation. The equation most often used for estimating k_a is based on a two-compartment model for the intestine and the rest of body (Fig. 4.7), according to which, the time course of $C_p(t)$ after oral administration of a drug can be described as in Eq. (4.24), which is known as the Bateman equation:

$$C_p(t) = \frac{k_a \cdot F \cdot D_{po}}{V \cdot (k_a - k)} \cdot (e^{-k \cdot t} - e^{-k_a \cdot t}) \tag{4.24}$$

where k is the elimination rate constant and V is the apparent volume of distribution of the compartment for the rest of the body; k_a can be estimated by fitting Eq. (4.24) to the exposure profiles of the drug after oral administration with F, V, and k estimated from intravenous studies using a nonlinear regression computer program. It is important to note that k_a estimated with Eq. (4.24) is valid only when the following assumptions are met: (1) first-order absorption and elimination rates, (2) homogeneous behaviors of the intestine for drug absorption, and (3) a one-compartment model for the rest of the body (a monoexponential decline of the plasma drug concentration–time profile on a semilog scale after intravenous bolus injection).

(ii) Method of residuals. The method of residuals is used to estimate k_a of a drug based on the assumption that its rate of absorption is much faster than its rate of elimination from the body, i.e., $k_a \gg k$. In this case, during the terminal phase after oral dosing, $e^{-k_a \cdot t}$ in Eq. (4.24) becomes much smaller than $e^{-k \cdot t}$, so that $C_p(t)$ during the terminal phase $[C_p^{Exp}(t)]$ can be approximated as

$$C_p^{Exp}(t) = \frac{k_a \cdot F \cdot D_{po}}{V \cdot (k_a - k)} \cdot e^{-k \cdot t} \tag{4.25}$$

Subtracting Eq. (4.24) from (4.25) yields

$$C_p^{Exp}(t) - C_p(t) = \frac{k_a \cdot F \cdot D_{po}}{V \cdot (k_a - k)} \cdot e^{-k_a \cdot t} \tag{4.26}$$

A plot of $C_p^{Exp}(t) - C_p(t)$, the "residuals" of $C_p^{Exp}(t)$ and $C_p(t)$ *vs.* time plots, becomes a straight line on a semilog scale and from its slope $(= -k_a/2.303)$, k_a can be estimated with curve-fitting as illustrated in Fig. 4.8.

The method of residuals is useful only when all the assumptions that were applied to the Bateman equation are satisfied and $k_a \gg k$. An estimate of k_a from the Bateman equation or the method of residuals is an apparent value reflecting the entire absorption process, including disintegration and dissolution rates from dosage forms (if the drug is not administered in solution) and transport rates passing through the intestine and the liver during absorption.

(iii) C_{max} *and* t_{max}. The highest drug concentration after oral administration (C_{max}) and t_{max}, the time at which C_{max} is observed, can be derived from the Bateman equation:

$$C_{max} = \frac{F \cdot D_{po}}{V} \cdot e^{-k \cdot t_{max}} \tag{4.27}$$

$$t_{max} = \frac{\ln(k_a/k)}{k_a - k} \tag{4.28}$$

As seen in Eqs. (4.27) and (4.28), both C_{max} and t_{max} are affected by k_a as well as by k. If two different formulations of the same drug (and hence, different k_a values, but the same k) are compared, the formulation with the faster absorption (a greater k_a) would produce a higher C_{max} with an earlier t_{max}.

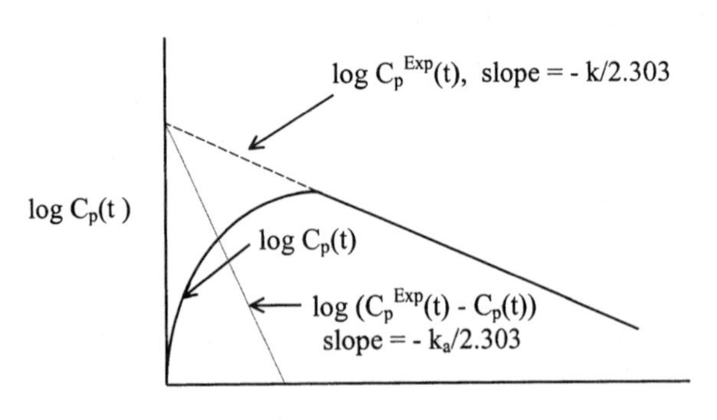

Figure 4.8. The method of residuals to estimate the absorption rate constant (k_a) of a drug after oral administration. A semilogarithmic plot of the difference between $C_p^{Exp}(t) - C_p(t)$ exhibits a straight line with a slope of $-k_a/2.303$. $C_p^{Exp}(t)$ represents a plasma drug concentration–time plot extrapolated from the terminal phase of $C_p(t)$ to the origin.

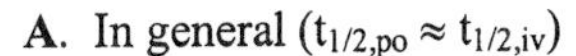

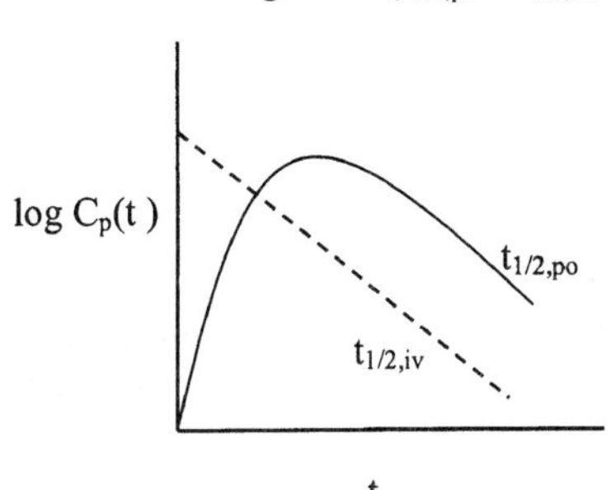

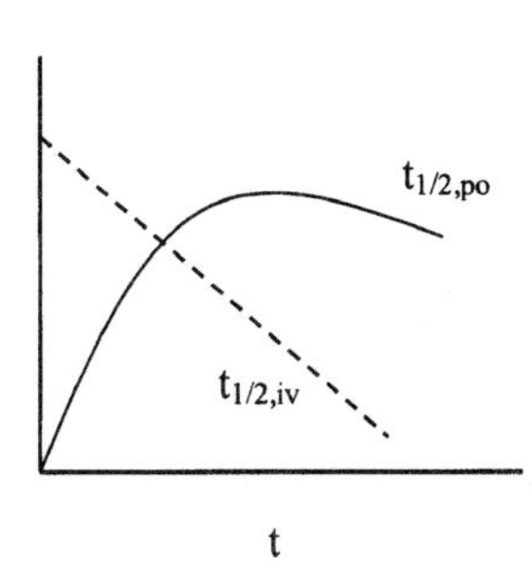

Figure 4.9. Plasma drug concentration *vs.* time profiles of hypothetical drugs after intravenous (----) or oral (——) administration on a semilogarithmic scale. In most cases, the absorption rate constant of drug after oral administration is much greater than the elimination rate constant evidenced by similar terminal half-lives between oral and intravenous administration (A). When the absorption rate constant is substantially smaller than the elimination rate constant, the terminal half-life of the drug after oral administration ($t_{1/2,po}$) becomes longer than that after intravenous administration ($t_{1/2,iv}$) (B).

(iv) Flip-flop kinetics. In a series of two consecutive, irreversible first-order rate processes such as absorption of a drug from the intestine and its subsequent systemic elimination, either step can be rate-limiting in the overall elimination process. In general, k_a of a drug after oral administration is greater than k so that elimination of the drug from the body after oral administration is governed primarily by how fast it can be removed once it enters the systemic circulation. In this case (e.g., $k_a > 3 \times k$), a plasma concentration–time profile after oral dosing exhibits a terminal half-life ($t_{1/2,po}$) similar to that after intravenous injection ($t_{1/2,iv}$). However, when k_a is much smaller than k (e.g., $k > 3 \times k_a$), drug disappearance from the body becomes governed by the rate of absorption rather than by the rate of elimination, and $t_{1/2,po}$ becomes longer than $t_{1/2,iv}$. This phenomenon is called "flip-flop kinetics" (Fig. 4.9).

(b) Moment Analysis.

(i) Mean absorption time. The value of k_a can also be estimated using moment analysis from the mean residence time (MRT), since the MRT of a drug after oral administration (MRT_{po}) includes the time required for absorption (mean absorption time, MAT) and MRT after intravenous administration (MRT_{iv}):

$$MRT_{po} = MAT + MRT_{iv} \tag{4.29}$$

and k_a is the reciprocal of MAT:

$$k_a = \frac{1}{MAT} = \frac{1}{MRT_{po} - MRT_{iv}} \tag{4.30}$$

and

$$\mathrm{MRT_{po}} = \frac{\mathrm{AUMC}_{0-\infty,\mathrm{po}}}{\mathrm{AUC}_{0-\infty,\mathrm{po}}} \quad \text{and} \quad \mathrm{MRT_{iv}} = \frac{\mathrm{AUMC}_{0-\infty,\mathrm{iv}}}{\mathrm{AUC}_{0-\infty,\mathrm{iv}}}$$

where $\mathrm{AUMC}_{0-\infty,\mathrm{iv}}$ and $\mathrm{AUMC}_{0-\infty,\mathrm{po}}$ are the areas under the first-moment curve of plasma drug concentration *vs.* time, i.e., AUC of the product of concentration and time *vs.* time profile from zero to infinity after intravenous and oral administration, respectively (see Chapter 2).

(ii) $\mathrm{MRT_{disint}}$, $\mathrm{MRT_{diss}}$, *and* $\mathrm{MRT_{abs}}$. Various MRT values for different steps of oral absorption of a drug can be calculated by moment analysis with exposure levels of the drug dosed in different formulations (Fig. 4.10). For instance, the difference in the MRT estimate after the administration of a tablet and of a suspension is the MRT for the disintegration process of the tablet ($\mathrm{MRT_{disint}}$) to particles in suspension. A difference in the MRT after administration of a suspension and a solution is the MRT for the dissolution process of the solid drug particles in suspension ($\mathrm{MRT_{diss}}$) to drug solution. A difference of MRT between an oral solution and an intravenous injection is the MRT for the absorption process of the drug molecules in solution ($\mathrm{MRT_{abs}}$) into the systemic circulation (Tanigawara *et al.*, 1982).

For example, $\mathrm{MRT_{po}}$ ($= \mathrm{AUMC}_{0-\infty,\mathrm{po}}/\mathrm{AUC}_{0-\infty,\mathrm{po}}$) of a drug determined after oral administration of a tablet is the sum of $\mathrm{MRT_{disint}}$, $\mathrm{MRT_{diss}}$, $\mathrm{MRT_{abs}}$, and $\mathrm{MRT_{iv}}$. Thus, MAT ($= \mathrm{MRT_{po}}$-$\mathrm{MRT_{iv}}$) is the sum of $\mathrm{MRT_{disint}}$, $\mathrm{MRT_{diss}}$, and

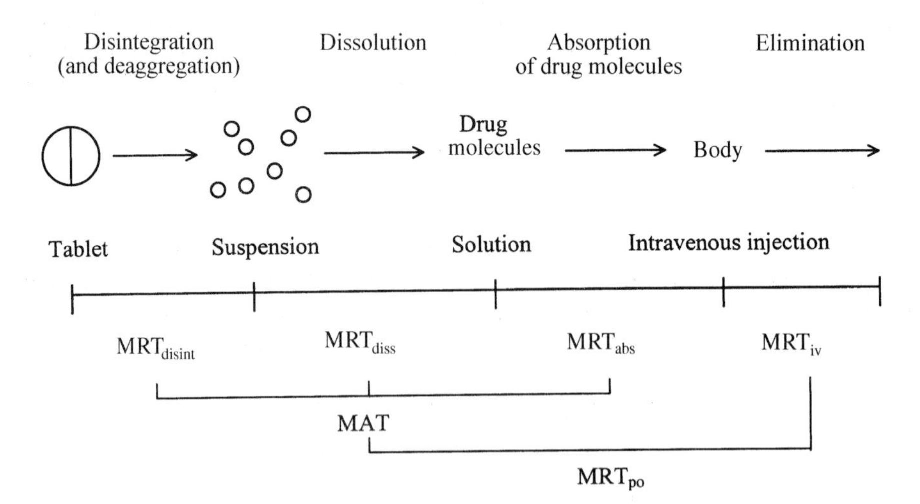

Figure 4.10. Relationships among different mean residence time (MRT) estimates reflecting various absorption processes after oral administration of drug in different oral dosage forms. MAT: mean absorption time of drug after oral administration, $\mathrm{MRT_{abs}}$: MRT for the absorption of the drug molecules in solution, $\mathrm{MRT_{disint}}$: MRT for the disintegration of the solid dosage form such as tablet, $\mathrm{MRT_{diss}}$: MRT for the dissolution of the solid drug particles in suspension, $\mathrm{MRT_{po}}$: MRT after oral administration, $\mathrm{MRT_{iv}}$: MRT after intravenous administration.

MRT_{abs}. MRT_{po} after oral administration of a suspension would include, therefore, MRT_{diss}, MRT_{abs}, and MRT_{iv}, and its MAT is then the sum of MRT_{diss} and MRT_{abs}.

4.3.4.2. In Situ or In Vitro Experiments

In general, most *in situ* or *in vitro* absorption or transport studies, such as intestinal perfusion or Caco-2 cell experiments, are performed with a drug solution rather than solid or suspension formulations. An estimate of the absorption rate constant obtained from *in situ* or *in vitro* studies, therefore, reflects only the membrane permeation process via enterocytes during absorption. It should be noted that the membrane permeation rate of a drug across the enterocytes becomes important in overall absorption only when disintegration and/or dissolution rates of dosage forms are significantly faster than the membrane permeation rate of drug molecules.

(a) Intestinal Perfusion. In situ intestinal perfusion studies are usually conducted with single-pass perfusion as opposed to recirculating perfusion of a drug solution through an isolated intestine segment under steady state conditions. Two different types of studies can be performed, depending on the site of sample collection. Samples can be collected from the inlet and outlet of perfusate only, in which case the rate and extent of the disappearance of the drug from the gut lumen can be determined. Blood samples in the mesenteric vein from the isolated intestine segment can be obtained in addition to perfusate samples. Analyses of these samples can provide information not only on drug disappearance from the gut lumen but also on drug appearance in the mesenteric vein, which is more relevant to actual drug absorption. As the studies are performed under steady state conditions, the effects of nonspecific adsorption of a drug to perfusion apparatus and tubes or intestinal membranes during perfusion on its drug disappearance from the lumen or its appearance in mesenteric vein can be ignored (Raoof *et al.*, 1998).

(i) Drug disappearance from the intestinal lumen. The rate and extent of drug disappearance from the lumen of the isolated intestine at steady state can be determined by measuring inlet and outlet drug concentrations of the perfusate. Drug disappearance from the perfusate can be due to transport of the drug into the enterocytes and/or metabolism by gut microflora inside the intestinal lumen. The apparent absorption rate constant ($k_{a,app}$), which reflects how fast drug molecules disappear from the perfusate flowing through the intestinal segment, can be estimated from the following equation:

$$k_{a,pp} = \frac{\overbrace{Q \cdot (C_{in,ss} - C_{out,ss})}^{\substack{\textit{Disappearance rate of drug} \\ \textit{in the intestinal lumen}}}}{C_{in,ss} \cdot V} \tag{4.31}$$

where $C_{in,ss}$ and $C_{out,ss}$ are the drug concentrations in a perfusate solution entering

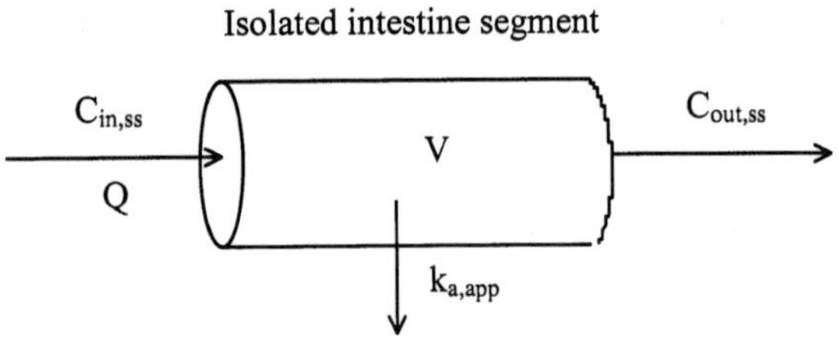

Figure 4.11. Schematic description of isolated intestine single-pass perfusion. $C_{in,ss}$: inlet drug concentration at steady state, $C_{out,ss}$: outlet drug concentration at steady state, $k_{a,app}$: apparent absorption rate constant, Q: perfusate flow rate, V: volume of the isolated intestine segment.

and leaving the isolated intestine segment at steady state, respectively, and Q and V are the perfusate flow rate and the volume of the intestine segment used in the experiment, respectively. A schematic description of an intestinal perfusion study is illustrated in Fig. 4.11.

If it is assumed that the disappearance of a drug from the intestine is a linear process, the relationship between steady state perfusate drug concentrations and effective permeability (P_{eff}) of the drug disappearing from the intestinal lumen can be expressed as

$$C_{out,ss}/C_{in,ss} = e^{-P_{eff} \cdot (2\pi \cdot r \cdot L)/Q} \tag{4.32}$$

where r and L are the radius and length of the gut lumen, respectively, P_{eff} is the effective permeability of a drug as it is transported from the intestinal lumen into the enterocytes, and may overestimate the true intestinal permeability of the compound *in vivo* owing to underestimation of the surface area of the gut lumen with $2\pi \cdot r \cdot L$.

The relationship between $k_{a,app}$ and P_{eff} is

$$k_{a,app} = \frac{Q(1 - e^{-P_{eff} \cdot (2\pi \cdot r \cdot L)/Q})}{V} \tag{4.33}$$

Unstirred water layer and permeability: The unstirred water layer (UWL), sometimes called the aqueous boundary layer, surrounds the surfaces of the brush border membranes of enterocytes. The permeability of a drug across the UWL can be affected by both the thickness of the boundary layer and the flow rate of the perfusate through the intestine. The rate of permeation of the compound in solution from the gut lumen into the splanchnic blood can be limited by the transport of drug molecules across the UWL or the enterocyte membranes (Fig. 4.12) (Amidon *et al.*, 1988; Zimmerman *et al.*, 1997). The effective permeability (P_{eff}) of the GI wall is considered to be a function of both the permeability of UWL (P_{aq}) and the permeability of the enterocyte membranes (P_m):

$$P_{eff} = \frac{P_{aq} \cdot P_m}{P_{aq} + P_m}$$

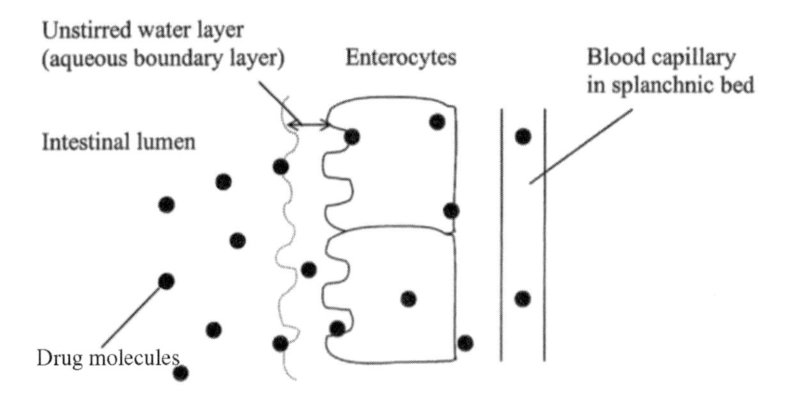

Figure 4.12. Schematic description of an unstirred water layer of enterocyte surfaces.

(ii) Drug appearance in the mesenteric vein. In addition to collection of inlet and outlet perfusate, blood samples taken from the mesenteric vein can provide an estimate of the absorption rate constant more relevant to actual drug absorption and information on the presystemic intestinal metabolism *in vivo*. To maintain a constant blood flow in the mesenteric vein and to avoid mixing of blood from the systemic circulation during sample collection, fresh blood is usually replenished into the mesenteric artery at a constant rate. A schematic description of sample collection from the mesenteric vein during intestinal perfusion of a drug solution is illustrated in Fig. 4.13. The equation describing mass balance during the study is as follows:

Amount of drug perfused into intestine ($Q \cdot C_{in,ss}$)

= Amount of drug leaving from the intestine ($Q \cdot C_{out,ss}$)

+ Amount of drug absorbed into the mesenteric vein ($Q_{mv} \cdot C'_{mv,ss}$)

+ Amount of drug eliminated by chemical instability in perfusate, intestinal microfloral metabolism, and/or first-pass metabolism by enterocytes

where $C_{in,ss}$ and $C_{out,ss}$ are the drug concentrations in a perfusate solution entering and leaving the intestine segment at steady state, respectively, $C_{mv,ss}$ is the blood drug concentration in the mesenteric vein at steady state, and Q and Q_{mv} are the perfusate and mesenteric blood flow rates, respectively.

Based on this relationship, the following information on absorption can be obtained:

FRACTION OF DRUG ABSORBED INTO MESENTERIC VEIN (F_{mv}):

$$F_{mv} = \frac{Q_{mv} \cdot C_{mv,ss}}{Q \cdot C_{in,ss}} \tag{4.35}$$

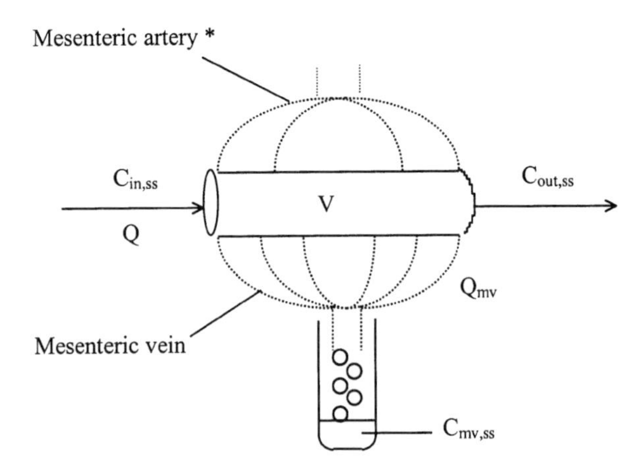

Figure 4.13. Schematic description of intestinal perfusion with blood collection from the mesenteric vein. $C_{in,ss}$: inlet drug concentration at steady state, $C_{out,ss}$: outlet drug concentration at steady state, $C_{mv,ss}$: blood drug concentration in mesenteric vein at steady state, Q: perfusate flow rate, Q_{mv}: mesenteric blood flow rate, V: volume of the isolated intestine, *In order to maintain blood flow in the mesenteric vein, fresh blood should be infused into the mesenteric artery at a constant rate.

APPARENT ABSORPTION RATE CONSTANT ($k_{a,app}$)

$$k_{a,app} = \frac{Q_{mv} \cdot C_{mv,ss}}{C_{i,ss} \cdot V} \tag{4.36}$$

where $C_{i,ss}$ is an average concentration within the intestine segment;

$$C_{i,ss} = (C_{in,ss} - C_{out,ss})/\ln (C_{in,ss}/C_{out,ss})$$

V is the volume of the intestine segment used in the experiment.

(b) Caco-2 Cells. Most drugs that are given orally are absorbed across the enterocytes primarily by passive diffusion. In order to transport from the intestinal lumen into the mesenteric vein, drug molecules must diffuse through a series of different physiological barriers, including the mucus gel layer (unstirred water layer), the intestinal epithelial cells, the laminal propia, and the endothelium of the intestinal capillary. Among these, it is the single layer of epithelial cells that has been recognized as the most significant barrier.

One of the most commonly used cells to investigate drug transport via enterocytes in humans is the Caco-2 cell line (Artursson, 1991). This cell line is derived from a human colon carcinoma and is distinguished from other cell lines of the same origin by its capability for spontaneous differentiation into monolayers of polarized enterocytes under conventional cell culture conditions (Artursson and Karlsson; 1991; Hidalgo *et al.*, 1989; Rubas *et al.*, 1993). The transport study can be performed from apical (luminal) to basal (blood) sides in an absorptive direction by

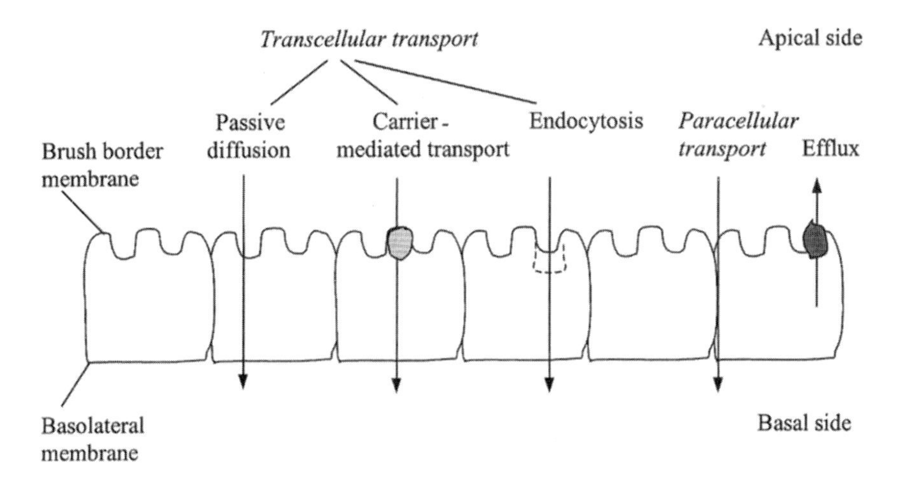

Figure 4.14. Various membrane transport mechanisms in enterocytes.

placing a drug solution in the apical side and collecting samples from the basal side at different incubation time points, or vice versa, depending on study needs (Fig. 4.14). Apparent membrane permeability (P_{app}) of a drug in Caco-2 cell experiments can be determined as follows:

$$P_{app} = \frac{\text{Amount of drug in acceptor side at time t}/\Delta t}{\text{Surface area of cell layer} \times \text{Concentration of drug in donor side at time 0}} \tag{4.37}$$

where Δt is an incubation period. In most cases, P_{app} is expressed in cm/sec.

(i) Direction of transport. The chambers where drug solution and blank buffer are placed are known as the donor and acceptor sides, respectively. When the apical side (A) of the Caco-2 cells is the donor side, the experiment is performed in an absorptive direction from the gut lumen to the mesenteric vein, whereas when the drug solution is placed in the basal side (B), drug efflux from the mesenteric vein to the gut lumen can be studied (Fig. 4.14). If transport of the drug is mediated solely by passive diffusion, P_{app} estimates must be the same regardless of the direction of the transport. However, when the drug is subject to active transport (Table 4.4) or efflux mechanisms such as P-glycoprotein (Gatmaitan and Arias, 1993; Leveque and Jehl, 1995) in the brush border membranes of the enterocytes, P_{app} values measured from A to B can be greater or smaller than those from B to A, respectively.

(ii) Validation and establishment of Caco-2 cell systems. When the Caco-2 cell line is newly established, a thorough validation of the cell integrity (confluence) and functionality (expression of active transporters and enzymes) is important because of a large variability in cell culture systems depending on study conditions. To assess the integrity of the monolayers, one or both of the following methods can be used.

Table 4.4. Carrier-Mediated Transport Systems in the Intestinal Epithelial Membranes[a]

Locations	Transporters	Substrates
Brush border membrane	Amino acid transporters	Amino acids and amino acid-mimetic compounds such as leucine, lysine, glutamate, L-dopa
	Oligopeptide transporters	Peptidomimetic compounds such as angiotensin converting enzyme (ACE) inhibitors, renin inhibitors, some β-lactam antibiotics
	Monocarboxylic acid transporter	Carboxylic acids such as salicylic acid, benzoic acid
	Glucose transporter	*p*-nitrophenyl-β-D-glucopyranoside
	Bile acid transporter	Taurocholic acid
	Phosphate transporter	Fosfomycin
	Membrane electric potential-dependent transport	Disopyramide, tyramine
	Proton antiporter	Tetraethylammonium, N-methylnicotine amide
	P-glycoprotein[b]	Cyclosporin-A, verapamil, vinblastine
Basolateral membrane	Amino acid transporters	
	Phosphate transporter	

[a]Data taken from Tsuji and Tomai (1996) and Zhang *et al.* (1998).
[b]P-glycoprotein, a multidrug resistance gene product, acts as a barrier to intestinal absorption of numerous xenobiotics by effluxing them out of cytoplasm of enterocytes and/or pumping right back into the intestinal lumen during their uptake into the cells (see Chapter 9, Gatmaitan and Arias, 1993, Hunter and Hurst, 1997; Leveque and Jehl 1995; Wacher *et al.*, 1996).

• Permeation of mannitol via the Caco-2 cell monolayers. Mannitol is known to transport via a paracellular pathway only, and a rate of flux from the donor to the acceptor side greater than 0.5%/hr in Caco-2 cells may indicate that the cells have been damaged and are not suitable for transport studies. P_{app} values of mannitol lower than 10^{-6} cm/sec indicate that the integrity of the monolayer of cells is well maintained. Propranolol is another control compound used for transcellular passive diffusion. P_{app} values of propranolol are usually greater than 10^{-5} cm/sec.

• Transepithelial electrical resistance. The development of a tight junction can be monitored by measuring transepithelial electrical resistance (TEER) across the Caco-2 cells. TEER values of intact Caco-2 cell monolayers range between 200 and 500 $\Omega \cdot cm^2$ (Hidalgo, 1996).

To assess the functionality of the cells, transport studies with known substrates for active transporters (Table 4.4) can be conducted. In general, it takes approximately 3 weeks after seeding for the full expression of active transporters in cell membranes. Other factors important in cell culture are the passage number of cells, material and surface area of inserts, compositions of incubation buffers, and amount of organic solvent used in studies. In general, less than 1% (v/v) of acetonitrile or methanol or 0.5% (v/v) of dimethylsulfoxide (DMSO) in incubation buffers can safely be used for study.

(iii) Relationship between P_{app} *values and the extent of absorption in humans in vivo.* General rules governing the relationship between P_{app} values of compounds determined from Caco-2 cell studies and the extent of absorption of compounds in humans (Artursson and Karlsson, 1991) are summarized in Table 4.5, provided that dissolution and/or intestinal metabolism of the compounds do not affect drug absorption to any significant extent.

(iv) Relationship between P_{app} *and* $k_{a,app}$. A pharmacokinetic relationship between P_{app} from Caco-2 cell studies and $k_{a,app}$ values from isolated intestinal perfusion experiments can be viewed as follows:

$$P_{app} = \frac{k_{a,app} \cdot V}{S} \tag{4.38}$$

where S and V are, respectively, the surface area and volume of the intestine available for absorption of the drug after oral administration. Equation (4.38) is an oversimplification of the true relationship between P_{app} and $k_{a,app}$ as it ignores experimental differences between *in vitro* and *in vivo* conditions; however, a linear relationship between P_{app} and $k_{a,app}$ has been reported among structurally similar compounds with similar absorption profiles (Cutler, 1991; Kim *et al.*, 1993).

(v) Membrane transport mechanisms in enterocytes. There are basically four different types of membrane transport processes across enterocytes including: (1) passive diffusion, (2) carrier-mediated transport (facilitated diffusion and active transport, (3) paracellular transport, and (4) endocytosis (pinocytosis). Figure 4.14 illustrates the various transport mechanisms in enterocytes.

• Passive diffusion. For most drugs, passive diffusion is a predominant membrane transport mechanism. Un-ionized, lipophilic molecules can diffuse better across membranes than ionized, hydrophilic molecules. Diffusion is a nonsaturable and concentration-gradient-dependent process, which does not require transport carriers or metabolic energy consumption.

• Carrier-mediated transport. Carrier-mediated transport can be divided into two different types, i.e., facilitated diffusion and active transport, which are carrier-mediated transport processes without and with ATP consumption, respectively.

Facilitated diffusion: Some compounds diffuse down electrochemical gradients across membranes more rapidly than expected from simple diffusion based on their physicochemical properties. This "facilitated" diffusion process mediated by carrier systems is saturable and stereospecific. Like simple diffusion, the facilitated diffusion is also a concentration-gradient-dependent process; i.e., once the concentrations between membranes reach equilibrium, apparent net transport of compounds via the facilitated diffusion ends. Facilitated diffusion is distinguished from active transport in that it does not require energy consumption.

Active transport: Active transport is saturable and differs from passive diffusion or facilitated diffusion in that solutes can be transported against thermodynamic equilibrium by consuming ATP. A few active transport systems identified at the

Table 4.5. Guidelines for the Relationship between P_{app} Values of Compounds Determined from Caco-2 Cell Studies and the Extent of Absorption of Compounds in Humans[a]

P_{app} of compounds in Caco-2 cells (cm/sec)	Extent of oral absorption in human
$>10^{-5}$	Well absorbed (>70%)
10^{-6}–10^{-5}	Moderately absorbed (20–70%)
$<10^{-6}$	Poorly absorbed (<20%)

[a]Data taken from Artursson and Karlsson (1991).

brush border and basolateral membranes of the intestinal epithelium are summarized in Table 4.5.

• Paracellular transport. In general, transport of small hydrophilic compounds with molecular weight smaller than 200 across the intestinal membrane occurs mainly via tight junctions between adjacent enterocytes. The extent of paracellular transport of compounds can be examined by performing transport studies in the absence and presence of divalent cations such as Ca^{+2}, which neutralize the inside-negatively charged paracellular channel. In general, paracellular transport is considered to be a minor absorption pathway.

• Endocytosis. Endocytosis is the process in which cells take up macromolecules such as proteins or polysaccharides by ingesting parts of their membranes to generate endocytotic vesicles enclosing a minute volume of extracellular fluid and its contents. There are two different types of this membrane engulfment process, i.e., phagocytosis and pinocytosis. Phagocytosis occurs only in specific cells such as macrophages and involves the ingestion of large particles such as viruses or cell debris, whereas pinocytosis occurs in all cell types and leads to the uptake of extracellular fluid and its contents.

(c) Other In Vitro Systems for Intestinal Drug Absorption. Various *in vitro* models utilizing isolated cells, membrane vesicles, cell culture systems other than Caco-2 such as HT-29, T84, MDCK, and excised tissues including isolated intestinal segments, Ussing chambers, everted sacs, intestinal rings, and stripped and unstripped mucosal sheets have been investigated for drug absorption to varying degrees (Hillgren, 1995).

4.4. ENTEROHEPATIC CIRCULATION

The liver secretes bile into the duodenum via the bile duct. Bile contains bile salts, which act as surfactants to promote the absorption of lipophilic substances including dietary components and drugs. Approximately 90% of the bile excreted into the intestine is reabsorbed and return to the liver for secretion. Drug can be excreted from hepatocytes into bile in unchanged and/or metabolized forms. Some metabolites excreted into the intestine via bile can be converted back to the parent drug by enzymatic or chemical reactions in the intestine. For instance, glucuronide conjugates of drugs can be deconjugated to the parent drug by β-glucuronidase produced by intestinal microflora. The unchanged drug or deconjugated metabolites back to the parent drug can be reabsorbed into the portal circulation, a part of which will reach the systemic circulation, and the rest become subject to further metabolism in the liver and/or subsequent biliary excretion (Tabata *et al.*, 1995). Nearly all drugs undergo enterohepatic circulation to a certain extent (Fig. 4.15).

4.4.1. Recognizing Enterohepatic Circulation

Enterohepatic circulation (EHC) of a drug is more apparent in animals having a gall bladder, such as mice, ferrets, dogs, monkeys, and humans, because of the distinctive "hump(s)," i.e., a transient increase, in drug exposure profiles around mealtimes. This is due to the pulsatile release of bile containing drug accumulated in the gall bladder into the duodenum upon food intake followed by subsequent reabsorption of the drug from the intestine. It is, however, common not to observe humps in an exposure profile even if a significant portion of the drug is subject to EHC, and this can be due in part to an insufficient number of data points. There are also examples in dogs where transient increases in exposure have been reported without EHC; e.g., delay in gastric emptying time or changes in the viscosity or pH of the GI tract can also cause the transient increases in exposure profiles (Mummaneni *et al.*, 1995; Reppas *et al.*, 1998).

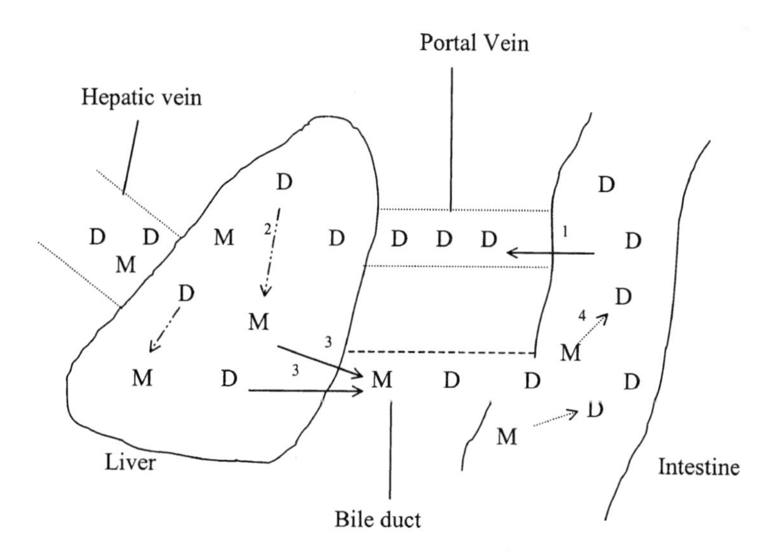

Figure 4.15. Schematic description of enterohepatic circulation of drug. D: drug, M: metabolite—Step 1: Absorption of drug molecules into the portal vein from the gut lumen. Step 2: Biotransformation of the drug to the metabolite by metabolizing enzymes. Step 3: Excretion of the drug and the metabolite into bile. Step 4: Conversion of the metabolite to the parent drug by intestinal microflora in some cases.

In animals without a gall bladder such as rats, the "hump(s)" in an exposure profile after meals may not be apparent for the drug subject to EHC owing to the continuous secretion of bile regardless of food intake. However, transient increases in plasma drug concentration profiles can be also observed in those animals. For example, when deconjugation of glucuronide conjugates of a drug excreted in bile by intestinal microflora occurs in a certain region of the intestine, subsequent reabsorption of the drug can produce a hump in the systemic exposure.

4.4.2. Pharmacokinetic Implications of Enterohepatic Circulation

1. Enterohepatic circulation should be viewed as a part of the distributional processes rather than the elimination processes.
2. In the presence of extensive EHC, the plasma exposure of a drug subject to EHC tends to be higher and is sustained longer than that of a drug with no EHC. As a result, a drug subject to EHC tends to exhibit a lower systemic clearance and a larger volume of distribution with a longer terminal half-life than one with no EHC.
3. It is often difficult to have an accurate estimate of oral bioavailability when the drug is subject to extensive EHC.
4. Biliary excretion of a drug seems to be a more important elimination pathway in laboratory animals such as the rat and the dog, than in the human. Therefore, care should be taken in extrapolating biliary excretion data of compounds obtained in animals to humans. Pharmacokinetic significance of EHC in the extent and duration of exposure of a drug is unclear in humans owing to the fact that there is limited information on the biliary excretion of drugs and their metabolites in humans *in vivo*.

4.4.3. Physicochemical Properties of Compounds for Biliary Excretion

Biliary excretion of compounds is thought to be mainly via carrier-mediated processes. Important physicochemical properties of compounds exhibiting relatively high biliary excretion are lipophilicity, molecular weight, and charge (Hirom *et al.*, 1974).

1. Lipophilicity ($\log P > -2$).
2. Molecular weight (>300 or 500 in rats or humans, respectively, for appreciable biliary excretion).
3. Charge: there appear to be separate biliary secretory mechanisms for acidic, basic, and neutral compounds.

4.4.4. Measuring Clearance in the Presence of Enterohepatic Circulation

Owing to the elevated exposure level of a drug from EHC, the systemic clearance (Cl_s) subject to EHC estimated by the intravenous dose (D_{iv}) divided by AUC after intravenous injection from zero to infinity ($AUC_{0-\infty,iv}$) may be lower than the sum of true individual organ clearances. In addition, a reliable determination of AUC from t_{last}, the last time point at which a quantifiable plasma drug concentration can be measured, to infinity ($AUC_{t_{last}-\infty}$) for the estimate of $AUC_{0-\infty,iv}$ using conventional curve-fitting may be difficult owing to the hump(s) in exposure caused by EHC. For a reliable estimate of $AUC_{t_{last}-\infty}$ of a drug subject to EHC after intravenous injection, renal clearance (Cl_s) is utilized.

1. Collect the first urine sample up to t_{last} and the second urine sample from t_{last} to time t in the same animals, at which excretion of a drug in the urine is considered to be completed.
2. Calculate Cl_r of the drug by dividing the amount of drug excreted unchanged in the urine from time 0 to t_{last} ($A_{e,0-t_{last}}$) by $AUC_{0-t_{last}}$. Since Cl_r is constant regardless of the shape of the plasma drug concentration–time profile, an estimation of $AUC_{t_{last}-\infty}$ can be obtained by dividing the amount of drug excreted from t_{last} to time t ($A_{e,t_{last}-t}$) in the urine by Cl_r:

$$Cl_r = A_{e,0-t_{last}}/AUC_{0-t_{last}}$$

   Therefore,

$$AUC_{t_{last}-\infty} = A_{e,t_{last}-t}/Cl_r$$

3. Cl_s can be estimated by dividing D_{iv} by $AUC_{0-\infty}$, which is the sum of $AUC_{0-t_{last}}$ and $AUC_{t_{last}-\infty}$.

4.4.5. Investigating Enterohepatic Circulation

Various experimental approaches to study the presence and extent of EHC of a drug in animals have been introduced:

1. Comparison of plasma exposure profiles between intact and bile-duct-cannulated animals: If exposure levels of a drug in intact animals are higher than those in bile-duct-cannulated animals, especially during the later stages of disposition, this may indicate that the drug undergoes EHC.
2. Donation of bile from one animal to another: The bile duct of one animal (donor animal) is surgically connected to the duodenum of another animal (recipient animal), and the bile duct of the recipient animal is cannulated for bile collection. After administration of a drug to the donor animal, blood and bile samples from the recipient animal are analyzed. If the drug is found in samples from the recipient animal, it is indicative of EHC.
3. Gut sterilization: When the conversion of conjugate metabolites to the parent drug by intestinal microflora and its subsequent absorption cause a transient increase in an exposure profile, the effects of microflora can be studied by sterilizing the gut lumen with pretreatment of animals with nonabsorbable antibiotics such as lincomycin.

4.5. FECAL EXCRETION OF DRUGS AND COPROPHAGY

Fecal excretion can be an important elimination pathway of xenobiotics from the body. Several important factors contributing to the extent of fecal elimination of drug are as follows:

1. Incomplete absorption: After oral administration of a drug, the nonabsorbed portion can be excreted in feces as unchanged. A significant portion of many macromolecules and ionized compounds at physiological pH may be excreted unchanged in feces after oral administration.
2. Biliary excretion: The biliary excretion of a drug is perhaps the most important factor for its fecal excretion.
3. Intestinal secretion: Drugs can be secreted from mesenteric blood into the intestinal lumen through enterocytes mainly by passive diffusion. The secretion into the intestinal lumen can become an important elimination pathway for certain lipophilic drugs only when other elimination processes are slow. Oral administration of activated charcoal or a fatty diet can facilitate the intestinal secretion of lipophilic drugs.

Most rodents such as rats and rabbits need to feed on their own feces as part of their normal diet, which is known as coprophagy. It is, therefore, sometimes necessary to restrict coprophagy by putting tail cups on the animals or conducting experiments in metabolic cages to separate feces, in order to avoid reintake of the drug in the feces.

4.6. LYMPHATIC ABSORPTION

The lymphatic system is an important vascular network for maintaining body water homeostasis in addition to blood circulation, and the whole GI tract is equipped with lymphatic vessels as well as blood vessels. The intestinal lymphatics are the major absorption pathway for many lipophilic nutrients including fats, lipid-soluble vitamins, and cholesterol. The orally dosed drug molecules have to be transported across enterocytes before entering either blood or lymphatic vessels. The lymph from the GI tract is collected into the thoracic lymph duct without passing through the liver before entering into the bloodstream. Therefore, the drug absorbed via lymphatic vessels in the GI tract can avoid hepatic first-pass effect, although it still has to undergo presystemic intestinal elimination. The lymphatic absorption of compounds is generally considered a minor absorption pathway even for highly lipophilic compounds (Muranishi, 1991).

REFERENCES

Amidon G. L. *et al.*, Estimating human oral fraction dose absorbed: a correlation using rat instestinal membrane permeability for passive and carrier-mediated compounds, *Pharm. Res.* **5**: 651–654, 1988.

Artursson P., Cell culture as models for drug absorption across the intestinal mucosa, *Crit. Rev. Ther. Drug Carrier Syst.* **8**: 305–330, 1991.

Artursson P. and Karlsson J., Correlation between oral drug absorption in humans and apparent drug permeability coefficients in human intestinal epithelial (Caco-2) cells, *Biochem. Biophys. Res. Comm.* **175**: 880–885, 1991.

Baijal P. K. and Fitzpatrick D. W., Effect of dietary protein on hepatic and extrahepatic phase I and phase II drug metabolizing enzymes, *Toxicol. Lett.* **89**: 99–106, 1996.

Cassidy M. K. and Houston J. B., *In vivo* assessment of extrahepatic conjugative metabolism in first pass effects using the model compound phenol, *J. Pharm. Pharmacol.* **32**: 57–59, 1980.

Cutler D., Assessment of rate and extent of drug absorption, *Pharmacol. Ther.* **14**: 123–160, 1991.

Dressman J. B., Comparsion of canine and human gastrointestinal physiology, *Pharm. Res.* **3**: 123–131, 1986.

Dressman J. B. *et al.*, Gastrointestinal parameters that influences oral medications, *J. Pharm. Sci.* **82**: 857–872, 1993.

Fujieda Y. *et al.*, Local absorption kinetics of levofloxacin from intestinal tract into portal vein in conscious rat using portal-venous concentration difference, *Pharm. Res.* **13**: 1201– 1204, 1996.

Gatmaitan Z. C. and Arias I. M., Structure and function of P-glycoprotein in normal liver and small intestine, *Adv. Pharmacol.* **24**: 77–97, 1993.

Hansch C. and Leo A., *Substituent Constants for Correlation Analysis in Chemistry and Biology*, Wiley (Interscience), New York, 1979.

Hidalgo I. J. *et al.*, Characterization of the human colon carcinoma cell line (Caco-2) as a model system for intestinal epithelial permeability, *Gastroenterology* **96**: 736–749, 1989.

Hidalgo I. J., Cultured intestinal epithelial cell models, in R. T. Borchardt, P. L. Smith and G. Wilson (eds.), *Models for Assessing Drug Absorption and Metabolism*, Plenum Press, New York, 1996, p. 39.

Hillgren K. M. *et al.*, *In vitro* systems for studying intestinal drug absorption, *Med. Res. Rev.*, **15**: 83–109, 1995.

Hirom P. C. *et al.*, The physicochemical factor required for the biliary excretion of organic cations in the bile of the rat, rabbit and guinea pig, *Biochem. Soc. Trans.* **3**: 327–330, 1974.

Hörter D. and Dressman J. B., Influence of physicochemical properties on dissolution of drugs in the gastrointestinal tract, *Adv. Drug Del. Rev.* **25**: 3–14, 1997.

Hunt S. M. and Groff J. L., The digestive system: mechanisms for nourishing the body, in S. M. Hunt and J. L. Groff (eds.), *Advanced Nutritional and Human Metabolism*, West, New York, 1990, p. 39.

Hunter J. and Hirst B. H., Intestinal secretion of drug: the role of P-glycoprotein and related drug efflux systems in limiting oral drug absorption, *Adv. Drug Del. Rev.* **25**: 129–157, 1997.

Kararli T. T., Comparison of the gastrointestinal anatomy, physiology, and biochemistry of humans and commonly used laboratory animals, *Biopharm. Drug Dispos.* **16**: 351–380, 1995.

Kim D. *et al.*, A correlation between the permeability characteristics of a series of peptides using an *in vitro* cell culture model (Caco-2) and those using an *in situ* perfused rat ileum model of the intestinal mucosa, *Pharm. Res.* **10**: 1710–1714, 1993.

Krarup L. H., Predicting drug absorption from molecular surface properties based on molecular dynamics simulations, *Pharm. Res.* **15**: 972–978, 1998.

Kwan K. C., Oral bioavailability and first-pass effects, *Drug Metab. Dispos.* **25**: 1329–136, 1998.

Kwon Y., Theoretical considerations on two equations for estimating the extent of absorption after oral administration of drugs, *Pharm. Res.* **13**: 566–569, 1996.

Lee C.-P. *et al.*, Selection of drug development candidates based on *in vitro* permeability measurements, *Adv. Drug Del. Rev.* **23**: 47–62, 1997.

Leveque D. and Jehl F., P-glycoprotein and pharmacokinetics, *Anticancer Res.* **15**: 331–336, 1995.

Lipinski C. A. *et al.*, Experimental and computational approaches to estimate solubility and permeability in drug discovery and development settings, *Adv. Drug Del. Rev.* **23**: 3–25, 1997.

Moriguchi I. *et al.*, Simple method of calculating octanol/water partition coefficient, *Chem. Pharm. Bull.* **40**: 127–130, 1992.

Mummaneni V. *et al.*, Gastric pH influences the appearance of double peaks in the plasma concentration–time profiles of cimetidine after oral administration in dogs, *Pharm. Res.* **12**: 780–786, 1995.

Muranishi S., Drug targeting towards the lymphatics, *Adv. Drug Res.* **21**: 1–37, 1991.

Palm K. *et al.*, Correlation of drug absorption with molecular surface properties, *J. Pharm. Sci.* **85**: 32–39, 1996.

Pang K. S., Metabolic first-pass effects, *J. Clin. Pharmacol.* **26**: 580–582, 1986.

Raoof A. A. *et al.*, Assessment of regional differences in intestinal fluid movement in the rat using a modified in situ single pass perfusion model, *Pharm. Res.* **15**: 1314–1316, 1998.

Reppas C. *et al.*, Effect of elevated viscosity in the upper gastrointestinal tract on drug absorption in dogs, *Eur. J. Pharm. Sci.* **6**: 131–139, 1998.

Rubas W. *et al.*, Comparison of the permeability characteristics of a human colonic epithelial (Caco-2) cell line to colon of rabbit, monkey, and dog intestine and human drug absorption, *Pharm. Res.* **10**: 113–118, 1993.

Smith D. A. *et al.*, Design of toxicokinetic studies, *Xenobiotica* **20**: 1185–1199, 1990.

Tabata K. *et al.*, Evaluation of intestinal absorption into the portal system in enterohepatic circulation by measuring the difference in portal-venous blood concentrations of diclofenac, *Pharm. Res.* **12**: 880–883, 1995.

Tanigawara Y. *et al.*, Moment analysis for the separation of mean *in vivo* disintegration, dissolution, absorption and disposition time of ampicillin products, *J. Pharm. Sci.*, **71**: 1129–1133, 1982.

Testa B. *et al.*, Lipophilicity in molecular modeling, *Pharm. Res.* **14**: 1332–1340, 1997.

Tsuji A. and Tamai I., Carrier-mediated intestinal transport of drugs, *Pharm. Res.* **13**: 963–977, 1996.

Wacher V. J. *et al.*, Active secretion and enterocytic drug metabolism barriers to drug absorption, *Adv. Drug Del. Rev.* **20**: 99–112, 1996.

Walter-Sack I., What is "fasting" drug administration? On the role of gastric motility in drug absorption, *Eur. J. Clin. Pharmacol.* **42**: 11–13, 1992.

Winne D., Influence of blood flow on intestinal absorption of xenobiotics, *Pharmacology*, **21**: 1–15, 1980.

Zhang L. *et al.*, Role of organic cation transporters in drug absorption and elimination, *Annu. Rev., Pharmacol. Toxicol.* **38**: 431–460, 1998.

Zhi J. *et al.*, Effects of dietary fat on drug absorption, *Clin. Pharmacol. Ther.*, **58**: 487–491, 1995.

Zimmerman C. L. *et al.*, Evaluation of gastrointestinal absorption and metabolism, *Drug Metab. Rev.* **29**: 957–975, 1997.

5

Distribution

Drug concentration in blood or plasma depends on the amount of the drug present in the body as well as how extensively it is distributed. The latter can be assessed from its volume of distribution.

5.1. DEFINITION

5.1.1. Proportionality Factor

The apparent volume of distribution (V) of a drug can be viewed simply as a proportionality factor between the total amount of drug present in the entire body and the drug concentration in the reference body fluid, usually the plasma, at any given time:

$$V(t) = A(t)/C_p(t) \tag{5.1}$$

A(t) is the amount of drug in the body at time t after administration, $C_p(t)$ is the drug concentration in the plasma at time t, and V(t) is the apparent volume of distribution with respect to the drug concentration in the plasma at time t.

There are several different volume terms depending on the reference fluids, where drug concentrations are measured, including blood, plasma, and plasma water, although the apparent volume of distribution based on the *plasma* drug concentration is the one most frequently determined. Relationships among different volume terms based on drug concentrations in different reference fluids are shown below:

$$V_b(t) \cdot C_b(t) = V(t) \cdot C_p(t) = V_u(t) \cdot C_u(t) \quad [=A(t)] \tag{5.2}$$

where $C_b(t)$, $C_p(t)$, and $C_u(t)$ are the drug concentration in blood, plasma, and plasma water, respectively, and $V_b(t)$, V(t), and $V_u(t)$ are the apparent volumes of distribution with respect to blood, plasma, and unbound drug concentrations, respectively. All the volume of distribution terms discussed hereafter will be the estimated volume of distribution based on the *plasma* drug concentrations, unless otherwise indicated.

Table 5.1. General Relationships among the Volume of Distribution at Steady State (V_{ss}), the Extent of Protein Binding, and the Distribution Characteristics of Drugs[a]

V_{ss}[b] (liter/kg)	Drug	General trend in protein binding	Distribution characteristics in the body
>1	Basic	More binding in tissue than in plasma	A drug may be concentrated in particular tissues in the body[c]
0.4–1	—	Similar binding between plasma and tissue	A drug distributes uniformly throughout the body
<0.4 (close to extracellular volume)	Acidic	More binding in plasma than in tissue	Drug molecules are confined mainly in the plasma pool with limited distribution to the tissues

[a]Data taken from Nau, 1986.
[b]Volume of distribution at steady state estimated based on drug concentration in plasma.
[c]The most common mechanisms for concentrating drug in tissues are binding to tissue macromolecules or partitioning into lipids within tissues. Another factor, which can lead to V_{ss} of the drug being apparently greater than total body volume, is reversible metabolism, i.e., metabolite(s) produced from drug converts back to the parent drug.

UNITS: Usually liters or liters/kg when normalized to kg body weight.

5.1.2. Pharmacokinetic Implications of the Volume of Distribution

5.1.2.1. Proportionality Factor between the Amount of a Drug in the Body and the Drug Concentrations

The amount of drug in the body at any given time can be estimated by multiplying V(t) by the plasma drug concentration at time t.

5.1.2.2. Extent of Distribution of Drug into Tissues

The volume of distribution is a direct measure of the extent of the distribution of a drug. Although it does not represent a real physiological volume, the volume of distribution at steady state (V_{ss}) can be used to assess the extent of distribution of a drug from the plasma into the tissues. Table 5.1 summarizes the general considerations concerning the extent of protein binding and distribution into tissues based on V_{ss}.

5.1.2.3. Determinants of Half-Life

The terminal half-life of a drug, which is indicative of the duration of drug exposure during the terminal phase of a plasma drug concentration–time profile, is affected by both the *volume of distribution at the terminal phase* (V_β), not V_{ss}, and the *systemic clearance* (Cl_s):

$$t_{1/2} = \frac{0.693 \cdot V_\beta}{Cl_s} \tag{5.3}$$

5.1.3. Summary of Characteristics of the Volume of Distribution

1. V(t) is an imaginary volume correlating the amount of a drug present in the body to the concentrations of the drug measured in the reference body fluid, usually plasma, at any given time t.
2. V(t) can vary depending on the reference fluids in which the drug concentrations are measured. In general, the volume of distribution is referred to the *plasma* drug concentrations, unless otherwise indicated.
3. As an independent parameter, V(t) is indicative of the extent of space or volume into which the drug is distributed, referred to drug concentration, not the rate of distribution. How fast the distribution occurs can be estimated by distributional clearance (Cl_d).

If a drug is confined to the plasma pool, its V(t) approaches the actual volume of plasma in the body, and may not change much with time. However, if drug molecules gradually diffuse from plasma into other tissues, V(t) of the drug changes with time upon administration and can be greater than the total plasma volume.

5.2. DIFFERENT VOLUME TERMS

When a plasma concentration–time profile after intravenous injection of the drug exhibits a biexponential decline and the body can be viewed as a two-compartment model (the central and the peripheral compartments, Fig. 5.1) there are three different volumes of distribution that can be considered: the apparent volume of distribution of the central compartment (V_c), the apparent volume of distribution at steady state (V_{ss}), and the apparent volume of distribution at pseudodistribution equilibrium (V_β).

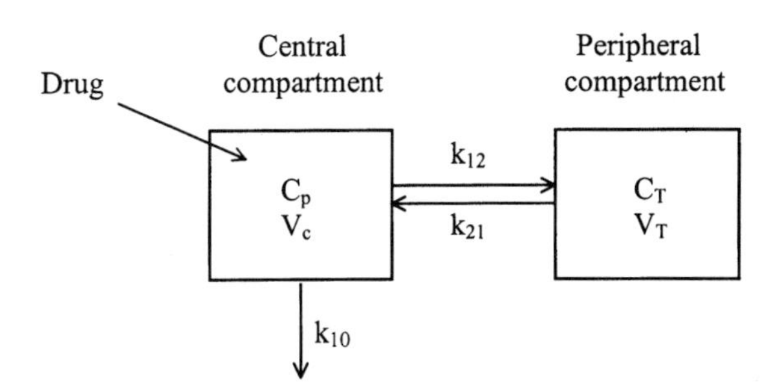

Figure 5.1. Two-compartment model of the body for drug disposition. Drug is administered and eliminated from the central compartment and distributes between the central and the peripheral compartments. C_p is the plasma drug concentration and C_T is a hypothetical average drug concentration in the peripheral compartment; k_{10} is the elimination rate constant, and k_{12} and k_{21} are the distribution rate constants; V_c and V_T are the apparent volumes of the central and the peripheral compartments, respectively.

5.2.1. Apparent Volume of Distribution of the Central Compartment

The apparent volume of distribution of the central compartment (V_c) is a proportionality factor between the amount of drug in the central compartment and the drug concentrations in the plasma. The central compartment may represent the plasma and highly perfused tissues and organs such as the liver, kidney, and spleen with which an instantaneous equilibrium of the drug in the plasma occurs.

Lower limit of V_c: After intravenous injection, drug molecules instantaneously disperse within the plasma pool and further distribute into blood cells and/or other tissues at a slower rate. Thus, V_c cannot be smaller than the actual volume of plasma in the body. The plasma volume of a human being (70 kg body weight) is approximately 3 liters (0.04 liter/kg) (Benet and Zia-Amirhosseini, 1995).

5.2.2. Volume of Distribution at Steady State

Two different approaches for estimating the volume of distribution at steady state (V_{ss}) will be discussed. V_{ss} based on a two-compartment system for the body will be described according to pharmacokinetic relationships with other volume terms. In addition, a physiologically more relevant description of V_{ss} will be discussed based on the extent of plasma and tissue protein binding and actual plasma and tissue volumes.

5.2.2.1. V_{ss} *Based on a Two-Compartment System*

By definition, V_{ss} is a ratio between the amount of drug in the body and its concentration in plasma *at steady state.* A steady state implies a condition in which the rate of change in the amount of drug [A(t)] in the body is zero, i.e., $dA(t)/dt = 0$, and can be achieved after continuous infusion of a drug when the rate of infusion equals the rate of elimination. V_{ss} can be also estimated after a single intravenous dosing. After a single intravenous bolus injection of a drug for which the disposition profile can be adequately described by a two-compartment system, the volume of distribution of that drug becomes V_{ss} at one time point, when its distribution between the central and the peripheral compartments reaches equilibrium. At this time point (i.e., distributional equilibrium), the rates of drug distribution from the central compartment into the peripheral compartment and vice versa become equal (Fig. 5.2). The pharmacokinetic expression of V_{ss} at this time point based on a two-compartment model is as follows:

$$V_{ss} = V_c \cdot (1 + k_{12}/k_{21}) \tag{5.4}$$

where k_{12} and k_{21} are the distribution rate constants from the central to the peripheral compartments and vice versa, respectively. V_c, k_{12}, and k_{21} can be calculated from the disposition parameters (A, B, α, and β) with curve-fitting of a biexponential equation [$C_p(t) = Ae^{-\alpha t} + Be^{-\beta t}$] to the plasma drug concentration–time profile (see Chapter 2).

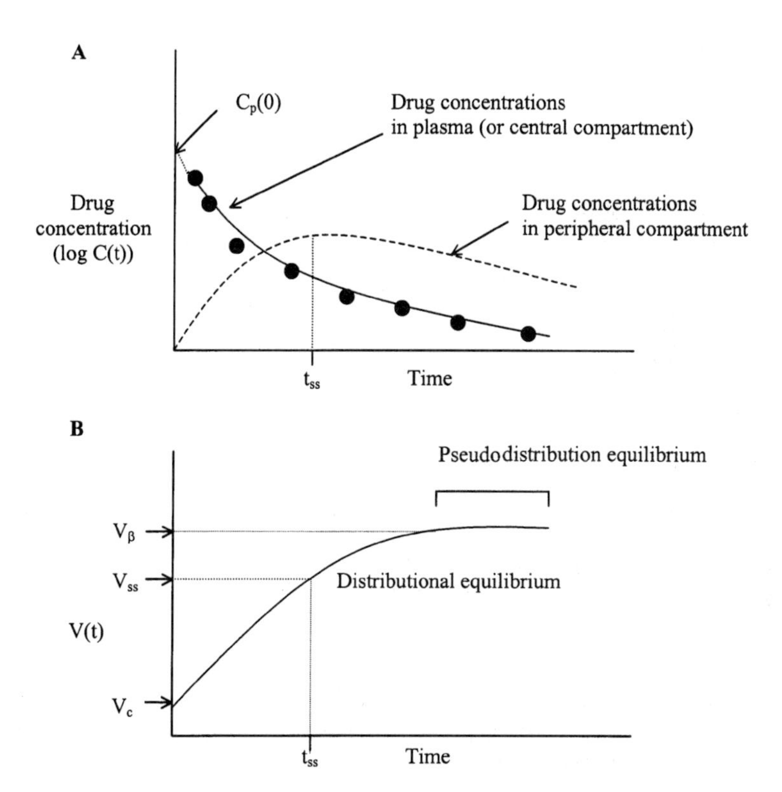

Figure 5.2. Semilogarithmic plots of measured drug concentrations (●) in plasma (central compartment, ——) and estimated drug concentrations in the peripheral compartment (– – – –) *vs.* time (A), and the corresponding changes in volume of distribution (B) with time, when the drug disposition profile can be best described with a two-compartment model. Only at time t_{ss} when V(t) becomes V_{ss}, is distributional equilibrium between the central (plasma) and the peripheral compartments attained.

5.2.2.2. V_{ss} *Based on Physiological Parameters*

Among several different volume of distribution terms, V_{ss} is especially important in that it has a certain physiological relevance that reflects the extent of drug distribution from the plasma pool into tissues and organs in the body. At equilibrium between the plasma and the tissues, the extent of distribution of the drug depends on its binding to plasma proteins, blood cells, and tissue components, and the actual total volumes of plasma and tissues:

$$V_{ss} \cdot C_{p,ss} = V_p \cdot C_{p,ss} + V_t \cdot C_{t,ss} \tag{5.5}$$

↓	↓	↓
Amount of drug in the body at steady state	*Amount of drug in plasma at steady state*	*Amount of drug in tissues at steady state*

where $C_{p,ss}$ and $C_{t,ss}$ are a plasma drug concentration and an average concentration

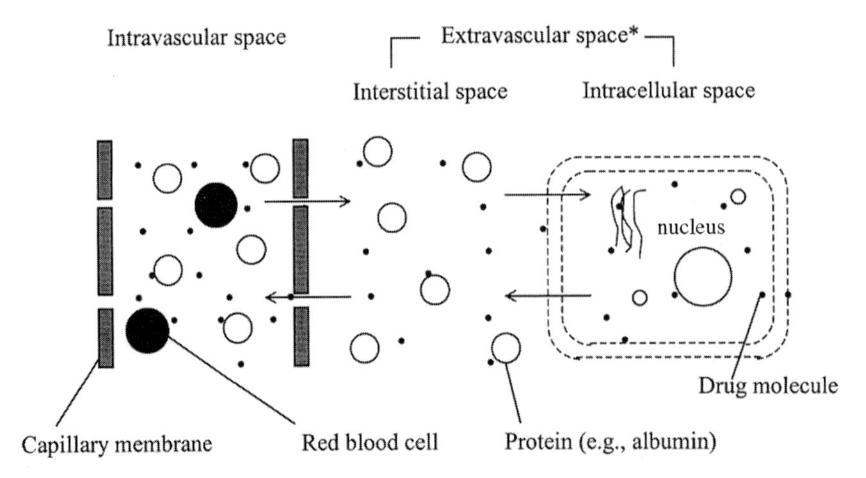

Figure 5.3. Schematic description of drug distribution. Note that only drug molecules (•) not bound to proteins (○) can transport from intravascular space into the rest of the total body water in extravascular and intracellular spaces. *Approximately 55–60% of the total extracellular (i.e., intravascular and interstitial) albumin and 40% of total α_1-acid glycoprotein are located in interstitial space.

of the drug in the tissues at steady state, and V_p and V_t are, respectively, the actual volume of the plasma and the extravascular space plus the erythrocyte volume into which the drug distributes. Therefore,

$$V_{ss} = V_p + \frac{C_{t,ss}}{C_{p,ss}} \cdot V_t \tag{5.6}$$

Drug molecules bound to a plasma protein such as albumin are not able to penetrate membrane lipid bilayers owing to the large sizes and charges of proteins. It is, therefore, assumed that only drug molecules not bound to (or free from) plasma proteins are capable of transporting between intravascular, extravascular, and intracellular spaces for distribution (free-drug hypothesis). Figures 5.3 illustrates unbound drug molecules transporting across different spaces in the body. Consequently, unbound drug concentrations in plasma and within tissues should be equal at steady state, i.e.,

$$\underbrace{f_u \cdot C_{p,ss}}_{\text{Unbound drug concentration in plasma}} = \underbrace{f_{u,t} \cdot C_{t,ss}}_{\text{Unbound drug concentration in extravascular space and red blood cells}}$$

and, hence,

$$C_{t,ss} = f_u \cdot C_{p,ss}/f_{u,t} \tag{5.7}$$

where f_u and $f_{u,t}$ are the ratios between unbound- and total-drug concentrations in plasma and tissues (extravascular space), respectively. Incorporating Eq. (5.7) into

Eq. (5.6) yields

$$V_{ss} = V_p + (f_u/f_{u,t}) \cdot V_t \tag{5.8}$$

As is clear from Eq. (5.8), the degree of protein binding in both the plasma and tissues can significantly affect V_{ss}. For instance, a drug with extensive binding to plasma proteins (small f_u) generally exhibits a small V_{ss} (Φie and Tozer, 1979). When the drug has a high affinity for tissue components (small $f_{u,t}$), V_{ss} can be much greater than the actual physiological volume of the body. Based on limited data, it has been suggested that tissue binding appears to be similar across species, whereas plasma protein binding can vary significantly and tends to be more extensive in larger animals. Unlike plasma protein binding, tissue binding of a drug cannot be readily measured, and it is thus difficult to assess the effects of changes in the latter on V_{ss}. Table 5.2 lists the physiological volumes of different body fluids.

5.2.2.3. Factors Affecting V_{ss}

Whenever physiological factors governing the volume of distribution are discussed, V_{ss} with respect to plasma drug concentrations should be considered. Basically four different factors affect the extent of V_{ss}, as was indicated in Eq. (5.8). (1) the size of entire plasma pool (V_P); (2) the actual body tissue volume (V_t); (3) the ratio between unbound- and total-drug concentrations in the plasma (f_u); (4) the ratio between unbound- and total-drug concentrations in the tissues ($f_{u,t}$). The magnitude of V_{ss} is affected not only by the physiological volume of reference fluids and tissues into which the drug distributes, but also by its physicochemical properties such as hydrophilicity and lipophilicity, which govern protein binding affinity.

5.2.3. Volume of Distribution at Pseudodistribution Equilibrium

The volume of distribution at pseudodistribution equilibrium (V_β) is simply a proportionality constant that relates the total amount of a drug present in the body

Table 5.2. The Physiological Volumes of Different Body Fluids and Tissues in an Average Adult (Body Weight 70 kg)[a]

	Volume	
Fluid	liters	liters/kg
Total body water	42	0.60
Plasma	3–3.3	0.04–0.05
Blood	5.5	0.08
Extravascular fluid	39	0.56
Interstitial fluid	11–12	0.16–0.17
Body solids and fat	20	0.28

[a]Data taken from Balant and Gex-Fabry (1990), Benet (1995), and Φie and Tozer (1979).

to the plasma drug concentrations during the pseudodistribution equilibrium phase (so-called, β-, postdistribution, elimination, or terminal phase). Once the pseudo-distribution equilibrium phase is attained, V_β multiplied by $C_p(t)$ during the phase provides a correct estimate of the amount of drug in the body at any time t thereafter.

5.3. ESTIMATING THE VOLUME OF DISTRIBUTION

The volume of distribution values can be estimated only after *intravenous bolus injection*. Equations (5.9), (5.10), and (5.11) for the measurement of different volume terms (V_c, V_{ss}, and V_β) should not be applied to data obtained after administration of a drug via other routes, even if complete bioavailability can be assumed. If a plasma drug concentration–time profile follows a monoexponential decline after intravenous injection, i.e., the body behaves like a single compartment, the three different volume terms described above become the same, and volume of distribution [V(t)] is independent of time. On the other hand, V(t) of a drug after intravenous administration changes from V_c, V_{ss} to V_β immediately after injection toward the terminal phase (Fig. 5.2), when the disposition of the drug in the body is better described with a two-compartment model, i.e., a biphasic profile of log $C_p(t)$ *vs.* time.

5.3.1. Apparent Volume of the Central Compartment

The apparent volume of the central compartment (V_c) can be estimated by dividing the intravenous dose (D_{iv}) by an estimated drug concentration in plasma at time 0, assuming that the drug injected into the systemic circulation instantaneously distributes into tissues and organs represented by the central compartment, but does not yet transport into tissues and organs represented by the peripheral compartment:

$$V_c = D_{iv}/C_p(0) \tag{5.9}$$

$C_p(0)$ is an estimated concentration at time 0 determined by backextrapolating the first two drug concentrations after intravenous injection (Fig. 5.2A), and V_c is sometimes called the initial dilution volume ($V_{extrapol}$).

5.3.2. Volume of Distribution at Steady State

The simplest way to estimate V_{ss} of a drug is to use moment analysis on the plasma drug concentration profiles after intravenous bolus injection:

$$V_{ss} = \underbrace{\frac{AUMC_{0-\infty,iv}}{AUC_{0-\infty,iv}}}_{} \cdot \underbrace{\frac{D_{iv}}{AUC_{0-\infty,iv}}}_{} = MRT_{iv} \cdot Cl_s \tag{5.10}$$

$AUC_{0-\infty,iv}$ and $AUMC_{0-\infty,iv}$ are, respectively, the area under the plasma drug concentration *vs.* time curve and the area under the first moment curve (plasma drug concentration × time *vs.* time curve) from time zero to infinity following intravenous bolus injection. Cl_s and MRT_{iv} are the systemic plasma clearance and mean residence time of a drug after intravenous injection.

5.3.3. Volume of Distribution at Pseudodistribution Equilibrium

The volume of distribution at pseudodistribution equilibrium (V_β) of a drug can be estimated by dividing Cl_s by the slope of the terminal phase (β) after intravenous administration (see Chapter 2):

$$V_\beta = \frac{Cl_s}{\beta} \tag{5.11}$$

5.3.4. Differences among V_c, V_{ss}, and V_β

V(t) of a drug with a two-compartmental disposition profile following intravenous injection increases from V_c immediately after injection, to V_{ss} when distributional equilibrium between the plasma and the peripheral compartment is achieved, and eventually to V_β when pseudodistribution equilibrium is attained (Gibaldi *et al.*, 1969). The trend of these volume changes with time is illustrated in Fig. 5.2.

Immediately after bolus injection of a drug, most of the drug molecules will remain momentarily in the plasma pool and highly perfused organs before distributing into other tissues. The drug dose divided by plasma concentration [$C_p(0)$] estimated at time zero yields the apparent volume of distribution corresponding to the central compartment, i.e., V_c, representing the volume of distribution term during this early phase.

Drug molecules will further distribute into the more slowly equilibrating tissues and/or organs (i.e., the peripheral compartment). During this initial stage of distribution, the rate of drug distribution from the plasma into those tissues is faster than that from the tissues to the plasma, simply because not much drug has yet been accumulated in the tissues. More drug molecules will be accumulated in the tissues with time, and thus the rate of drug distribution from the tissues to the plasma increases. At a certain time point (t_{ss}), the rate of drug distribution from the plasma into the tissues becomes equal to the rate of drug distribution back from the tissues to the plasma, i.e., the rate of change of the amount of the drug in the tissues becomes zero. At t_{ss}, an average drug concentration in the tissues (the peripheral compartment) reaches its highest point. The apparent volume of distribution of the drug at t_{ss} becomes V_{ss}.

After V_{ss} is attained, V(t) continues to increase as time goes on until the pseudodistribution equilibrium (terminal phase) between the plasma drug concentrations and those in all the tissues is achieved. At the pseudodistribution equilibrium stage, the ratio of the amounts of a drug between the plasma (central compartment) and the tissues (peripheral compartment) remains constant. The apparent volume of distribution of a drug during this phase is called V_β.

It is noteworthy that V_{ss} multiplied by $C_p(t)$ equals the amount of a drug remaining in the body only at one time point, t_{ss}, after intravenous bolus injection. Before or after t_{ss}, respectively, the amount of the drug in the body estimated by multiplying V_{ss} by $C_p(t)$ is either an over- or an underestimate of the true value.

5.3.5. Relationships among V_c, V_{ss}, V_β, Cl_s, and Cl_d

When a log $C_p(t)$ *vs.* time plot exhibits biphasic behavior, i.e., the slope of the plot during the initial distribution phase is much steeper than that of the terminal phase, V_β can be approximated with respect to V_c, V_{ss}, and Cl_s and distributional clearance (Cl_d) as follows (Jusko and Gibaldi, 1972; Kwon, 1996):

$$V_\beta = V_{ss} + (V_{ss} - V_c) \cdot (Cl_s/Cl_d) \tag{5.12}$$

REFERENCES

Benet L. Z. and Zia-Amirhosseini P., Basic principles of pharmacokinetics, *Toxicol. Pathol.* **23**: 115–123, 1995.

Balant L. P. and Gex-Fabry M., Physiological pharmacokinetic modelling, *Xenobiotica* **20**: 1241–1257, 1990.

Gibaldi M. *et al.*, Relationship between drug concentration in plasma or serum and amount of drug in the body, *J. Pharm. Sci.* **58**: 193–197, 1969.

Jusko W. J. and Gibaldi M., Effects of change in elimination on various parameters of the two-compartment open model, *J. Pharm. Sci.* **61**: 1270-1273, 1972.

Kwon Y., Volume of distribution at pseudo-distribution equilibrium: relationship between physiologically based pharmacokinetic parameters and terminal half-life of drug, *Pharm. Sci.* **2**: 387–388, 1996.

Nau H., Species differences in pharmacokinetics and drug teratogenesis, *Environ. Health Perspect.* **70**: 113–129, 1986.

Φie S. and Tozer T. N., Effect of altered plasma protein binding on apparent volume of distribution, *J. Pharm. Sci.* **68**: 1203–1205, 1979.

6

Clearance

Clearance is a measure of the ability of the body or an organ to eliminate a drug from the blood circulation (Gibaldi, 1986; Tozer 1981; Wilkinson, 1987). Systemic (or total body) clearance is a measure of the ability of the entire body to eliminate the drug, whereas organ clearance such as hepatic or renal clearance is a measure of the ability of a particular organ to eliminate the drug.

6.1. DEFINITION

6.1.1. Proportionality Factor

The most general definition of clearance is a proportionality factor between the rate of elimination of a drug from the entire body (systemic clearance) or an organ (organ clearance), and its concentration at the site of measurement, i.e., reference body fluid such as blood or plasma. For instance, when drug concentrations are measured in the plasma, systemic *plasma* clearance (Cl_s) can be defined as

$$\text{Systemic plasma clearance} = \frac{\text{Elimination rate of drug from the entire body}}{\text{Drug concentration in plasma}}$$

$$Cl_s = \frac{-dA(t)/dt}{C_p(t)} \tag{6.1}$$

A(t) is the amount of drug in the body and $C_p(t)$ is the plasma drug concentration at time t.

6.1.2. Apparent Volume of Reference Fluid Cleared of a Drug per Unit Time

A physiologically more meaningful definition of systemic clearance is the *apparent* volume of reference fluid such as plasma (or blood) cleared of drug per unit time. It is important to note that the clearance terms depend on where the drug concentrations are measured. The estimated clearance based on drug concentrations in blood, plasma, and plasma water (plasma without proteins) are referred to as blood clearance (Cl_b), plasma clearance (Cl_p), and clearance based on the unbound drug concentration (Cl_u), respectively. Systemic *blood* or *unbound-drug* clearance can be obtained by replacing $C_p(t)$ in Eq. (6.1) with the drug concentration in the blood [$C_b(t)$] or unbound drug concentration [Cu(t)], respectively.

NOTE: PHARMACOKINETIC IMPLICATIONS OF CLEARANCE. Clearance is not an indicator of how much of a drug is being eliminated per unit time, but rather *how much apparent volume of reference fluid is cleared of a drug per unit time.* The amount of a drug eliminated from the body over a certain period of time depends on both the extent of clearance and the concentrations of the drug in the reference fluid. For a better understanding, let us assume that there is a tank filled with water that is being well circulated and continuously filtered by a pump (Fig. 6.1). If a drop of blue dye is added to the tank, then dye molecules will be instantaneously and evenly distributed in the water by the continuous circulation caused by the pump. As the water passes through the filter, which can remove the dye molecules, its color will gradually fade. The tank and the water can be viewed as a whole body and a reference fluid, respectively. The pump can be viewed as the heart and the filter can be considered the elimination mechanisms of the body such as metabolism or renal elimination. A drop of blue dye is an intravenous dose of drug and thus blue dye molecules can be viewed as drug molecules in the body. The changes in the intensity of the color of the water with time can be viewed as a concentration *vs.* time profile of a drug in a reference fluid in the body after intravenous bolus injection.

The efficiency of the filter in removing dye molecules from the water depends on how much water passes through the filter and how efficiently the filter can remove the dye molecules from the water. The volume of the water completely cleared of the dye molecules per unit time is equivalent to the clearance of the drug in the body. For instance, the amount of dye removed from the water (the amount of drug eliminated from the body) per unit time can be calculated by multiplying the volume of water cleared of the dye per unit time (systemic clearance) by the concentration of

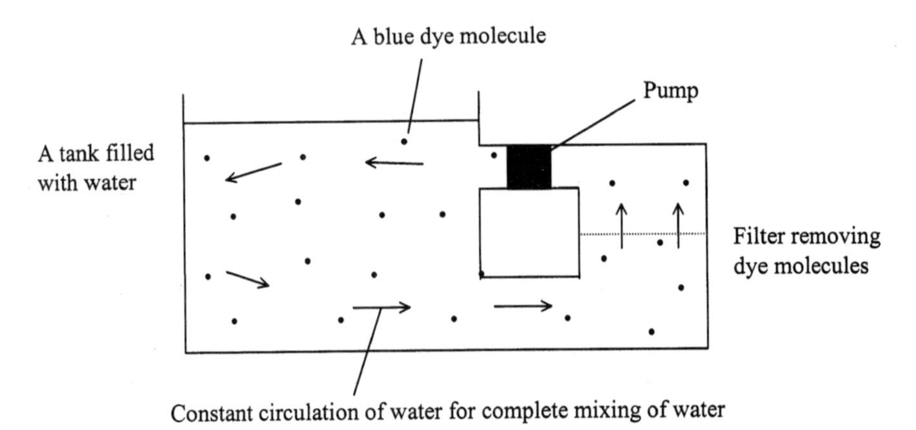

Figure 6.1. A tank filled with water being continuously circulated by a pump and filtered.

Table 6.1. Important Pharmacokinetic Implications of Clearance

Parameters	Relationship with clearance
Amount of a drug eliminated per unit time	Systemic clearance × concentration of a drug in reference fluid at time t
Amount of a drug eliminated	Systemic clearance × $AUC_{t_1-t_2}$ from time t_1 to t_2
Terminal half-life of a drug	Proportional to the apparent volume of reference fluid divided by the systemic clearance
Steady state drug concentration after continuous infusion	Infusion rate divided by systemic clearance

dye in the water (drug concentration in a reference fluid). Also, how fast the dye molecules disappear from the water (the terminal half-life of the drug) is governed by both the volume of the water completely cleared of the dye per unit time (clearance) and the total volume of the water containing the dye molecules in the tank (the apparent volume of distribution of the reference fluid). The important pharmacokinetic implications of clearance are summarized in Table 6.1.

UNITS: The unit of clearance is the same as flow rate (i.e., volume/time). For instance, ml/min or ml/min·kg (i.e., ml/min/kg) when normalized to body weight (kg).

6.2. SYSTEMIC (PLASMA) CLEARANCE

6.2.1. Estimation

In general, the systemic plasma clearance (Cl_s) of a drug can be estimated from the plasma drug concentration–time profile after intravenous bolus injection. Systemic clearance can also be measured following administration via routes other than intravenous injection as long as bioavailability is known and the total absence of administration-route-dependent differences in kinetics can be assumed. The integral of the numerator and denominator of Eq. (6.1) with respect to time yields

$$Cl_s = \frac{-\int_0^\infty [dA(t)/dt]\,dt}{\int_0^\infty C_p(t)\,dt} = \frac{A(0) - A(\infty)}{AUC_{0-\infty,iv}} \qquad (6.2)$$

$A(0)$ and $A(\infty)$ are the total amount of drug present in the body after intravenous bolus injection at time zero and infinity, respectively. $A(0)$ is, therefore, equal to the intravenous dose (D_{iv}), and $A(\infty)$ is zero because at infinite time there should be no drug left in the body. $AUC_{0-\infty,iv}$ is the area under the plasma drug concentration *vs.* time curve from time zero to infinity after intravenous injection.

The systemic plasma clearance of the drug can then be estimated by dividing D_{iv} by $AUC_{0-\infty,iv}$:

(6.3) $$\boxed{\text{Systemic plasma clearance} = D_{iv}/AUC_{0-\infty,iv}}$$

For example, if $AUC_{0-\infty,iv}$ of a drug in the plasma is 5 $\mu g \cdot hr/ml$ after intravenous administration of 10 mg/kg body weight in rats, systemic plasma clearance of this drug in rats would be 33.3 ml/min/kg as shown below:

$$\text{Systemic plasma clearance} = \frac{\overset{D_{iv}}{\downarrow}\ 10\ mg/kg}{(5\ \mu g \cdot hr)/ml\ \underset{AUC_{0-\infty,iv}}{\uparrow}} = \frac{(10 \cdot 1000\ \mu g)/kg}{(5\ \mu g \cdot 60\ min)/ml}$$

$$= 33.3\ ml/min/kg$$

In most cases, drug concentrations are measured in *plasma* rather than in *blood* because sample preparation and drug analysis in plasma are simpler than in blood, and thus systemic clearance implies systemic *plasma* clearance, unless otherwise indicated.

6.2.2. Relationship between Systemic Clearance and the Volume of Distribution

The clearance and the volume of distribution at steady state are somewhat related to each other in that the extent of plasma protein binding of a drug can affect both parameters (see Chapter 7). In most other cases, however, the clearance and the volume of distribution can be assumed to be independent of one another.

6.2.3. Relationship between Systemic Clearance and the Terminal Half-Life

Terminal half-life ($t_{1/2}$) of a drug after intravenous injection is a function of both the systemic clearance and the volume of distribution at pseudodistribution equilibrium (V_β).

(6.4) $$\boxed{t_{1/2} = \frac{0.693 \cdot V_\beta}{Cl_s}}$$

6.2.4. Amount of Drug Eliminated from the Body

The amount of a drug eliminated from the body up to time t after administration can be estimated by multiplying Cl_s by AUC from time 0 to t (AUC_{0-t}):

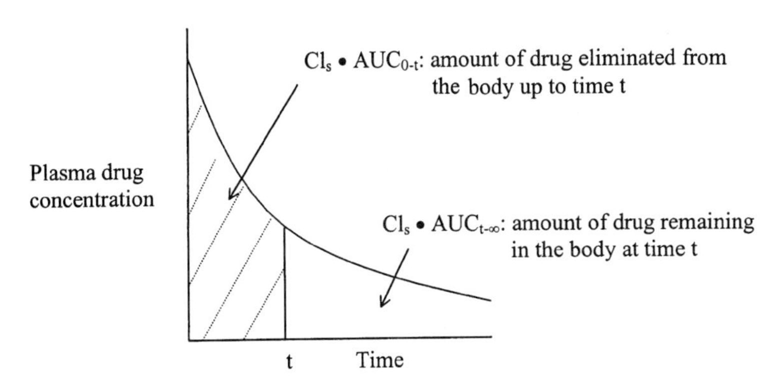

Figure 6.2. Plasma concentration–time curve of a hypothetical drug after intravenous administration on a linear scale. If the area up to time t (AUC_{0-t}) is 30% of the total AUC (i.e., $AUC_{0-\infty}$), then the indications are that 30% of the intravenous dose has been eliminated by time t. The remaining 70% of $AUC_{0-\infty}$ accounting for AUC from time t to ∞ represents the fraction of the dose remaining in the body at time t.

$$\text{Amount of drug eliminated from the body up to time t after administration} = Cl_s \cdot AUC_{0-t} \quad (6.5)$$

Equation (6.5) holds true at any time after drug administration, regardless of the route of administration. Obviously, the amount of drug remaining in the body at time t can be calculated by subtracting $Cl_s \cdot AUC_{0-t}$ from the total amount of drug administered into the systemic circulation. Figure 6.2 illustrates the relationship between the amount of drug eliminated from the body up to time t and the amount of drug remaining in the body at time t after intravenous bolus injection.

6.3. ORGAN CLEARANCE

Organ clearance reflects the ability of the eliminating organ to remove a drug from the *blood*. A pharmacokinetically more meaningful interpretation of organ clearance would be the real physiological volume of blood cleared of drug per unit time. This is because organ clearance is estimated based on the drug concentration in the *blood*, not in the *plasma*, and therefore, when organ clearance is referred to, it generally implies *blood clearance*. It is important to note that whenever physiological factors governing the clearance of drug are considered, blood clearance by the organ has to be used.

Like systemic clearance, organ clearance can also be viewed as a proportionality constant relating the elimination rate of a drug from blood by the organ to the drug concentration in blood perfusing through the organ. This concept can be best illustrated by considering elimination processes of a drug in a single organ using isolated organ perfusion under a steady state condition (Fig 6.3).

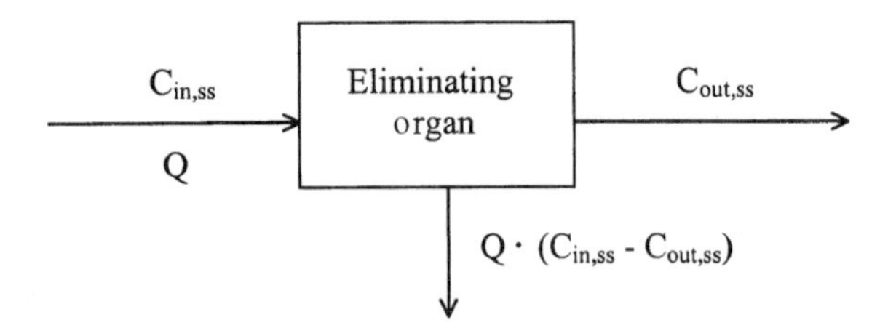

Figure 6.3. Schematic representation of organ perfusion under steady state conditions: $C_{in,ss}$: drug concentration in blood entering the organ at steady state; $C_{out,ss}$: drug concentration in blood leaving the organ at steady state; and Q: blood flow rate.

Based on mass balance at steady state, the rate of elimination of a drug by the organ is equal to the difference between the input and output rates of the drug through the organ:

$$\text{Rate of elimination at steady state} = \text{Input rate at steady state} - \text{Output rate at steady state} = Q \cdot C_{in,ss} - Q \cdot C_{out,ss}$$

$C_{in,ss}$ and $C_{out,ss}$ are, respectively, the drug concentrations in the blood entering and leaving the organ at steady state, and Q is the blood flow rate. According to the definition of clearance, organ clearance can be also described as a proportionality constant between the rate of elimination of the drug by the organ and the drug concentration in the blood, i.e., inlet blood drug concentration, at steady state:

$$\text{Organ clearance} = \frac{\text{Rate of drug elimination by the organ at steady state}}{\text{Inlet drug concentration in blood at steady state}}$$

$$\boxed{\text{Organ clearance} = \frac{Q \cdot (C_{in,ss} - C_{out,ss})}{C_{in,ss}}} \quad (6.6)$$

$$\text{Organ clearance} = Q \cdot E \quad (6.7)$$

$$E = \frac{C_{in,ss} - C_{out,ss}}{C_{in,ss}} \quad (6.8)$$

As in Eq. (6.7), organ clearance is often expressed as the blood flow rate multiplied by the extraction ratio [Eq. (6.8)]. The extraction ratio (E) represents the fraction of the amount (or concentration) of drug entering the organ that is extracted by the organ during perfusion. E is dimensionless and ranges between 0 and 1 (sometimes expressed as a percent). $E = 0$ means that the organ does not remove drug at all during perfusion, whereas $E = 1$ indicates the complete elimination of a drug from the blood by the organ during perfusion. In other words, E reflects the organ's efficiency in removing drug from the blood stream.

Let us assume that a rat liver is perfused with blood containing 10 μg/ml of a drug at 12 ml/min via the portal vein, and that the blood concentration of the drug in the hepatic vein at steady state is 2 μg/ml. The hepatic clearance of this drug is, then, 9.6 ml/min, and the extraction ratio of the liver for the drug is 0.8, as described below:

$$\text{Hepatic clearance} = \frac{\underset{\substack{\text{Flow rate}\\ \text{(ml/min)}}}{12} \times (\underset{\substack{C_{in,ss}\\ (\mu g/ml)}}{10} - \underset{\substack{C_{out,ss}\\ (\mu g/ml)}}{2})}{\underset{C_{in,ss}\ (\mu g/ml)}{10}} = 12 \times \underset{E}{0.8} = 9.6\ \text{ml/min}$$

Note: Extraction Ratio. In general, when E is greater than 0.7, between 0.3 and 0.7, or smaller than 0.3, organ clearance is considered to be high, moderate, or low, respectively. Availability of a drug after it passes through the eliminating organ can be expressed as 1 – E, which is the fraction of the amount of drug that entered the organ and was not cleared or the fraction of the volume of blood containing the drug that passed through the organ without being cleared.

Upper limit of organ clearance: As indicated in Eq. (6.6), maximum organ clearance, which can be achieved with complete removal of the drug by the organ during single-pass perfusion, i.e., $C_{out,ss} = 0$, cannot be greater than the blood flow rate perfusing through the organ.

6.3.1. Hepatic Clearance

The liver is the most important organ in the elimination of a drug from the body. It is highly perfused and under normal conditions receives approximately 75% of its blood supply from the portal vein and 25% from the hepatic artery. The highly branched capillary system and fenestrated endothelium enable direct contact between the blood components and all cell types within the organ, including, e.g., hepatocytes, Kupfer cells, and fat storage cells. Hepatocytes, the principal cell type in the liver, contain various metabolizing enzymes such as cytochrome P450 and uridine diphosphate glucuronyl transferase (UDPGT) and are well equipped with active transporters for efficient uptake of drug and excretion into the bile. In general, hepatic drug clearance implies clearance via both metabolism and biliary excretion.

Although Eq. (6.6) elucidates the concept of organ clearance rather intuitively, it has a limited applicability in estimating organ clearance because of the specific experimental conditions required, i.e., *in situ* organ perfusion at steady state, or in characterizing physiological factors governing organ clearance *in vivo*. Several pharmacokinetic models have been developed to enable estimates of organ clearance without the need for organ perfusion studies or an understanding of the physiological factors that govern organ clearance. The most well-known hepatic clearance models include “well-stirred (or venous equilibrium),” “parallel-tube (or sinusoidal perfusion),” and “dispersion” models. Pharmacokinetic differences among the models are due mainly to the differences in assumptions made on the *anatomical structure of the liver* and the *extent of blood mixing within the liver*.

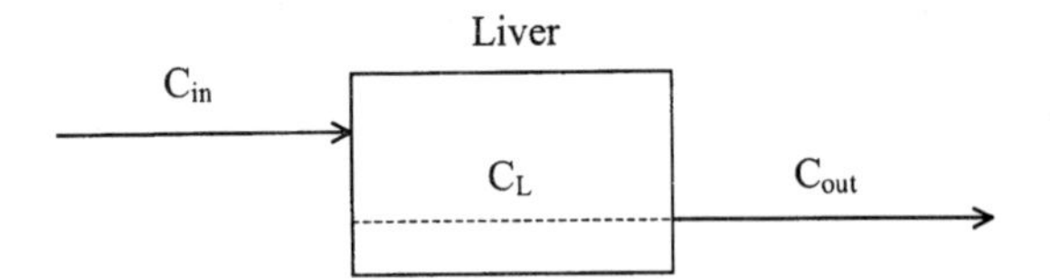

Figure 6.4. The well-stirred model: C_{in}: drug concentration in portal vein blood or the systemic circulation; C_L: drug concentration within the liver; and C_{out}: drug concentration in hepatic vein blood (i.e., emergent venous blood).

6.3.1.1. Well-Stirred (Venous Equilibrium) Model

The well-stirred model, which is the most popular one for hepatic clearance (Fig. 6.4), assumes that the entire liver, i.e., both the liver tissues including hepatocytes and the blood in the sinusoid, is well mixed, so that drug molecules are distributed instantaneously and homogeneously within the liver upon entering. As a result, the drug concentration within the liver is assumed to be equal throughout the organ (Pang and Rowland, 1977). In other words, the well-stirred model views the liver as a single compartment (*anatomy of the liver*) with a complete mixing of blood (*extent of blood mixing*). Important assumptions for the well-stirred model for hepatic clearance include: (a) only unbound drug in blood is subject to elimination (metabolism and/or biliary excretion), (b) no membrane transport barrier, (c) no concentration gradient of the drug within the liver, (d) the concentration of the drug within the liver is equal to that in emergent venous blood, and (e) linear kinetics.

Hepatic clearance (Cl_h) based on the well-stirred model is described as follows:

$$Cl_h = \frac{Q_h \cdot f_{u,b} \cdot Cl_{i,h}}{Q_h + f_{u,b} \cdot Cl_{i,h}} \tag{6.9}$$

$Cl_h = Q \cdot E$; therefore

$$E = \frac{f_{u,b} \cdot Cl_{i,h}}{Q_h + f_{u,b} \cdot Cl_{i,h}} \tag{6.10}$$

$Cl_{i,h}$ is the intrinsic hepatic clearance, Q_h is the hepatic *blood* flow rate, and $f_{u,b}$ is the ratio between unbound and total (both bound and unbound) drug concentrations in *blood*, not plasma. Since in most studies, the fraction of a drug unbound in plasma, f_u, is more frequently measured than $f_{u,b}$, it is useful to note the relationship between $f_{u,b}$ and f_u:

$$f_{u,b} = \frac{f_u \cdot C_p}{C_b} \tag{6.11}$$

since $f_{u,b} \cdot C_b = f_u \cdot C_p$, where C_b and C_p are the drug concentrations in blood and plasma, respectively. Equation (6.9) is one of the most important equations in pharmacokinetics with various applications to understanding fundamental concepts of hepatic clearance.

6.3.1.2. Parallel-Tube (Sinusoidal Perfusion) Model

The parallel-tube model (Fig. 6.5) views the liver as a group of identical tubes (*anatomy of the liver*) arranged in parallel, with metabolizing enzymes and biliary excretion functions evenly distributed around the tubes with a bulk flow of blood (*extent of blood mixing*) passing through them (Pang and Rowland, 1977). The parallel-tube model produces a concentration gradient of a drug within the liver along the blood flow path from the periportal (portal vein) to the perivenous (hepatic vein) regions. Important assumptions for the parallel-tube model are: (a) only the drug not bound to blood components is subject to elimination (metabolism and/or biliary excretion), (b) no membrane transport barrier, (c) a concentration gradient of the drug within the liver ranged between inlet and outlet drug concentrations, and (d) linear kinetics.

Hepatic clearance based on the parallel-tube model is

$$Cl_h = Q_h \cdot (1 - e^{-f_{u,b} \cdot Cl_{i,h}/Q_h}) \tag{6.12}$$

The average drug concentration ($C_{L,avg}$) within the liver is

$$C_{L,avg} = \frac{C_{in} - C_{out}}{\ln(C_{in}/C_{out})} \tag{6.13}$$

The differences between the well-stirred and the parallel-tube models are summarized in Table 6.2.

6.3.1.3. Dispersion Model

The well-stirred and the parallel-tube models represent the two extreme cases of the anatomical features of the liver and hepatic blood flow patterns. The dispersion model views the liver somewhere between these two extremes. It considers the liver to be a meshed organ (*anatomy of the liver*) with internal blood dispersion, the degree of which can be reflected by the so-called "dispersion number (D_N, *extent of blood mixing*)" (Roberts and Rowland, 1985, 1986).

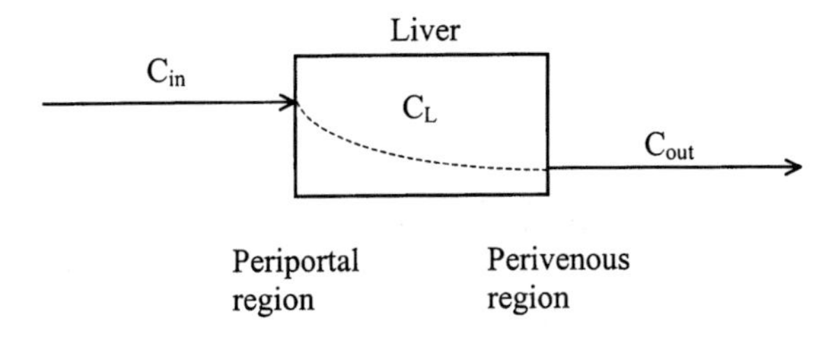

Figure 6.5. The parallel-tube model (see Fig. 6.4 for abbreviations).

Table 6.2. Differences between the Well-Stirred and the Parallel-Tube Models of Hepatic Clearance

	Well-stirred model	Parallel tube model
Liver anatomy	A single homogeneous compartment	A group of identical tubes
Blood flow	Complete mixing	Bulk flow
Drug concentration in the liver	Constant and equal to that of emergent venous blood	$(C_{in} - C_{out})/[\ln(C_{in}/C_{out})]$, decreases from periportal to perivenous regions
Hepatic clearance	$(Q_h \cdot f_{u,b} \cdot Cl_{i,h})/(Q_h + f_{u,b} \cdot Cl_{i,h})$	$Q_h \cdot (1 - e^{-f_{u,b} \cdot Cl_{i,h}/Q_h})$
Extraction ratio[a]	$f_{u,b} \cdot Cl_{i,h}/(Q_h + f_{u,b} \cdot Cl_{i,h})$	$1 - e^{-f_{u,b} \cdot Cl_{i,h}/Q_h}$

[a]The difference in hepatic clearance of the same drug between the models is not significant when its clearance is low ($E < 0.3$). When the clearance of a drug is high ($E > 0.7$), hepatic clearance estimated based on the parallel-tube model tends to be slightly higher than that using the well-stirred model. Thus, the selection of the models becomes important only for drugs showing extensive clearance.

Hepatic clearance based on the dispersion model is expressed as

$$Cl_h = Q_h \cdot (1 - F_h) \tag{6.14}$$

and

$$F_h = \frac{4a}{(1 + a)^2 \cdot \exp[(a - 1)/2D_N] - (1 - a)^2 \cdot \exp[-(a + 1)/2D_N]}$$

where, $a = (1 + 4R_N \cdot D_N)^{1/2}$ and $R_N = f_u \cdot Cl_{i,h}/Q$. D_N ranges between 0 and 1, and the well-stirred and the parallel-tube models are the two extreme cases of the dispersion model with $D_N = 1$ (complete dispersion of blood) and 0 (no dispersion of blood) within the liver, respectively. It is likely that a D_N value somewhere between these two extremes would be more reasonable for describing the degree of blood mixing inside the liver. It appears that the estimates of D_N depend heavily on the experimental conditions and substrates studied.

NOTE: INTRINSIC HEPATIC CLEARANCE $Cl_{i,h}$ reflects the inherent ability of the liver to eliminate the drug not bound in blood components (to be exact, drug molecules not bound to tissue components within the hepatocytes) and is governed solely by

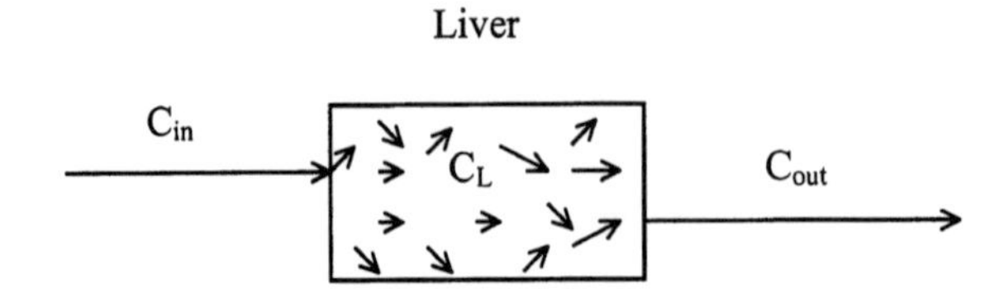

Figure 6.6. The dispersion model (see Fig. 6.4 for abbreviations).

the activities of metabolizing enzymes and/or biliary excretion (both active and passive transport of a drug into the bile). In theory, it is the same as the hepatic clearance of a drug when there are no limitations in drug delivery to the liver (sufficient and fast blood flow), protein binding (no protein binding), and other factors such as cofactor availability when needed. Physiological factors governing $Cl_{i,h}$ are as follows:

$$Cl_{i,h} = \frac{V_{max}}{K_m + C_{L,u}} \tag{6.15}$$

V_{max} is the maximum rate of metabolizing enzyme(s) and/or biliary excretion, K_m is the apparent Michaelis–Menten constant of metabolizing enzyme(s) and/or biliary excretion activity, and $C_{L,u}$ is the concentration of unbound drug within the hepatocytes available for metabolizing enzyme(s) and/or biliary excretion. If $C_{L,u}$ is much smaller than $K_m (C_{L,u} < 0.1 \cdot K_m)$, $Cl_{i,h}$ becomes simply a ratio between V_{max} and K_m. At low $C_{L,u}$, therefore, $Cl_{i,h}$ is a concentration-independent constant:

$$Cl_{i,h} = V_{max}/K_m \tag{6.16}$$

(a) Factors Affecting Hepatic Clearance. According to Eq. (6.9), there are basically three different physiological factors governing hepatic clearance: (1) *hepatic blood flow rate*, indicating how fast the drug can be delivered to the liver, (2) *fraction of a drug not bound to blood components*, reflecting what portion of a drug in the blood is available for clearance, (3) *intrinsic hepatic clearance*, representing the intrinsic ability of the liver to remove a drug from the blood by metabolism and/or biliary excretion, when there are no restrictions in drug delivery (blood flow), protein binding, and cofactor availability.

(b) Applications of Hepatic Clearance Models. The most important value of hepatic clearance models is probably their ability to elucidate how the three different factors described above that affect hepatic clearance, i.e., blood flow, protein binding, and intrinsic clearance, are related each other. In some models, the effects of the degree of blood dispersion inside the liver on clearance can also be taken into account.

(i) Estimation of intrinsic hepatic clearance. $Cl_{i,h}$ of a drug can be estimated according to Eq. (6.9), if Cl_h, Q_h, and $f_{u,b}$ are known. Cl_h can be estimated either by measuring the extraction ratio from *in situ* perfusion studies under steady state conditions [Eq. (6.6)] or by determining the *in vivo* systemic clearance. If there is experimental evidence suggesting that the systemic clearance of a drug is solely via hepatic elimination, systemic blood clearance of the drug determined based on blood drug concentrations can be considered to equal Cl_h. For Q_h, published values (Table 6.3) are usually used; $f_{u,b}$ can be measured experimentally *in vitro* (see Chapter 7).

Table 6.3. Hepatic Blood Flow Rate of Laboratory Animals and Humans[a]

Species	Body weight (kg)	Hepatic blood flow rate (ml/min/kg)
Mouse	0.02	90
Rat	0.25	47.2, 81
Monkey	5	43.6
Dog	10	30.9
Human	70	20.7

[a]Data taken from Davies and Morris (1993) and Houston (1994).

(ii) Estimating hepatic clearance from in vitro experiments. It is possible to predict Cl_h if $Cl_{i,h}$ can be estimated from experimentally measured V_{max} and K_m values of metabolizing enzymes *in vitro* (see Chapter 12).

(c) Limits of Hepatic Clearance. Cl_h of a drug cannot be greater than Q_h, which is the upper limit of Cl_h. Let us think about the following two extreme cases of Cl_h based on the well-stirred model [Eq. 6.9)].

IF $f_{u,b} \cdot Cl_{i,h}$ IS MUCH GREATER THAN Q_h (HIGH CLEARANCE):

Cl_h approaches Q_h:

$$Cl_h = \frac{Q_h \cdot f_{u,b} \cdot Cl_{i,h}}{Q_h + f_{u,b} \cdot Cl_{i,h}} \approx \frac{Q_h \cdot \cancel{f_{u,b}} \cdot \cancel{Cl_{i,h}}}{\cancel{f_{u,b}} \cdot \cancel{Cl_{i,h}}} = Q_h$$

When Cl_h is greater than 70% of Q_h (in other words, $E \geqslant 0.7$), it is generally considered high extraction or high hepatic clearance for the drug. In this case, changes in $f_{u,b}$ and/or $Cl_{i,h}$ do not affect Cl_h to any significant extent, whereas alterations in Q_h can have substantial effects.

IF $f_{u,b} \cdot Cl_{i,h}$ IS MUCH SMALLER THAN Q_h (LOW CLEARANCE):

Cl_h becomes limited by $f_{u,b} \cdot Cl_{i,h}$:

$$Cl_h = \frac{Q \cdot f_{u,b} \cdot Cl_{i,h}}{Q + f_{u,b} \cdot Cl_{i,h}} \approx \frac{\cancel{Q} \cdot f_{u,b} \cdot Cl_{i,h}}{\cancel{Q}} = f_{u,b} \cdot Cl_{i,h}$$

When Cl_h of a drug is lower than 30% of Q_h ($E \leqslant 0.3$), it is generally considered low extraction or low hepatic clearance. In this case, Cl_h of a drug is not much affected by changes in Q_h, whereas alterations in $f_{u,b}$ and/or $Cl_{i,h}$ can have significant effects.

6.3.2. Biliary Clearance

Once a drug molecule gets into the hepatocytes, it can be subject to both metabolism and biliary excretion. Excretion of a drug into the bile occurs through canalicular membranes surrounding the bile canaliculi of the hepatocytes. Hepatic clearance (Cl_h) of a drug is the sum of its clearances via both metabolism (Cl_m) and biliary excretion (Cl_{bl}):

$$Cl_h = Cl_m + Cl_{bl} \tag{6.17}$$

By definition, Cl_{bl} of a drug is a proportionality constant relating the rate of biliary excretion of unchanged drug to drug concentration in the blood:

$$Cl_{bl} = \frac{(\text{Bile flow rate}) \cdot (\text{Drug concentration in bile})}{\text{Drug concentration in blood}} \tag{6.18}$$

As the bile flow is relatively slow, approximately 0.06 and 0.008 ml/min/kg in rats and humans, respectively, Cl_{bl} would be small, unless the drug concentration in the bile is substantially higher than that in the blood.

Enterohepatic circulation: A drug excreted in the bile passes into the intestine. Some or most of the secreted drug may be reabsorbed from the intestine, undergoing enterohepatic circulation (EHC) and the rest of it is excreted in the feces. This may be repeated many times until the drug is ultimately eliminated from the body by other elimination processes such as metabolism or renal or fecal excretion. In this way, EHC may increase the persistence of a drug in the body. If some of the drug excreted in the bile is subject to EHC, the biliary excretion process is no longer solely an elimination process, but rather contains components of both elimination and distribution processes.

6.3.3. Renal Clearance

The renal clearance (Cl_r) of a drug consists of four different processes—glomerular filtration (Cl_f), active secretion (Cl_{rs}), passive reabsorption (F_r), and renal metabolism (Cl_{rm}):

$$Cl_r = (Cl_f + Cl_{rs}) \cdot (1 - F_r) + Cl_{rm} \tag{6.19}$$

$$Cl_f = f_{u,b} \cdot GFR \tag{6.20}$$

$$Cl_{rs} = \frac{Q_r \cdot f_{u,b} \cdot Cl_{i,s}}{Q_r + f_{u,b} \cdot Cl_{i,s}} \tag{6.21}$$

Cl_f, Cl_{rs}, and Cl_{rm} are clearances representing glomerular filtration, active secretion, and renal metabolism. $Cl_{i,s}$ and Q_r are intrinsic renal tubular secretion clearance by active transporter(s) and renal blood flow rate, respectively. F_r is the fraction of the drug reabsorbed into the blood from the urine after excretion. GFR is the glomerular filtration rate, at which plasma water is filtered through the glomerulus, e.g., approximately 125 ml/min in a 70-kg man.

6.3.3.1. Glomerular Filtration

In the Bowman capsule (glomerulus), drug molecules not bound to blood components are physically filtered through the glomerular capillary because of the renal artery blood pressure (Fig. 6.7). For instance, molecules smaller than 15 Å in

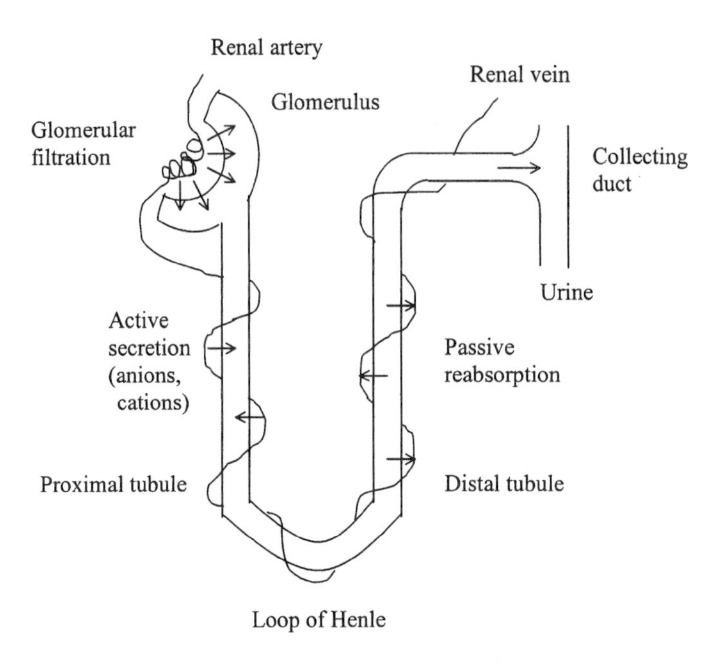

Figure 6.7. A schematic description of renal clearance processes.

diameter can readily pass through the glomeruli with approximately 125 ml of plasma/min in a healthy adult [glomerular filtration rate (GFR)], which is less than 20% of the total renal blood flow rate of 650 to 750 ml/min. This physical filtration process is represented by the Cl_f of a drug. As only unbound drug in the blood can be filtered, Cl_f is a function of both $f_{u,b}$ and GFR [Eq. (6.20)]. Creatinine (endogenous substance) or inulin can be used to assess GFR because of their negligible protein binding, tubular secretion, and reabsorption.

6.3.3.2. Active Secretion

Active secretion of a drug occurs mainly at the proximal tubule by transporters located in the tubular membranes. It has been reported that there are several distinct transporters for various cationic or anionic substrates (Giacomini, 1997). Equation (6.21), which describes Cl_{rs} is similar in principle to the well-stirred hepatic clearance model. The difference is that in the kidney, active secretion clearance is mediated by membrane transporters, whereas hepatic clearance is governed by both metabolizing enzyme and biliary excretion activities. Cl_{rs} is affected by Q_r, $f_{u,b}$, and $Cl_{i,s}$. $Cl_{i,s}$ can be described as T_{max}/K_m, where T_{max} is the maximum capacity of transporter(s) and K_m is the apparent Michaelis–Menten constant under linear conditions.

6.3.3.3. Reabsorption

In general, reabsorption of drug molecules into the renal venous blood occurs mainly at the distal tubule. Owing to reabsorption of the drug from the urine before it is excreted into the bladder, renal clearance has to be corrected for the fraction of drug (F_r) excreted in urine that is reabsorbed before release into the bladder. The extent of F_r depends on the lipophilicity and ionizability of a drug. As a rule of thumb, reabsorption can be nearly complete for a drug with log D > 0. Because this process is considered to be passive and diffusional, urinary pH can also play an important role for the reabsorption of weak acidic or basic drugs, as drug molecules un-ionized at urinary pH are more readily reabsorbed than their ionized counterparts. On an average, urinary pH is close to 6.3. Diet, drugs, and various disease states can alter urine pH. Under forced acidification and alkalinization, urinary pH can vary from approximately 4.4 to 8.2.

6.3.3.4. Renal Metabolism

Although metabolism in the kidney is a minor elimination pathway for most compounds, renal metabolism, especially glucuronidation and amino acid conjugation, appears to play an important role in renal clearance of several drugs, such as zidovudine (Lohr *et al.*, 1998).

NOTE: ESTIMATING RENAL CLEARANCE. Equations (6.19)–(6.21) are useful for understanding physiological mechanisms of renal drug elimination; however, they have limited practical value in estimating Cl_r, owing to experimental difficulties in measuring $Cl_{i,s}$ and F_r *in vivo*. The Cl_r of a drug can be determined experimentally *in vivo* without organ perfusion simply by measuring drug concentrations in blood and urine simultaneously. By definition, Cl_r is a proportionality ratio between renal elimination rate of a drug and its concentration in blood.

$$Cl_r = \frac{(\text{Urine flow rate}) \cdot (\text{Drug concentration in urine})}{\text{Drug concentration in blood}} \tag{6.22}$$

One drawback of this approach is that it needs measurement of the *in vivo* urine flow rate, which is difficult to measure accurately. Alternatively, Cl_r can be estimated by dividing the amount of unchanged drug excreted into the urine over an extended period of time (usually over 24 hr after drug administration in small laboratory animals) by AUC_{0-t} in blood, regardless of the route of administration [Eq. (6.23)]. The amount of unchanged drug excreted into the urine can be obtained by the urinary drug concentration multiplied by the volume of urine collected, and, in general, $AUC_{0-t} > 90\%\ AUC_{0-\infty}$ is desirable for a reliable estimate of Cl_r:

$$\boxed{Cl_r = \frac{\text{Amount of unchanged drug excreted in the urine up to time t}}{AUC_{0-t}}} \tag{6.23}$$

The difference between systemic clearance (Cl_s) and Cl_r, both experimentally measured, is called nonrenal clearance (Cl_{nr}), which can be assumed to be equal to Cl_h when the liver is the major clearance organ:

$$Cl_s = Cl_r + Cl_{nr} \tag{6.24}$$

Creatinine clearance and "intact nephron hypothesis": Renal clearance of creatinine, an endogenous end product of muscle metabolism with a narrow range of constant concentrations in the blood, is widely recognized as a reliable indicator of kidney impairment. Plasma protein binding of creatinine is negligible, and its tubular active secretion and reabsorption are also considered minimal. Owing to these characteristics of creatinine, its renal clearance is close to the actual GFR, which ranges from 100 to 125 ml/min in man. It has been observed that creatinine clearance correlates well with the entire kidney function; e.g., creatinine clearance lower than 50 ml/min is indicative of moderate to severe renal failure. These empirical observations suggest that renal impairments do not selectively affect any particular kidney function or cell type, but rather affect the entire nephron. This relationship between creatinine clearance and the functions of the entire nephron is known as "the intact nephron hypothesis."

6.4. RELATIONSHIP BETWEEN SYSTEMIC BLOOD AND ORGAN CLEARANCES

The systemic blood clearance (Cl_b) can equal the sum of the organ clearances by each of the eliminating organs when the following conditions are met: (1) Elimination of the drug from the body occurs solely via the eliminating organs (in a case where the drug is unstable in the blood, systemic blood clearance might be greater than the sum of organ clearances). (2) The eliminating organs except the liver and intestine are anatomically arranged independently of each other:

$$Cl_b = Cl_r + Cl_{h,app} + Cl_{others} \tag{6.25}$$

where Cl_r is the renal clearance, $Cl_{h,app}$ is the (apparent) hepatic clearance, and Cl_{others} is the sum of clearances via other organs, including, e.g., the lung, brain, and muscle (but not the intestine).

For instance, the liver and the kidney neither share a blood vessel nor are they connected to each other, so that the extent of drug clearance by one of them does not affect that by the other. The intestine and the liver are, however, directly connected to each other by a single vein, i.e., the portal vein, which makes most of the blood perfusing through the intestine, except blood from the lower part of rectum, flow into the liver as well. Because of this anatomical arrangement of the organs, the extent of apparent hepatic clearance of a drug *in vivo* is also affected by intestinal elimination. In general, it is assumed that the systemic intestinal metabolism, i.e., the elimination of a drug in the systemic circulation by the intestine, is negligible for most compounds owing to the limited diffusibility of compounds across the basement membranes between the splanchnic blood and the enterocytes. When drug molecules in blood are not subject to intestinal metabolism, $Cl_{h,app}$ in Eq. (6.25) becomes the true hepatic clearance of the drug. For those compounds subject to both systemic intestinal and hepatic clearances *in vivo*, $Cl_{h,app}$ becomes the apparent clearance, reflecting both intestinal and hepatic elimination [Eq. (6.26)]. This is because the liver is connected to the intestine via a single vein, which makes both organs act as one eliminating organ in a kinetic sense (Kwon, 1997):

$$Cl_{h,app} = Cl_g + (1 - E_g) \cdot Cl_h \tag{6.26}$$

where Cl_g is intestinal clearance, E_g is the intestinal extraction ratio of the drug, and Cl_h is hepatic clearance.

6.5. APPARENT CLEARANCE FOLLOWING ORAL DOSING

Apparent clearance following oral dosing (Cl_{po}) of a drug is simply the ratio between an oral dose (D_{po}) and AUC from time zero to infinity after oral administration ($AUC_{po,0-\infty}$), and is sometimes referred to as "oral clearance":

$$Cl_{po} = D_{po}/AUC_{po,0-\infty} \tag{6.27}$$

The Cl_s of a drug is D_{iv}/AUC_{iv}, and if D_{po} and D_{iv} are the same, then

$$Cl_{po} = \frac{AUC_{iv,0-\infty} \cdot Cl_s}{AUC_{po,0-\infty}} = \frac{Cl_s}{F} \tag{6.28}$$

where F is the oral bioavailability. Cl_{po} does not have any particular pharmacokinetic implications other than as the ratio between the systemic clearance and the oral bioavailability, except when the following conditions are met: (a) complete absorption of a drug after oral administration, (b) only hepatic clearance for drug elimination, (c) linear kinetics, and (d) blood clearance.

Under the above conditions, F becomes equal to the drug availability $(1 - E)$ of hepatic clearance, and thus Cl_{po} becomes the same to $f_{u,b} \cdot Cl_{i,h}$ by substituting $1 - E$ for F. This relationship according to the well-stirred model for hepatic clearance is as follows:

$$Cl_s = \frac{Q \cdot f_{u,b} \cdot Cl_{i,h}}{Q + f_{u,b} \cdot Cl_{i,h}} \quad \text{and} \quad 1 - E = \frac{Q}{Q + f_{u,b} \cdot Cl_{i,h}}$$

Therefore,

$$Cl_{po} - Cl_s/F - f_{u,b} \cdot Cl_{i,h} \tag{6.29}$$

6.6. DISTRIBUTIONAL CLEARANCE

One of the least explored parameters in pharmacokinetics may be distributional clearance (Cl_d). Cl_d reflects the ability of the body to distribute drug molecules from the plasma into the organs or tissues and vice versa (Jusko, 1986), and is a function of drug permeability (P) across the membrane and the surface area of the membrane (S):

$$Cl_d(cm^3/sec) = P(cm/sec) \cdot S(cm^2) \tag{6.30}$$

Cl_d has the same unit of flow rate, usually cm^3/sec, as any other clearance term.

6.7. BLOOD *VS.* PLASMA CLEARANCES

Estimates of clearance can vary depending on the reference body fluids, such as blood or plasma, for which drug concentrations are measured. For instance, the systemic clearance of a drug, which is referred to as drug concentration in *blood*, is "systemic *blood* clearance," whereas if drug concentrations in *plasma* are used, the clearance becomes "systemic *plasma* clearance."

6.7.1. Blood Clearance

Blood clearance can be viewed as the *actual* volume of blood cleared of a drug per unit time from the entire blood pool in the body (systemic blood clearance) or from the blood pool passing through the eliminating organs (organ blood clearance). The systemic blood clearance is the sum of organ blood clearances.

6.7.2. Plasma Clearance

Plasma clearance does not represent the actual volume of plasma cleared of a drug. It is rather the *apparent* volume of plasma cleared per unit time, reflecting simply the ratio between the rate of drug elimination from the entire body (or the organs) and the drug concentrations in the plasma. Plasma clearance is more widely used than blood clearance, however, because sample preparation and analysis are easier in plasma than in blood. If one wishes to estimate the extraction ratio via the eliminating organ, the plasma clearance value has to be converted to the blood clearance value based on the concentration ratios between the plasma and the blood.

6.7.3. Relationship between Blood and Plasma Clearances

For most pharmacokinetic applications, it does not matter much whether clearance measurements are based on blood or plasma drug concentrations. However, this does become important when clearance estimates are compared directly with an organ blood flow rate to obtain the extraction ratio of the organ clearance. In this case, blood clearance must be used because the organ clearance relates to organ *blood* flow rate, not *plasma* flow rate, and the extraction ratio is measured based on differences in drug concentrations in *blood*, not plasma, between entering and leaving the organ. A direct comparison of a plasma clearance value with the blood flow rate of the eliminating organ to assess the organ's drug extraction ratio is, therefore, incorrect, unless both the plasma and the blood drug concentrations are the same. In general, blood clearance is considered a more appropriate measure of organ function than plasma clearance whenever physiological implications of clearance are considered. The relationships between blood clearance (Cl_b) and plasma clearance (Cl_p) can be obtained from the definition of clearance, which is the ratio between the rate of drug elimination (dA/dt) and drug concentrations in the reference fluid, blood (C_b), or plasma (C_p):

$$Cl_b = \frac{dA/dt}{C_b(t)} \quad \text{and} \quad Cl_p = \frac{dA/dt}{C_p(t)}$$

Therefore,

$$Cl_b \cdot C_b = Cl_p \cdot C_p \tag{6.31}$$

6.7.4. Relationship between Blood and Plasma Concentrations

Relationships between blood and plasma drug concentrations and unbound drug concentration are illustrated in Fig. 6.8 and can be described by the following equation:

$$C_b = Hct \cdot C_r + (1 - Hct) \cdot C_p \tag{6.32}$$

where C_b, C_p, and C_r are the drug concentrations in the blood, plasma, and red blood cells, respectively, and Hct is the hematocrit (a ratio of the volume of erythrocytes to the volume of blood, which is usually 0.4–0.5).

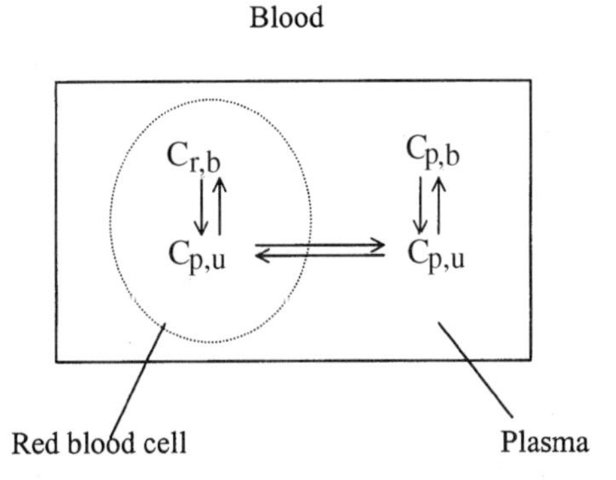

Figure 6.8. Diagram showing drug concentrations in different blood components. The unbound drug concentration in plasma is the same as that inside the red blood cells, if drug transport across the red blood cell membranes occurs via passive dffusion: $C_{p,b}$: concentration of drug bound to plasma proteins; $C_{p,u}$: concentration of drug not bound to plasma proteins; and $C_{r,b}$: concentration of drug bound to red blood cell components.

If drug binding is more extensive to red cells than to plasma proteins, $C_b/C_p > 1$, and if there is stronger binding to plasma proteins, $C_b/C_p < 1$. Plasma clearance is often found to be greater than the blood flow rate of the eliminating organ(s) such as the liver, which is the upper limit of the organ clearance. This may indicate that the drug partitions more to red blood cells than to plasma, and drug molecules bound to red blood cells are also available for extraction by the organ. C_b/C_p ratios of most drugs are between 0.8–1.2. It is, therefore, reasonable to assume that Cl_p of drugs is generally similar to Cl_b, and the extent of binding to plasma proteins or red blood cells for most drugs within the therapeutic window is independent of drug concentration.

6.7.5. Clearance Based on Unbound Drug Concentration in Plasma

Occasionally, it is observed that the plasma clearance increases with the drug concentration. This apparent concentration-dependent clearance may be simply due to nonlinear protein binding. In this case, clearance measurements based on unbound drug concentration in plasma (Cl_u) should remain unchanged, regardless of the drug concentration. The relationship between Cl_p and Cl_u is

$$Cl_p \cdot C_p = Cl_u \cdot C_u \tag{6.33}$$

Since C_u equals $f_u \cdot C_p$ and f_u is a ratio between the unbound and total drug concentrations in plasma, Cl_p equals $Cl_u \cdot f_u$. It is important to note that Cl_u does not have any particular physiological meaning and should be viewed simply as a proportionality constant between the elimination rate of a drug from the body and the unbound drug concentrations in plasma.

6.7.6. Relationship among Blood, Plasma, and Unbound Drug Clearances

From Eqs. (6.31) and (6.33) (Tozer, 1981), the following relationship can be established:

$$Cl_b \cdot C_b = Cl_p \cdot C_p = Cl_u \cdot C_u \tag{6.34}$$

For example, Cl_p of 33.3 ml/min/kg does not mean that 33.3 ml of plasma is cleared of drug per min per kg of body weight. Rather, it shows that 33.3 ml of *apparent* volume of plasma, which is equal to Cl_b, i.e., the *actual* volume of blood cleared of drug per minute in the entire body per kg body weight, multiplied by C_b/C_b, a ratio of drug concentrations between the blood and the plasma.

REFERENCES

Davies B. and Morris T., Physiological parameters in laboratory animals and humans, *Pharm. Res.* **10**: 1093–1095, 1993.

Giacomini K. M., Membrane transporters in drug disposition, *J. Pharmacokinet. Biopharm.* **25**: 731–741, 1997.

Gibaldi M., The basic concept: clearance, *J. Clin. Pharmacol.* **26**: 330–331, 1986.

Houston J. B., Utility of *in vitro* drug metabolism data in predicting *in vivo* metabolic clearance, *Biochem. Pharmacol.* **47**: 1469–1479, 1994.

Jusko W. J., Guidelines for collection and analysis of pharmacokinetic data, in W. E. Evans, J. J. Schentag, and W. J. Jusko (eds.), *Applied Pharmacokinetics: Principles of Therapeutic Drug Monitoring*, 2nd ed., Applied Therapeutics, Washington, 1986, pp. 19–37, 1986.

Kwon Y., Effects of diffusional barriers on the extent of presystemic and systemic intestinal elimination of drugs, *Arch. Pharm. Res.* **20**: 24-28, 1997.

Lohr J. W. *et al.*, Renal drug metabolism, *Pharmacol. Rev.* **50**: 107–141, 1998.

Pang K. S. and Rowland M., Hepatic clearance of drugs: I. Theoretical considerations of a "well-stirred" model and a "parallel tube" model. Influence of hepatic blood flow rate, plasma and blood cell binding, and the hepatocellular enzymatic activity on hepatic drug clearance, *J. Pharmacokinet. Biopharm.* **5**: 625–654, 1977.

Roberts M. S. and Rowland M., Hepatic elimination-dispersion model, *J. Pharm. Sci.* **74**: 585–587, 1985.

Roberts M. S. and Rowland M., A dispersion model of hepatic elimination: 1. Formulation of the model and bolus consideration, *J. Pharmacokinet. Biopharm.* **14**: 227–260, 1986.

Tozer T. N., Concepts basic to pharmacokinetics, *Pharmacol. Ther.* **12**: 109–131, 1981.

Wilkinson G. R., Clearance approaches in pharmacology, *Pharmacol. Rev.* **39**: 1–47, 1987.

7

Protein Binding

Upon entering into plasma, most drugs bind rapidly to blood constituents. When the phenomenon of protein binding of a drug is considered, it is usually the protein binding of drug molecules to blood components, including blood cells, albumin, and α_1-acid glycoprotein. The extent of binding can vary with both drug and protein concentrations. Binding of a drug to plasma and tissue proteins is a saturable process, and is generally considered reversible with rapid equilibrium within milliseconds. For most drugs, protein binding at physiologically relevant concentrations seems to be concentration-independent.

It is generally assumed that only unbound drug is able to transport across membranes and become subject to absorption, distribution, metabolism, and excretion processes. Characterization of protein binding of a drug and the effects of various pathophysiological conditions, such as disease states and concomitant medications, on protein binding are important for an understanding of the pharmacokinetic behavior of a drug. Furthermore, as only unbound drug is considered to be able to interact with pharmacological receptors, the proper integration of drug pharmacokinetics and pharmacodynamics should be based on a thorough understanding of the nature and extent of drug protein binding in both plasma and tissues.

7.1. DEFINITION

Protein binding indicates how much of the total amount of a drug in plasma or tissue is bound to plasma or tissue proteins.

Plasma proteins: Important physiological functions of plasma proteins include maintenance of the osmotic pressure of the blood and transport capacity for numerous endogenous and exogenous substrates through specific and/or nonspecific binding. Albumin and α_1-acid glycoprotein are the two major proteins in plasma, with albumin being by far the most abundant [approximately 4% (w/v) of plasma] (Table 7.1). Albumin in plasma is approximately 40% of the total albumin in the body with the rest being found mainly in interstitial fluid (Fig. 7.1). The distribution pattern of albumin in the human body is summarized in Table 7.2. In general, albumin appears to have higher binding affinity for acidic compounds. The content of α_1-acid glycoprotein in plasma is less than 0.1% (w/v), and basic compounds tend

Table 7.1. Plasma Proteins in Humans

Plasma protein	Molecular weight (kDa)[a]	Concentration in plasma (g/dl)	Drugs binding to the protein[b]
Albumin	69	3.5–5.0	Acidic
α_1-acid glycoprotein	44	0.04–0.1	Basic
Lipoproteins	200–3400	Variable	Basic
Globulins	140	2.5	—
Steroid binding globulin (transcortin)	53	0.003–0.007	Steroids (cortisol)
Fibrinogen	400	0.3	—

[a]kDa = 1000.
[b]Preferable characteristics of drugs binding to the corresponding protein.

to bind more to α_1-acid glycoprotein, although they do bind to albumin to a significant extent as well. In general, the binding affinity for α_1-acid glycoprotein, which is referred as the acute phase reactant protein, is much higher than for albumin. The primary physiological role of lipoproteins is the synthesis and transport of endogenous fatty acids such as triglycerides, phospholipids, and cholesterol. It has also been found that lipoproteins can be important in the binding of very lipophilic and/or basic compounds. Various specific proteins are responsible for the plasma binding and transport of certain endogenous compounds, including hormones. In addition, the red and white blood cells as well as the platelets can bind drugs, especially basic drugs, although such binding is usually minor (Wilkinson, 1983).

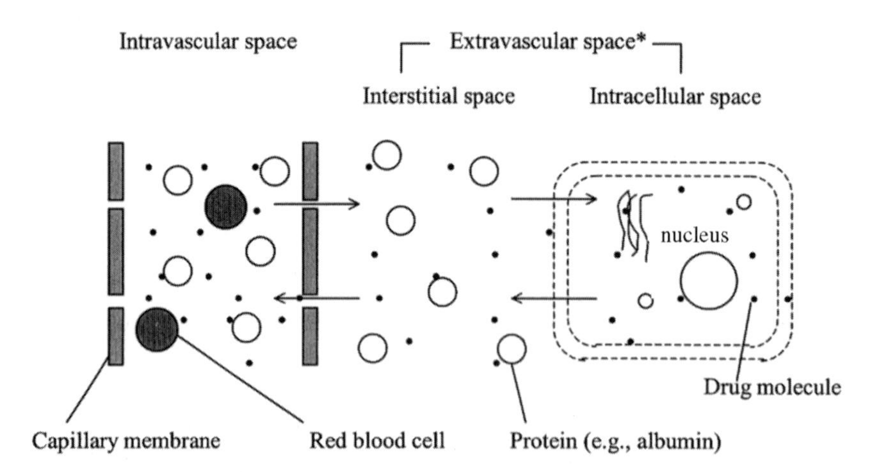

Figure 7.1. Schematic description of drug distribution. Note that only drug molecules not bound (•) to proteins (○) can distribute from intravascular space into interstitial and intracellular spaces throughout the rest of the total body water. *Approximately, 55–60% of the total extracellular (intravascular and interstitial) albumin is found in interstitial space (after Φie and Tozer, 1979).

Table 7.2. Distribution of Albumin in Human Plasma and Tissue[a]

Organ	Concentration (mg/g organ)	Amount (g/70 kg man)	Total albumin in the body (ca. %)
Intravascular			
Plasma[b]	43	140	40
Extravascular[c]			
Skin[d]	12	60	18
Muscle[e]	1.7	50	15
Intestine	4	8	2
Liver	1.4	2	1
Heart, kidney, lungs, spleen	—	11	3
Other tissues	3	79	21
Total	—	210	60
Total in the body		350	100

[a]Data taken from Øie and Tozer (1979) and Rothschild *et al.* (1955).
[b]The plasma volume is about 5% of body weight.
[c]Extravascular space consists of interstitial and intracellular space, and interstitial fluid volume is about 17% of body weight in a normal 70-kg man. Albumin is located primarily in the interstitial fluid.
[d]Skin mass is about 6–7% of the body weight and about 60% of the skin weight is interstitial fluid.
[e]Muscle constitutes about 40–45% of the body weight and about 10–16% of muscle is interstitial fluid.

7.2. ESTIMATING THE EXTENT OF PROTEIN BINDING

The binding of a drug to proteins can be viewed as a reversible and rapid equilibrium process. In the simplest case, assuming only one reversible-binding site in protein for a drug molecule, the binding equilibrium between a drug and proteins can be described in the following scheme:

$$[D] + [P] \underset{k_{-1}}{\overset{k_1}{\rightleftharpoons}} [DP] \tag{7.1}$$

where [D] is the unbound-(free)-drug concentration and [P] is the concentration of protein that is free of drug; [DP] is the concentration of the drug–protein complex, i.e., the concentration of drug bound to protein; k_1 and k_{-1} are the association and dissociation rate constants, respectively. At equilibrium,

$$k_1 \cdot [D] \cdot [P] = k_{-1} \cdot [DP] \tag{7.2}$$

Rearranging Eq. (7.2) yields the association constant (K_a):

$$K_a = \frac{k_1}{k_{-1}} = \frac{[DP]}{[D] \cdot [P]} \tag{7.3}$$

The ratio of unbound and total plasma drug concentrations (f_u) is

$$f_u = \frac{[D]}{[D] + [DP]} \tag{7.4}$$

Table 7.3. Advantages and Disadvantages of Equilibrium Dialysis and Ultrafiltration for Measuring the Extent of Protein Binding of Drug in Plasma[a]

	Equilibrium dialysis	Ultrafiltration
Advantages	Considered as standard method	Needs small amount of sample (<1 ml)
	Temperature controlled	Fast (takes ~30 min)
	Thermodynamically sound	No buffer needed
		Commercially available kits
		Disposable device (easy cleanup)
		Small changes in drug concentration during filtration
Disadvantages	Long time to reach equilibrium[b]	Nonspecific binding of drug to plastic tube or ultrafiltration membrane
	Need of buffer	Volume of ultrafiltrate may not be sufficient for drug assay
	Degradation of unstable compounds	Usually not temperature controlled
	Dilution of drug[c]	Constriction of membrane pores during ultrafiltration
	Donnan ion effect	
	Volume shift[d]	Donnan ion effect
	pH changes	
	Nonspecific binding to dialysis device and membrane	
Applications	More suitable for highly (>98%) protein bound drugs	Suitable for fast screening when nonspecific binding is less than 10%.
		More applicable for highly concentrated protein solutions or tissue homogenates

[a]Information taken from Bower *et al.* (1984) and Pacifici and Viani (1992).
[b]Using a commercial dialysis membrane, a dialysis of 4 hr at 37°C appears to be optimal for most drugs. Owing to the long incubation time, the possible degradation of proteins and the chemical or enzymatic stability of the drug should be taken into consideration.
[c]The initial concentration of drug in plasma decreases during incubcation as the plasma and the buffer equilibrate. Equilibrium dialysis may be inappropriate when there are significant changes in the extent of protein binding of a drug resulting from its dilution in plasma with buffer during equilibrium.
[d]Owing to the osmotic pressure difference between plasma (high) and buffer (low), water molecules from the buffer side are continuously moving into the plasma side during incubation, causing an increase in plasma volume and a decrease in buffer volume as compared to the original values.

There are several *in vitro* methods for measuring the unbound drug concentration in plasma, including equilibrium dialysis, ultrafiltration, ultracentifugation, gel filtration, and albumin column (Oravcova *et al.*, 1996). Among them, equilibrium dialysis and ultrafiltration are the two most commonly used for determining the unbound drug concentrations in plasma, serum, or diluted tissue homogenate. In general, equilibrium dialysis is considered to be the standard method for protein binding measurements; however, ultrafiltration can be adopted as the initial method for conducting protein binding studies, because it is less time-consuming and involves simpler sample preparation. Neither of these methods is free of biological, chemical, and physical artifacts, and the advantages and disadvantages of each are summarized in Table 7.3.

7.2.1. Equilibrium Dialysis

Equilibrium dialysis is based on the establishment of an equilibrium state between plasma containing a drug and a buffer after a period of incubation, usually longer than 2 hr, at a fixed temperature (e.g., 37°C). The equilibrium dialysis chambers for plasma containing a drug and a buffer free of drug are separated by a semipermeable membrane, which allows only low-molecular-weight ligands, such as drug molecules, to transport between the two chambers (Fig. 7.2). Sodium or potassium phosphate buffers at pH 7.4 are the ones most commonly used, although for some compounds others are required owing to the formation of insoluble salts or interactions with drug binding sites in protein molecules.

As illustrated in Fig. 7.2, water molecules from the buffer side are continuously moving into the plasma side during incubation because of the difference in osmotic pressure between the plasma and the buffer and/or the Donnan ion effect. This phenomenon is called the "volume shift," i.e., an increase in plasma volume and a decrease in buffer volume compared to their initial values. The ratio of unbound and total drug concentrations in plasma (f_u) can be estimated after equilibrium dialysis using the following equation with a correction factor for the volume shift, which is usually about 15–20%:

$$f_u = \frac{C_{be}}{(C_{pe} - C_{be}) \cdot (V_{pe}/V_{pb}) + C_{be}} \quad (7.5)$$

C_{pe} and C_{be} are the concentrations of the drug in the plasma and buffer sides of the

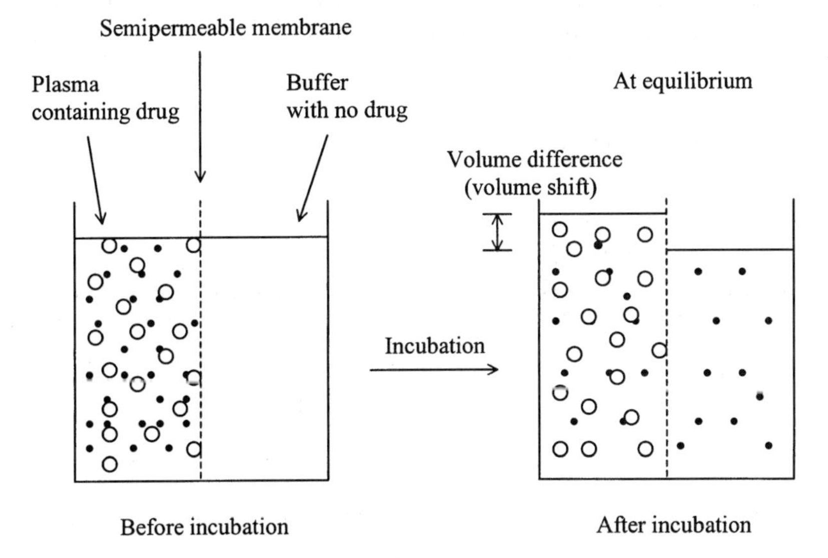

Figure 7.2. Schematic description of the equilibrium dialysis process between drug molecules not (•) bound to proteins (○) in plasma and a buffer with no drug molecules via a semipermeable membrane. There is a volume shift from the buffer side to the plasma side, while reaching an equilibrium of free drug molecules between the two sides owing to the higher osmotic pressure of the plasma containing the drug.

equilibrium chambers at equilibrium after incubation, respectively, and V_{pb} and V_{pe} are the original volume of the plasma before incubation and the volume of the plasma at equilibrium after incubation, respectively. It should be stressed that maintenance of physiological temperature and pH during the experiment is important for accurate assessment of protein binding of drugs (Boudinot and Jusko, 1980; McNamara and Bogardus, 1982; Tozer *et al*., 1983).

7.2.2. Ultrafiltration

Ultrafiltration is based on physical separation of free drug molecules in plasma water (plasma without proteins) from drug bound to plasma proteins by filtering plasma samples through a semipermeable membrane under a positive pressure generated by centrifugation (Judd and Pesce, 1982). Concentration of a drug in an ultrafiltrate is an unbound drug concentration at the particular plasma drug concentration examined. There are several advantages to the ultrafiltration method over equilibrium dialysis. Since ultrafiltrate is free of proteins, small fraction sample preparation for drug assay is relatively simple. Ultrafiltration takes about 30 min, which is significantly faster than equilibrium dialysis, and as the ultrafiltration device is disposable, cleanup after experiments is easy. A major drawback of this method is the potential nonspecific binding of a drug to the plastic tube and ultrafiltration membrane. Although the nonspecific drug binding to the ultrafiltration apparatus can be corrected by performing separate studies with plasma water spiked with the drug, equilibrium dialysis is considered to be more reliable for protein binding measurements, if nonspecific binding to ultrafiltration apparatus is greater than 20%.

In order to correct nonspecific binding of a drug to the centrifuge device, plasma water (PW), can be spiked with a known amount of the drug and ultrafiltrated. A correction factor for nonspecific binding, i.e., the ratio between drug concentration in the original PW (C_{pw}) and concentration of the drug in PW ($C_{pw,f}$) after

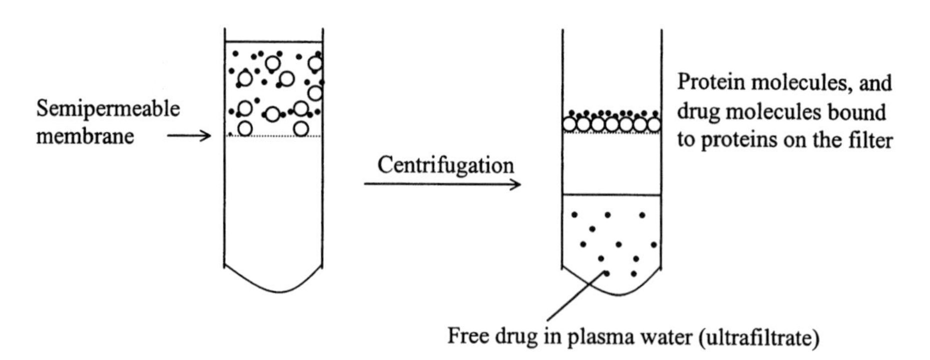

Figure 7.3. Schematic description of ultrafiltration. Drug molecules (•) not bound to protein (○) and plasma water can pass through a semipermeable filter with a molecular-weight cutoff against plasma protein, whereas drug molecules bound to proteins remain on the top of the filter.

ultrafiltration, can be incorporated into the estimate of f_u using Eq. (7.6):

Correction factor for nonspecific binding measured using PW spiked with drug

$$f_u = \frac{C_f \cdot (C_{pw}/C_{pw,f})}{C_p} \tag{7.6}$$

C_f is the drug concentration to be measured in the ultrafiltrate *after* centrifugation of the plasma containing the drug and C_p is the concentration of the drug in the plasma *before* centrifugation.

NOTE: EFFECTS OF pH CHANGES IN PLASMA ON PROTEIN BINDING. The physiological pH of blood (or plasma) is tightly controlled *in vivo* within a range between 7.2 and 7.6. Immediately upon withdrawal from animals, the pH of blood starts to rise, up to 8.2 owing to the escape of blood CO_2 into the atmosphere, which may affect the nature and extent of protein binding of a drug. To avoid this problem, plasma has to be treated with a small volume of concentrated phosphoric acid prior to protein binding experiments to adjust the pH to 7.4.

7.2.3. Microdialysis

Microdialysis has been used to estimate concentrations of unbound drug present in the extracellular fluids in various tissues or organs or in blood *in vivo* (Fig. 7.4). For microdialysis sampling, a dialysis capillary with a typical outer diameter of 500 μm is implanted in tissues or within blood vessels, and the capillary is then

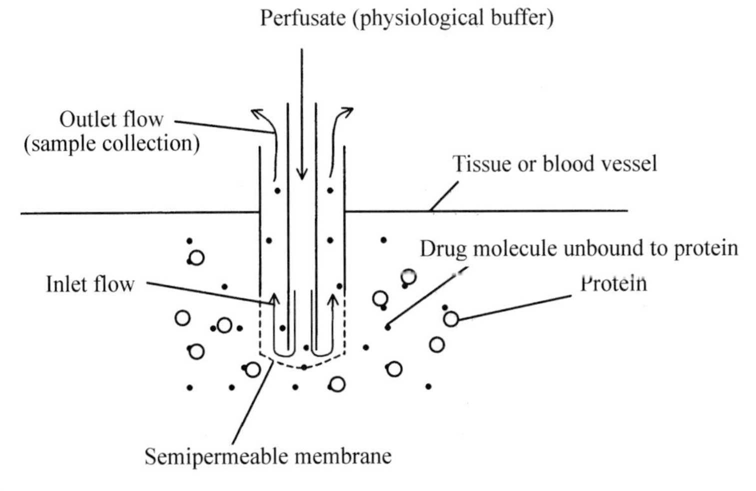

Figure 7.4. Schematic diagram describing a microdialysis probe in tissues or blood vessels.

perfused with a physiological buffer such as Ringer's solution at a low flow rate (usually <2 μl/min). Owing to the semipermeable membrane of microdialysis with a typical molecular weight cutoff of 20 kDa, only low-molecular-weight compounds transport across the membrane into the perfusate. Concentrations of a drug in perfusate collected from microdialysis, therefore, are concentrations of drug that is not bound to blood or tissue components (Elmquist and Sauchuk, 1997). Microdialysis allows continuous sampling from awake, freely moving animals, along with simple sample preparation for assay (Telting-Diaz *et al.*, 1992). Owing to the relatively small volume of samples (typically 0.1–10 μl/min for 1–5 min), the major limitation of microdialysis is the sensitivity of the assay.

Assessment of microdialysis probe recovery: Since microdialysis operates under nonequilibrium conditions, the concentration of analyte in the dialysate tends to be lower than that in the extracellular fluid surrounding the dialysis probe. The ratio between these concentrations is defined as a relative recovery of analyte. Relative recovery decreases as the flow rate of the perfusate increases. It is important to assess recovery efficiency *in vivo* rather than *in vitro*, since diffusion coefficients determined *in vitro* can be substantially different from those measured *in vivo*.

7.3. PHARMACOKINETIC AND PHARMACODYNAMIC IMPLICATIONS OF PROTEIN BINDING

It is assumed that only unbound drug is available for uptake into tissue, subject to the intrinsic elimination activities, and able to bind to pharmacological target sites or tissue proteins governing the onset, duration, and intensity of its pharmacological efficacy [free hormone hypothesis (Mendel, 1989)]. Bound drug must dissociate from the drug–protein complex before it becomes available for these processes. The extent of protein binding in plasma is, therefore, considered one of the important physiological factors affecting disposition profiles as well as the pharmacological efficacy of a drug.

7.3.1. Effects on Clearance

Clearance of a drug can be significantly affected by the extent of its protein binding. In general, a high-protein-bound drug is considered to have lower systemic clearance than a low-protein-binding drug, when other pharmacokinetic conditions are similar.

7.3.1.1. Hepatic Clearance

To be eliminated via hepatic metabolism and/or renal excretion, the drug molecules in blood must dissociate from the plasma proteins, after partitioning out of the red blood cells in case they are bound to those cells. For a better understanding of the relationship between protein binding in plasma (in this case, *blood* to be exact) and hepatic clearance (Cl_h), let us consider the well-stirred hepatic clearance model (see Chapter 6):

(7.7)
$$\boxed{Cl_h = \frac{Q_h \cdot f_{u,b} \cdot Cl_{i,h}}{Q_h + f_{u,b} \cdot Cl_{i,h}}}$$

$f_{u,b}$, $Cl_{i,h}$, and Q_h are the ratio of the unbound and total drug concentrations in blood, the intrinsic hepatic clearance, and the hepatic blood flow rate, respectively. For a drug with a low extraction ratio, i.e., $f_{u,b} \cdot Cl_{i,h} \ll Q_h$, Cl_h becomes proportionally related to $f_{u,b}$ and $Cl_{i,h}$. For a drug with a high extraction ratio, i.e., $f_{u,b} \cdot Cl_{i,h} \gg Q_h$, Cl_h becomes similar to Q_h, and independent of $f_{u,b}$ and $Cl_{i,h}$ (Balant and Gex-Fabry, 1990). The relationship between $f_{u,b}$ and Cl_h can be better understood after Eq. (7.7) is rearranged in terms of $f_{u,b}$ and Cl_h:

(7.8)
$$\frac{1}{f_{u,b}} = Cl_{i,h} \cdot \left(\frac{1}{Cl_h}\right) - \frac{Cl_{i,h}}{Q_h}$$

As indicated in Eq. (7.8), the reciprocal of $f_{u,b}$ is positively related to the reciprocal of Cl_h. The extensive protein binding of a drug in blood can, therefore, result in a low hepatic clearance and vice versa. It is important to note that $f_{u,b}$ is the ratio of unbound and total drug concentrations estimated under *equilibrium* conditions between association and dissociation of drug molecules with *blood components*. An estimate of hepatic clearance of a drug based on the conventional hepatic clearance models such as the well-stirred model (see Chapter 6) is valid only when it can be assumed that the binding equilibrium between unbound and bound drug molecules exists within the sinusoids of the liver.

Basically, three different processes determine the fate of drug bound to proteins in the capillary bed (or blood) during perfusion through the eliminating organ: the rate of dissociation of drug from its protein complex, the permeability of unbound drug across the capillary membrane, and the mean transit time of drug molecules passing through the organ. Binding of a drug to plasma proteins is considered to be more rapid than the dissociation of the drug–protein complex. For some drugs, dissociation from the drug–protein complex may take longer than the transit time along the capillary bed during perfusion, and, thus, there may not be binding equilibrium between unbound and bound drugs in the capillary bed in the case of rapid cellular uptake of a drug from the blood. In this case, the rapid cellular uptake and subsequent elimination of unbound drug from the blood can result in the dissociation of the drug from proteins becoming the rate-limiting process in overall cellular uptake and elimination (Weisiger, 1985). On the other hand, rapid dissociation of a drug in addition to fast membrane permeation and subsequent elimination can make the extent of dissociation of the complex and cellular uptake of a supplementary amount of free drug *in vivo* greater than expected from *in vitro* $f_{u,b}$ measurements. This phenomenon becomes more apparent when the capillary transit time of a drug in the organ is long (Tillement *et al.*, 1988).

Relationship between unbound fractions of drug in blood and plasma: In general, the free drug concentration in red blood cells is considered to be the same as that in plasma, unless there is an active transport system(s) in the membranes of red blood cells:

$$\underbrace{f_{u,b} \cdot C_b}_{\text{Unbound drug concentration in blood}} = \underbrace{f_u \cdot C_p}_{\text{Unbound drug concentration in plasma}} \tag{7.9}$$

C_b and C_p are the (total) drug concentrations in blood and plasma, and $f_{u,b}$ and f_u are the ratios of unbound and total drug concentrations in the blood and plasma, respectively.

7.3.1. Renal Clearance

Protein binding is especially important in the urinary excretion of unchanged drug, as renal clearance (Cl_r) is closely related to $f_{u,b}$:

$$Cl_r = f_{u,b} \cdot \left(GFR + \cdot \frac{Q_r \cdot Cl_{i,s}}{Q_r + f_{u,b} \cdot Cl_{i,s}} \right) \cdot (1 - F_r) + Cl_{rm} \tag{7.10}$$

$Cl_{i,s}$ and Q_r are the intrinsic clearance for renal tubular secretion by active transporter(s) and renal blood flow rate, respectively; Cl_{rm} is the renal metabolism; F_r is the fraction of the drug reabsorbed back into the blood from the urine after excretion; and GFR is the glomerular filtration rate. As indicated in Eq. (7.10), protein binding of a drug can affect its glomerular filtration, active secretion, and renal metabolism. Clearance by glomerular filtration, in particular, is directly related to $f_{u,b}$, as this is a physical filtration process with a molecular-weight cutoff (Balant and Gex-Fabry, 1990). Like Cl_h, Cl_r of a high-protein-bound drug tends to be lower than that of a less extensive protein-binding drug, when other conditions are similar.

7.3.2. Effects on the Volume of Distribution

The volume of distribution of a drug at steady state (V_{ss}) referred to drug concentration in plasma is affected by the extent of binding to both plasma proteins and tissue components [Eq. (7.11)], because only the unbound drug is considered to be capable of transporting across the membranes:

$$\boxed{V_{ss} = V_P + \frac{f_u}{f_{u,t}} \cdot V_t} \tag{7.11}$$

f_u and $f_{u,t}$ are the ratios of unbound and total drug concentrations in plasma and tissues, respectively; and V_P and V_t are the actual physiological plasma and extravascular volumes into which the drug distributes. As indicated in Eq. (7.11), the degree of protein binding in both plasma and tissue can significantly affect the extent of V_{ss}. For instance, a drug with extensive binding to plasma proteins (small f_u) generally exhibits a small V_{ss}. On the other hand, when a drug has a high affinity for tissue components (small $f_{u,t}$), V_{ss} can be much greater than the actual volume of the body. It is, however, difficult to reliably assess the effects of tissue protein binding on V_{ss}, because unlike protein binding in plasma, it is difficult to obtain an accurate estimate of the extent of drug binding in tissues (Benet and Zia-Amirhusseini, 1995).

7.3.3. Effects on Half-Life

Owing to the fact that the terminal half-life of a drug ($t_{1/2}$) is dependent on both the volume of distribution in the pseudodistribution equilibrium phase (V_β) and the systemic clearance, it is difficult to reliably predict the effects of changes in protein binding on $t_{1/2}$:

$$t_{1/2} = \frac{0.693 \cdot V_\beta}{Cl} \tag{7.12}$$

7.3.4. Effects on Pharmacological Efficacy

The blood protein binding of drugs targeting receptors inside cells can have a significant effect on their efficacy since only unbound drug is available for interaction with receptors (du Souich *et al.*, 1993). In fact, *in vitro* potency of a drug can be drastically reduced in the presence of plasma protein in an incubation buffer, compared to that in the absence of protein. In most clinical cases, however, the changes in the extent of protein binding of a drug alone may have only a limited effect on its efficacy because the changes in protein binding are usually small *in vivo*, and thus may not alter the unbound drug concentrations to any significant extent.

7.3.5. Effects on Drug–Drug Interaction

Displacement from plasma binding of one drug by another *in vivo* may not lead to an increase in the concentration of unbound drug in plasma, because the drug molecules released from the plasma proteins can further distribute into tissues and the transient increase in the unbound drug concentration will be buffered by tissue binding. As indicated in Table 7.4, displacement of a drug from plasma proteins by

Table 7.4. Potential Effects of a Decrease in Plasma Protein Binding of One Drug by Another on Systemic Clearance, Volume of Distribution at Steady State, Terminal Half-Life, and Unbound Concentrations in Plasma

	Effects of an increase in f_u of drug A by drug B[b]				
Clearance of drug A[a]	Cl	V_{ss}	$t_{1/2}$	C_p	C_u
High ($\approx Q_h$)	$\leftrightarrow$	$\uparrow$	$\uparrow$	$\leftrightarrow^c$, $\downarrow^d$	$\uparrow^c$, $\leftrightarrow^d$
Low ($\approx f_u \cdot Cl_{i,h}$)	$\uparrow$	$\uparrow$	$\downarrow$	$\downarrow$	$\leftrightarrow$

[a] Assuming systemic elimination of drug A occurs primarily via hepatic clearance.
[b] C_u: unbound concentration of A in plasma ($C_u = f_u \cdot C_p$); $Cl_{i,h}$: intrinsic hepatic clearance; f_u: ratio of unbound and total concentrations of drug A in plasma; and Q_h: hepatic blood flow rate. Note that C_u is not affected by an increase in f_u, except after parental administration of the drug.
[c] After parenteral administration of drug A.
[d] After oral administration of drug A. Complete oral absorption is assumed. C_p: total concentration of drug A in plasma.

other drugs *in vivo* does not significantly alter its unbound drug concentrations (C_u), so plasma protein binding interactions between different drugs are rarely of clinical significance. Exceptions can be found for a drug with a high systemic clearance after parenteral administration, in which case its C_u increases when it is displaced in proteins by other drug(s).

7.4. FACTORS AFFECTING PROTEIN BINDING

In general, basic compounds tend to bind more to α_1-acid glycoprotein, whereas albumin seems to have higher binding affinity for acidic compounds. Various endogenous ligands such as bilirubin, free fatty acids, heparin, pregnancy factors, uremic middle molecules, and uremic peptides can inhibit protein binding of drugs. Studies of drug binding to a particular protein(s) in plasma are useful for defining the major binding site(s) for a particular drug; however, physiological or clinical implications are rather limited. Concentrations of albumin and α_1-acid glycoprotein in plasma can be altered under various pathophysiological conditions (Table 7.5).

Table 7.5. Various Physiological or Pathological Conditions Altering Protein Concentrations in Plasma[a]

Changes	Albumin	α_1-acid glycoprotein	Lipoprotein
Increase	Benign tumor, exercise, hypothyroidism, gynecological disorder, myalgia, neurological diseases, psychosis, schizophrenia	Age (geriatric) acute illness, burns,[b] chronic pain syndrome, enzyme induction by phenobarbital (in dogs), infection, inflammations,[b] morbid, obesity,[b] myocardiac infarction, neoplastic diseases, renal failure, rheumatoid arthritis, smoking, stress, surgery,[b] trauma[b]	Diabetes, hypothyroidism, liver disease, nephrotic syndrome
Decrease	Injury, acute infection, age (neonates, geriatric), bone fractures, burns,[c] chronic bronchitis, cystic fibrosis, freezing, gastrointestinal diseases, leprosy, liver diseases,[c] malnutrition, neoplasms, pregnancy, renal failure,[c] surgery, trauma	Age (neonates), nephrotic syndrome, oral contraceptives, pregnancy (?)	Hyperthyroidism, trauma

[a]Data taken from Jusko and Gretch (1976), Notarianni (1990), Φie (1986), and Verbeeck *et al.* (1984).
[b]Factors causing more than 50% increase in α_1-acid glycoprotein, compared to the normal value.
[c]Factors causing a larger decrease in albumin concentration than the others.

7.5. NONLINEARITY OF PLASMA PROTEIN BINDING

The extent of protein binding of a drug can be drug- or protein-concentration-dependent, based on the affinity and capacity of the plasma protein. It is desirable to measure the extent of protein binding *in vitro* at more than two different concentrations, one close to the therapeutic concentration and another close to a toxic concentration level. For drugs with multiple protein binding sites, the fraction of a drug free in plasma can vary with drug concentrations such that linear protein binding can occur over multiple ranges of drug concentrations, as illustrated in Fig. 7.5 (Boudinot and Jusko, 1980).

7.6. PLASMA *VS.* SERUM AND *IN VITRO VS. EX VIVO* PROTEIN BINDING MEASUREMENTS

In general, the extent of protein binding of drugs to plasma proteins is based in "drug-spiked serum or plasma" *in vitro*. Between serum and plasma, serum is preferable for protein binding studies because of the possible interference of anticoagulants, such as heparin, EDTA, or citric acid, when plasma is used. The extent of plasma protein binding of a drug can be affected by qualitative and/or quantitative changes in the plasma proteins under various pathophysiological conditions (Table 7.4). Because of the possible interactions of metabolites of drugs copresent in plasma, for a highprotein-bound drug, it is more desirable to measure *ex vivo* protein binding in blood samples obtained from animals or individuals exposed to the drug than otherwise.

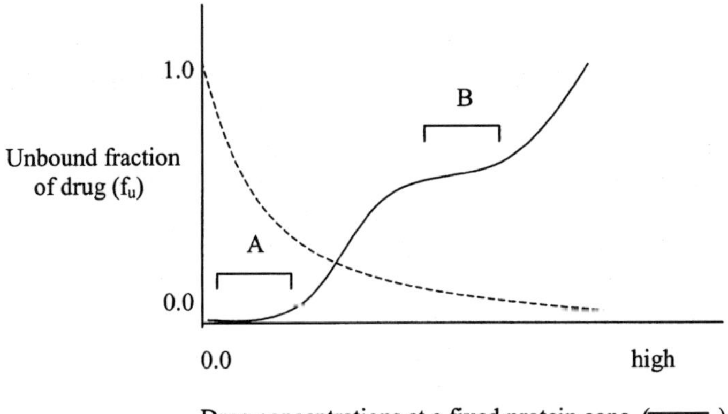

Figure 7.5. Drug- or protein-concentration-dependent changes in protein binding of a hypothetical drug with two discrete binding sites of protein molecules. A, B: Linear protein binding region over the ranges of drug concentrations, reflecting high-affinity/low-capacity and low-affinity/high-capacity binding sites of the protein, respectively.

7.7. PROTEIN BINDING IN TISSUES

Determining the "true" unbound and total drug concentrations in pharmacological target tissues at the sampling time is of utmost pharmacological importance in understanding the relationships among *in vitro* potency, *in vivo* efficacy, and toxicological responses of a drug. In addition, the degree of protein binding to the tissues can have a more pronounced effect on the extent of drug distribution because the total amount of protein in the tissues exceeds by several times the quantity of plasma protein. Not much is known about tissue binding of a drug compared with plasma protein binding, owing to the fact that the reliable *in vitro* estimation of drug binding to tissue components *in vivo* is experimentally much more difficult than in plasma. In addition to tissue proteins, there are also membrane lipids and other macromolecular constituents involved in drug binding, all of which make protein-binding study in tissues more complicated.

Several methods using isolated organ perfusion, tissue slices, tissue homogenate, or isolated subcellular organelles have been investigated to estimate the extent of drug binding to or within tissues. All the techniques, however, suffer from various methodological problems that confine the determination of tissue binding to relatively simple experimental approaches, such as ultrafiltration of diluted tissue homogenates.

7.7.1. General Trends in Drug Binding to (Muscle) Tissues

Not much is known concerning the characteristics of protein binding of drugs in tissues. The following is a summary of some experimental findings, but their general applicability to a large number of compounds has not been thoroughly examined (Fichtl *et al.*, 1991; Kurz and Fichtl, 1983).

- For many drugs, the extent of binding to intact muscle tissues *in vivo* can be extrapolated from *in vitro* data obtained using tissue homogenates. This is because the extent of protein binding in tissue homogenates is found to be almost linearly related to the protein concentrations within them.
- Ultrafiltration appears to be more suitable than equilibrium dialysis for measuring the extent of protein binding in tissue homogenates. In general, tissue homogenate diluted with saline or phosphate buffer [1:1–1:4 (w/v)] is used for ultrafiltration studies. Equilibrium dialysis of tissue homogenate tends to take much longer (sometimes up to 48 hr) than ultrafiltration.
- Unlike in plasma protein binding, there appear to be no species-dependent differences in the extent of binding of drugs to muscle.
- For acidic compounds, there tends to be more binding to plasma proteins than to muscle.
- For basic compounds, there tends to be more binding to tissue than to plasma proteins.
- There is a good positive correlation between plasma and muscle tissue binding of structurally related compounds.

- Muscle tissue binding seems to be linear over a wide range of drug concentrations of many drugs.
- In general, drug binding to different tissues decreases in the rank order of liver > kidney > lung > muscle.

7.7.2. Pharmacokinetic Implications of Tissue Binding

The following is the summary of important effects of changes in tissue binding on drug disposition profiles:

- *Clearance:* Clearance of a drug is not affected by the extent of the tissue binding [Eq. (7.7)].
- *Volume of distribution:* Drugs with extensive tissue binding [Eq. (7.11)] tend to have larger volumes of distribution at steady state compared to those with limited tissue binding.
- *Terminal half-life:* Drugs with extensive tissue binding tend to have a longer terminal half-life than those with limited tissue binding, when other parameters are similar.
- *Drug–drug interaction:* If a displacement of one drug by another takes place only at tissue binding sites and does not occur in plasma, the total concentration of the drug may decrease, but unbound or total drug concentrations in plasma and unbound drug concentration in tissues are not changed because drug clearance is not affected by changes in tissue binding.

7.8. SPECIES DIFFERENCES IN PROTEIN BINDING

The extent of drug binding to plasma proteins can vary considerably among different species. There are slight differences in the amino acid sequences in protein molecules such as albumin among different species, despite their structural and functional homologies (approximately 590 amino acids), and similar concentrations (e.g., 500–600 μM of albumin in plasma in rats and humans). This may cause the differences in binding affinity and/or number of binding sites of drugs in protein molecules among different species. In general, more extensive plasma protein binding is expected in larger animals, including humans, than in small laboratory animals. However, a similar extent of tissue binding of drugs has been found among different species with a few exceptions (Fichtl and Schulmann, 1986; Fichtl *et al.*, 1991; Sawada *et al.*, 1984).

REFERENCES

Balant L. P. and Gex-Fabry M., Physiological pharmacokinetic modeling, *Xenobiotica*, **20**: 1241–1257, 1990.

Benet L. Z. and Zia-Amirhosseini P., Basic principles of pharmacokinetics, *Toxicol. Pathol.* **23**: 115–123, 1995.

Boudinot F. D. and Jusko W. J., Fluid shifts and other factors affecting plasma protein binding of prednisolone by equilibrium dialysis, *J. Pharm. Sci.* **73**: 774–780, 1980.

Bowers W. F. *et al.*, Ultrafiltration vs. equilibrium dialysis for determination of free fraction, *Clin. Pharmacokinet.* **9**(Suppl. 1): 49–60, 1984.

du Souich P. *et al.*, Plasma protein binding and pharmacological response, *Clin. Pharmacokinet.* **24**: 435–440, 1993.

Elmquist W. F. and Sawchuk R. J., Application of microdialysis in pharmacokinetic studies, *Pharm. Res.* **14**: 267–288, 1997.

Fichtl B. and Schulmann G., Relationships between plasma and tissue binding of drugs, in J. P. Tillement and E. Lindenlaub (eds.), *Protein Binding and Drug Transport*, Schattauer, New York, 1986, pp. 255–271.

Fichtl B. *et al.*, Tissue binding versus plasma binding of drugs: general principles and pharmacokinetic consequences, in B. Test (ed.), *Advances in Drug Research, Vol. 20*, Academic Press, London, 1991, pp. 117–166.

Judd R. L. and Pesce A. J., Free drug concentrations are constant in serial fractions of ultrafiltrate, *Clin. Chem.* **28**: 1726–1727, 1982.

Jusko W. J. and Gretch M., Plasma and tissue protein binding of drugs in pharmacokientics, *Drug Metab. Rev.* **5**: 43–140, 1976.

Kurz H. and Fichtl B., Bindings of drugs to tissues, *Drug Metab. Rev.* **14**: 467–510, 1983.

McNamara P. J. and Bogardus J. B., Effect of initial conditions and drug-protein binding on the time to equilibrium in dialysis systems, *J. Pharm. Sci.* **71**: 1066–1068, 1982.

Mendel C. M., The free hormone hypothesis: a physiologically based mathematical model, *Endocrine Rev.* **10**: 232–274, 1989.

Notarianni L. J., Plasma protein binding of drugs in pregnancy and in neonates, *Clin. Pharmacokinet.* **18**: 20–36, 1990.

Oravcova J. *et al.*, Drug-protein binding studies: new trends in analytical and experimental methodology, *J. Chromatogr. B. Biomed. Sci. Appl.* **677**: 1–28, 1996.

Pacifici G. M. and Viani A., Methods of determining plasma and tissue binding of drugs, Pharmacokinetic consequences, *Clin. Pharmacokinet.* **23**: 449–468, 1992.

Φie S., Drug distribution and binding, *J. Clin. Pharmacol.* **26**: 583–586, 1986.

Φie S. and Tozer T. N., Effect of altered plasma protein binding on apparent volume of distribution, *J. Pharm. Sci.* **68**: 1203–1205, 1979.

Rothschild M. A. *et al.*, Tissue distribution of I131 labeled human serum albumin following intravenous administration, *J. Clin. Invest.* **34**: 1354–1358, 1955.

Sawada Y. *et al.*, Prediction of the volumes of distribution of basic drugs in humans based on data from animals, *J. Pharmacokinet. Biopharm.* **12**: 587–596, 1984.

Telting-Diaz M. *et al.*, Intravenous microdialysis sampling in awake, freely-moving rats, *Anal. Chem.* **64**: 806–810, 1992.

Tillement J. P. *et al.*, Blood binding and tissue uptake of drugs. Recent advances and perspectives, *Fundam. Clin. Pharmacol.* **2**: 223–238, 1988.

Tozer T. N. *et al.*, Volume shifts and protein binding estimates using equilibrium dialysis: application to prednisolone binding in humans, *J. Pharm. Sci.* **72**: 1442–1446, 1983.

Verbeeck R. K. *et al.*, Effects of age and sex on the plasma binding of acidic and basic drugs, *Eur. J. Clin. Pharmacol.* **27**: 91–94, 1984.

Weisiger R., Dissociation from albumin: a potentially rate-limiting step in the clearance of substances by the liver, *Proc. Natl. Acad. Sci.* **82**: 1563–1567, 1985.

Wilkinson G. R., Plasma and tissue binding considerations in drug disposition, *Drug Metab. Dispos.* **14**: 427–465, 1983.

8

Metabolism

8.1. INTRODUCTION

Metabolism (biotransformation) is the major elimination pathway for most drugs. A thorough understanding of the metabolic pathways and profiles of a drug is important in improving its pharmacokinetic profiles and addressing potential metabolism-related issues such as toxic metabolites, metabolic interactions, and polymorphic metabolism. In general, enzymatic metabolism transforms lipophilic parent drugs to more hydrophilic metabolites, which can be readily excreted into bile or urine. Drug metabolism in the body can be divided into two different types of reactions: phase I and phase II metabolism. Phase I metabolism generally results in the introduction of a functional group into molecules or the exposure of new functional groups of molecules, whereas phase II metabolism involves conjugation of functional groups of molecules with hydrophilic endogenous substrates (Caldwell *et al.*, 1995; Parkinson, 1996*a*).

$$\text{Drugs (lipophilic and less polar)} \xrightarrow{\text{Metabolism}} \text{Metabolites (hydrophilic and more polar)}$$

Phase I: addition (or revealing) of hydrophilic moieties
Phase II: conjugation with hydrophilic endogenous substrates

Important characteristics of phase I and phase II metabolism and metabolizing enzymes are summarized below.

8.1.1. Phase I Metabolism

Phase I metabolism is sometimes called a "functionalization reaction," since it results in the introduction of new hydrophilic functional groups to compounds.

1. Function: introduction (or unveiling) of functional group(s) such as $-OH$, $-NH_2$, $-SH$, $-COOH$ into the compounds.
2. Reaction types: oxidation, reduction, and hydrolysis
3. Enzymes:
   3.1. Oxygenases and oxidases: Cytochrome P450 (P450 or CYP), flavin-containing monooxygenase (FMO), peroxidase, monoamine oxidase

(MAO), alcohol dehydrogenase, aldehyde dehydrogenase, and xanthine oxidase.
3.2. Reductase: Aldo-keto reductase and quinone reductase.
3.3. Hydrolytic enzymes: esterase, amidase, aldehyde oxidase, and alkylhydrazine oxidase.
3.4. Enzymes that scavenge reduced oxygen: Superoxide dismutases, catalase, glutathione peroxidase, epoxide hydrolase, γ-glutamyl transferase, dipeptidase, and cysteine conjugate β-lyase.

Examples of phase I metabolism:

1. Oxidation
   1.1. Oxidation by cytochrome P450 isozymes (microsomal mixed-function oxidases).
   1.2. Oxidation by enzymes other than cytochrome P450s—most of these enzymes are involved primarily in oxidation of endogenous substrates: (a) oxidation of alcohol by alcohol dehydrogenase, (b) oxidation of aldehyde by aldehyde dehydrogenase, and (c) N-dealkylation by monoamine oxidase.
2. Reduction
   Enzymes responsible for reduction of xenobiotics require NADPH as a cofactor. Substrates for reductive reactions include azo- or nitrocompounds, epoxides, heterocyclic compounds, and halogenated hydrocarbons: (a) azo- or nitroreduction by cytochrome P450; (b) carbonyl (aldehyde or ketone) reduction by aldehyde reductase, aldose reductase, carbonyl reductase, quinone reductase; and (c) other reductions including disulfide reduction, sulfoxide reduction, and reductive dehalogenation.
3. Hydrolysis
   Esters, amides, hydrazides, and carbamates can be hydrolyzed by various enzymes.

8.1.2. Phase II Metabolism

Phase II metabolism includes what are known as conjugation reactions. In general, the conjugation reaction with endogenous substrates occurs on the metabolite(s) of the parent compound after phase I metabolism; however, in some cases, the parent compound itself can be subject to phase II metabolism.

1. Function: conjugation (or derivatization) of functional groups of a compound or its metabolite(s) with endogenous substrates.
2. Reaction types: glucuronidation, sulfation, glutathione-conjugation, N-acetylation, methylation and conjugation with amino acids (e.g., glycine, taurine, glutamic acid).
3. Enzymes: Uridine diphosphate-glucuronosyltransferase (UDPGT): sulfotransferase (ST), N-acetyltransferase, glutathione S-transferase (GST), methyl transferase, and amino acid conjugating enzymes.

Examples of phase II metabolism:

1. Glucuronidation by uridine diphosphate-glucuronosyltransferase
2. Sulfation by sulfotransferase
3. Acetylation by N-acetyltransferase
4. Glutathione conjugation by glutathione S-transferase
5. Methylation by methyl transferase
6. Amino acid conjugation

8.1.3. Subcellular Locations of Metabolizing Enzymes

ENDOPLASMIC RETICULUM (microsomes): the primary location for the metabolizing enzymes. (a) Phase I: cytochrome P450, flavin-containing monooxygenase, aldehyde oxidase, carboxylesterase, epoxide hydrolase, prostaglandin synthase, esterase. (b) Phase II: uridine diphosphate-glucuronosyltransferase, glutathione S-transferase, amino acid conjugating enzymes.

CYTOSOL (the soluble fraction of the cytoplasm): many water-soluble enzymes. (a) Phase I: alcohol dehydrogenase, aldehyde reductase, aldehyde dehydrogenase, epoxide hydrolase, esterase. (b) Phase II: sulfotransferase, glutathione S-transferase, N-acetyl transferase, catechol O-methyl transferase, amino acid conjugating enzymes.

MITOCHONDRIA. (a) Phase I: monoamine oxidase, aldehyde dehydrogenase, cytochrome P450. (b) Phase II: N-acetyl transferase, amino acid conjugating enzymes.

LYSOSOMES. Phase I: peptidase.

NUCLEUS. Phase II: uridine diphosphate-glucuronosyltransferase (nuclear membrane of enterocytes).

Brief descriptions of the metabolic implications and biological importance of several metabolizing enzymes in humans are described below.

8.2. PHASE I ENZYMES

8.2.1. Cytochrome P450 Monooxygenase (Cytochrome P450, P450, or CYP)

METABOLIC IMPLICATIONS. Cytochrome P450 monooxygenases (Cytochrome P450, P400, or CYP) play a major role in the biosynthesis and catabolism of various endogenous compounds such as steroid hormones, bile acids, fat-soluble vitamins, and fatty acids. P450 enzymes are also considered the most important metabolizing enzymes for xenobiotics ($>85\%$ of the drugs in the market are metabolized by P450s) (Rendic and DiCarlo, 1997).

REACTION TYPE. Mainly oxidation in the presence of oxygen or reduction under low oxygen tension. A basic reaction scheme for the oxidation of a substrate by P450 is

$$\text{RH (substrate)} + O_2 \xrightarrow[\text{(2H}^+\text{2e}^-\text{)}]{\text{Cytochrome P450}} \text{ROH (metabolite)} + H_2O$$

NADPH, H^+ → NADP +
NADPH-cytochrome P450 reductase

1. Oxidation
   (a) Aromatic hydroxylation: $R{-}C_6H_5 \rightarrow R{-}C_6H_4OH$
   (b) Aliphatic hydroxylation: $R{-}CH_3 \rightarrow R{-}CH_2 \rightarrow OH$
   (c) N, O, S-dealkylation:
   $R{-}NH\ (O, S){-}CH_3 \rightarrow R{-}NH_2\ (OH, SH) + HCHO$
   (d) N-oxidation: $R{-}C_6H_4{-}NH_2 \rightarrow R{-}C_6H_4{-}N(OH)(H)$
   (e) S-oxidation: $R{-}S{-}CH_3 \rightarrow R{-}SO{-}CH_3$
   (f) Epoxidation: $R_1{-}C{=}C{-}R_2 \rightarrow R_1{-}C{-}C{-}R_2$ (with O bridging the two C atoms)
   (g) Dehalogenation: $R_1R_2CH{-}X \rightarrow R_1R_2C{=}O$ (X: Cl, Br)
2. Reduction
   (a) Azo- or nitroreduction: $R_1{-}N{=}N{-}R_2$ (or $R{-}NO_2$) $\rightarrow R_1{-}NH_2 + R_2{-}NH_2$ (or $R{-}NH_2$)
   (b) Dehalogenation: $CCl_4 \rightarrow CHCl_3$
3. Others
   (a) Hydrolysis of esters
   (b) Dehydrogenation

COFACTORS. Nicotinamide-adenine-dinucleotide-phosphate (NADPH) and O_2.

TISSUE DISTRIBUTION. Ubiquitous, especially in liver, intestine, kidney, lung and skin.

SUBCELLULAR LOCALIZATION. Endoplasmic reticulum (microsomes), mitochondria (P450 enzymes in mitochondria are involved mainly in steroid biosynthesis and vitamin D metabolism).

ISOZYMES. There are at least eight mammalian P450 gene families. In humans, at least seventeen liver cytochrome P450 isoforms have been characterized.

POLYMORPHISM. Polymorphic metabolism by CYP2C18, CYP2C19, CYP2D6, and CYP3A5 in humans.

NADP:

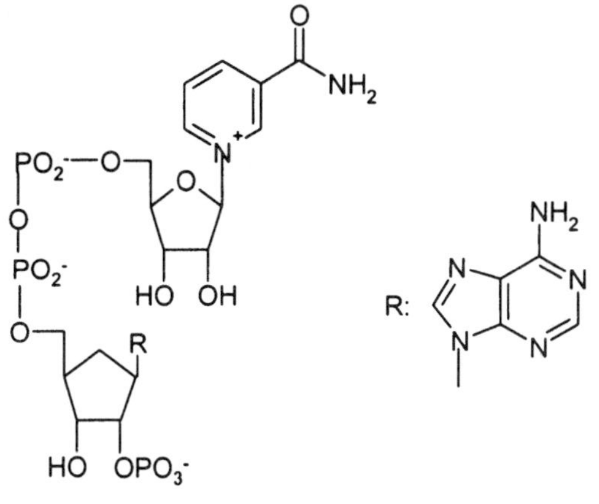

R:

8.2.1.1. Cytochrome P450 Enzyme System

The P450 system consists of three protein components embedded in the phospholipid environment of the endoplasmic reticulum, including: (a) Cytochrome P450, a membrane-bound hemoprotein, which binds directly to substrates and molecular oxygen. (b) NADPH-cytochrome P450 reductase, a membrane-bound flavoprotein, containing 1 mole each of flavin adenine dinucleotide (FAD) and flavin mononucleotide (FMN), which transfers electrons from NADPH to the cytochrome-P450–substrate complex; cytochrome P450 and NADPH–cytochrome-P450 reductase are embedded in the phospholipid bilayer of the endoplasmic reticulum, which facilitates their interaction. (c) Cytochrome b_5, a membrane-bound hemoprotein, which enhances the catalytic efficiency (rate) of some P450 isoforms by donating the second of the two electrons required for cytochrome P450 reactions.

(a) Relationship between P450 and NADPH-Cytochrome P450 Reductase/ Cytochrome b_5. There are various cytochrome P450 isoforms, but only one form of NADPH-cytochrome P450 reductase and cytochrome b_5 in liver microsomes. Approximately 10–20 molecules of cytochrome P450s and 5–10 molecules of cytochrome b_5 surround each molecule of NADPH-cytochrome P450 reductase in liver microsomes.

(b) Nomenclature.

(i) Cytochrome P450. When the heme iron [usually in ferric (Fe^{+3}) state] in cytochrome P450 is reduced to ferrous (Fe^{+2}) state, cytochrome P450 can bind ligands such as O_2 and CO. The name cytochrome P450 was derived from the findings that the complex between ferrous cytochrome P450 and CO absorbs light maximally between 447 and 452 nm (an average of 450 nm).

(ii) Cytochrome P420. When the thiolate bond in the fifth ligand (cysteine–thiolate) to the heme moiety of cytochrome is disrupted, cytochrome P450 is

converted to a catalytically inactive form called cytochrome P420, which absorbs light maximally at 420 nm upon binding CO.

(iii) Cytochrome P450 isoforms. Classification of cytochrome P450 depends on the extent of amino acid sequence identity of different P450 enzymes, not on catalytic activities or substrate specificity (Nelson, 1996).

- Gene families (e.g., CYP1, CYP2, CYP3, etc). P450 enzymes with less than 40% amino acid sequence identity are assigned to different gene families. There are at least eight mammalian P450 gene families.

- Subfamilies (e.g., CYP2A, CYP2B, CYP2C, etc). P450 enzymes with 40–55% amino acid sequence identity are assigned to different subfamilies.

- Isoforms (e.g., CYP2C8, CYP2C9, etc). P450 enzymes with more than 55% amino acid sequence identity are classified as members of the same subfamily.

Most cytochrome P450s are named irrespective of species. CYP1A1, CYP1A2, and CYP2E1 are present in all mammalian species. Similar isoform names of cytochrome P450 imply similar structure, but not necessarily similar catalytic functions, since slight changes in enzyme structure can cause marked differences in metabolic activities.

(c) Human Cytochrome P450 Isoforms. Currently, four P450 gene families, eight subfamilies, and at least seventeen human liver cytochrome P450 isoforms involved in drug metabolism have been characterized to varying degrees. CYP1, CYP2, and CYP3 are the three main P450 gene families in the human liver. Metabolically important isoforms include CYP1A2, 2A6, 2B6, 2C8, 2C9, 2C19, 2D6, 2E1, and 3A4. Table 8.1 lists the cytochrome P450 families, subfamilies, and isoforms known in humans, and Table 8.2 summarizes the average content of different P450 isoforms in the human liver. The relative amount of individual P450 isoforms present in the liver is 3A subfamily (mainly 3A4) > 2C subfamily > 1A2 > 2E1 > 2A6 > 2D6 > 2B6 (Fig. 8.1).

Table 8.1. Human Cytochrome P450 Isoforms

Category	Cytochrome P450 isoforms			
Family	CYP1	CYP2	CYP3	CYP4
	↓	↓	↓	↓
Subfamily	1A	2A, B, C, D, E	3A	4A
	↓	↓	↓	↓
Isoforms	1A1, 1A2	2A6 2B6 2C8, 2C9, 2C10, 2C18, 2C19 2D6 2E1	3A3, 3A4, 3A5, 3A7	4A9, 4A11

Table 8.2. The Average Content of Cytochrome P450 Enzymes in Human Liver Microsomes[a]

P450 enzymes	Average content: pmol/mg microsomal protein	Average content: nmol/g liver[b]
Total P450 (determined spectrally)	344	18.1
Total P450 (determined immunochemically)	240	12.6
CYP isoforms		
1A2	42	2.2
2A6	14	0.7
2B6	1	0.05
2C[c]	60	3.2
2D6	5	0.3
2E1	22	1.2
3A[d]	96	5.0

[a]Data taken from Shimada *et al.* (1994).
[b]The values were converted to nmol/g liver, considering the microsomal protein content of 52.5 mg/g liver.
[c]Sum of CYP2C8, 2C9, 2C10, 2C18, and 2C19.
[d]Sum of CYP3A3, 3A4, 3A5, and 3A7.

(d) Important Human Cytochrome P450 Isoforms in Drug Metabolism. Most of the drugs on the market today are metabolized by CYP3A4 and 2D6, followed by CYP2C9, 2C19, 1A2, and 2E1 (Fig. 8.2). The discrepancy between the relative abundance of P450 isoforms in the liver and the extent of their contribution to the overall metabolism of xenobiotics is due to differences in the affinity (K_m values) of

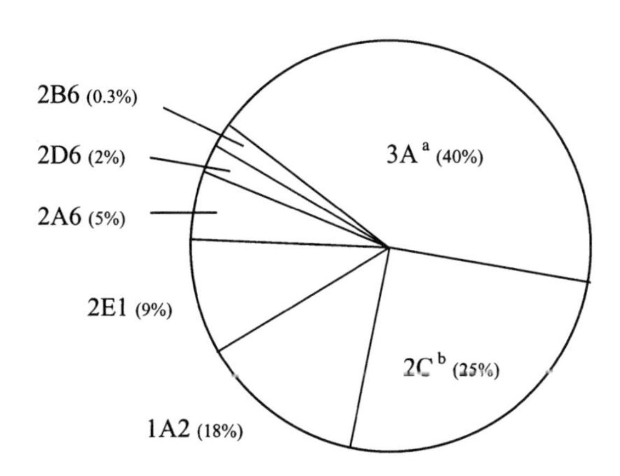

a Sum of CYP3A3, 3A4, 3A5 and 3A7

b Sum of CYP2C8, 2C9, 2C10, 2C18 and 2C19

Figure 8.1. Percent amount of individual P450 isoforms in the total P450 determined by immunochemical methods in human liver microsomes (Shimada, 1994).

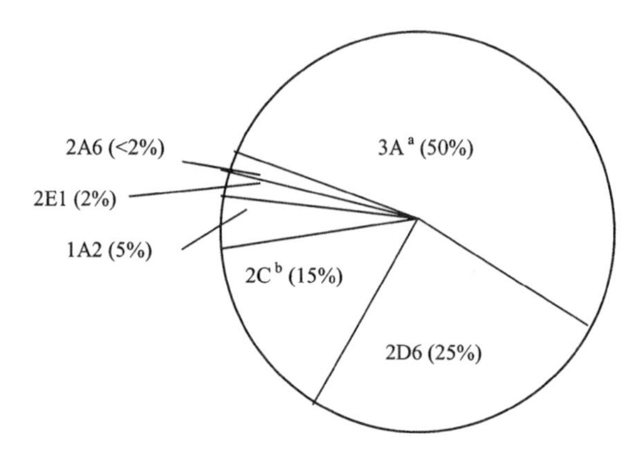

[a] Sum of CYP3A4 and 3A5

[b] Sum of CYP2C8, 2C9, 2C18 and 2C19

Figure 8.2. Percent of drugs on the market metabolized by various cytochrome P450 isoforms.

the enzymes to substrates. Since the rate of metabolism by an enzyme is determined by the amount (capacity) of the enzyme (V_{max}) as well as its affinity (K_m) to substrates, the enzyme with high affinity can exhibit a fast turnover rate despite its smaller quantity compared to other enzymes. For instance, CYP2D6 metabolizes more drugs than CYP2C9 owing to the higher affinity (lower K_m values) of CYP2D6, although the absolute amount of CYP2D6 present in the liver is substantially lower (smaller V_{max}) than the amount of CYP2C9. This becomes especially important for the metabolism of substrates at low concentrations. As a rule of thumb, $K_m > 100\ \mu M$ is usually considered "low affinity," whereas $K_m < 20\ \mu M$ can be viewed as "high affinity" of a particular enzyme for a substrate of interest.

(e) Characteristics of Cytochrome P450 Reactions.

1. Cytochrome P450-mediated metabolism is responsible for the metabolism of more than 85% of the drugs on the market.
2. Cytochrome P450 metabolism often precedes phase II metabolism and is generally slower. As a result, in many cases, cytochrome P450–mediated metabolism is the rate-limiting step for the biotransformation of drugs.
3. The same compound can be a substrate for many different P450 isoforms with markedly different affinities, and also one P450 isoform can metabolize the same compound in multiple sites at different rates.
4. There is a large variability in P450 functions among different mammalian species.
5. Genetic and environmental factors have significant effects on P450 expression, which leads to substantial interindividual and species variability in P450 mediated–metabolism.

Typical substrates and inhibitors of human cytochrome P450 isoforms, and their tissue locations and inducibility are listed in Table 8.3.

NOTE: SUICIDE INHIBITOR (MECHANISM-BASED INHIBITOR). A suicide inhibitor (sometimes called a mechanism-based inhibitor) is a compound that inhibits the activity of a metabolizing enzyme such as cytochrome P450 by forming a covalent bond(s) with the enzyme as a result of its own metabolism by the enzyme. For instance, 1-aminobenzotriazole (ABT) is a suicide inhibitor of various cytochrome P450 enzymes. To be activated, ABT undergoes a P450-catalyzed oxidation to form benzyne, a reactive intermediate, which covalently binds to the prosthetic heme group of cytochrome P450 and thereby causes the irreversible loss of its enzymatic activity (Mugford *et al.*, 1992).

(f) Substrate Structure Specificity. Certain structural properties of substrates for several cytochrome P450 isozymes have been characterized, although there are some exceptions (DeGroot and Vermeuler, 1997; Smith and Jones, 1992; Smith, 1994*a*,*b*).

CYP1A: Most substrates are planar and aromatic [so-called, polycyclic aromatic hydrocarbon (PAH)].

CYP2C9: Most substrates are lipophilic, and neutral or acidic with strong hydrogen-bonding and/or ion-pairing ability. (b) Oxidation of substrates usually occurs 5–20 Å from a proton-donor heteroatom in substrates.

CYP2D6: Most substrates are arylalkylamines, which are basic (cationic) at physiological pH. (b) Oxidation of substrates usually occurs 5 to 7 Å from a protonated nitrogen in the substrates, which interacts with an anionic residue (Glu^{301}) of the enzyme.

CYP3A4: No apparent selectivity in terms of overall structure of substrates. (b) Most substrates are lipophilic, and neutral or basic at physiological pH. (c) Oxidation often occurs on a basic nitrogen atom (N-dealkylation) or an allylic position in a substrate. (d) Metabolism can occur at multiple sites in a single molecule. (e) K_m values of substrates for CYP3A4 are usually higher than those for CYP2D6.

8.2.2. Flavin-Containing Monooxygenase (FMO)

METABOLIC IMPLICATIONS. In addition to the cytochrome P450, hepatic microsomes contain a second class of monooxygenase, the flavin-containing monooxygenase (FMO). Although the FMOs are considered to be important for metabolizing heteroatom (N, S, Se or P)-containing compounds rather than direct oxidation at a carbon atom, the quantitative role of FMOs in hepatic drug metabolism in humans is limited (Cashman, 1995).

Table 8.3. Typical Model Substrates and Inhibitors, Tissue Distribution, and Inducibility of Human Cytochrome P450s[a]

CYP isoforms	Model substrates	Inhibitors	Notes
1A1	Acetoaminophen Caffeine 7-Ethoxyresorufin[b] (polycyclic hydrocarbons)	α-Naphthoflavone	Lung, liver, placenta Inducible by smoking, charcoal-broiled meat, and cruciferous vegetables
1A2	Acetoaminophen Caffeine[c] Theophylline R-Warfarin (heterocyclic amines)	Furafylline α-Naphthoflavone	Liver only (large interindividual variability in levels) Inducible by smoking, charcoal-broiled meat, cruciferous vegetables, omeprazol, and lansoprazole
2A6	Coumarin[d]	Pilocarpine	Hepatic (a large interindividual variability in levels) Inducible by barbiturates, dexamethasone, and rifampicin
2B6	7-Ethoxy-4-trifluoromethyl-coumarin[e]		Hepatic Constitutive Neonatal form (?) Inducible by rifampicin
2C8, 9, 10	Phenytoin Taxol[f] Tolbutamide[g] S-Warfarin (many NSAIDs)	Sulfaphenazole (2C9, 10)	Hepatic, renal (2C8) Inducible by barbiturates and rifampicin Rare genetic variation [poor metabolizer ~0.2% (2C8)]
2C18, 19	Diazepam S-Mephenytoin[h]	Tranylcypromine	Hepatic Inducible by rifampicin, and omeprazole Polymorphic [poor metabolizer: 3% in Caucasians (2C18, 2C19) and 20% in Asians (2C19)]
2D6	Bufarolol Codein Debrisoquine[i] Dextromethorphan[j] Sparteine (amines)	Quinidine Yohimbine	Hepatic Constitutive Not inducible Polymorphic (poor metabolizer 5–10% in Caucasians and $<2\%$ in Asians)
2E1	Chlorzoxazone[k] p-nitrophenol Nitrosamine Ethanol Vinyls (small molecules)	Diethyldithiocarbamate Disulfiram 4-Methylpyrazole	Hepatic and extrahepatic Constitutive Inducible by acetone, ethanol and isoniazid Rare genetic variation ($<0.3\%$)

Table 8.3. Continued

CYP isoforms	Model substrates	Inhibitors	Notes
3A3, 4	Caffeine[l] Cortisol[m] Cyclosporin Erythromycin[n] Lidocaine Midazolam[o] Nifedipine Tamoxifen Testosterone[p]	Gestodene Ketoconazole Naringenin (grapefruit juice) Quercetin Troleandomycin	Hepatic and extrahepatic (intestine) Constitutive Inducible by barbiturate, carbamazepine, dexamethason, phenytoin, rifampicin, and troleandomycin
3A5	Cyclosporin Midazolam[q] Nifedipine Testosterone	Gestodene Troleandomycin	Hepatic and extrahepatic Not inducible Polymorphism (expressed 10–30% and 80% in all human livers and kidneys, respectively)
3A7	Dehydroepiandrosteron (DHEA)		Hepatic Fetal enzyme (also found in adult liver)
4A9, 11	Lauric acid[r]		Hepatic Not inducible

[a]Data taken from Birkett *et al.* (1993), Bourrié *et al.* (1996), Gonzalez (1992), Halpert *et al.* (1994), Meyer (1996), Newton *et al.* (1995), Thummel (1994), Wrighton and Stevens (1992), and Wrighton *et al.* (1993). NSAID stands for nonsteroidal antiinflammatory drug. Footnotes b–r refer to the metabolic reactions of common marker substrates used for phenotyping the *in vivo* catalytic activities of P450 isozymes in human.
[b]7-Ethoxyresorufin *O*-dealkylation.
[c]Caffeine-3-demethylation.
[d]Coumarin 7-hydroxylation.
[e]7-Ethoxy-4-trifluoromethyl coumarin *O*-dealkylation.
[f]Taxol 6α-hydroxylation.
[g]Tolbutamide methyl-hydroxylation.
[h]S-Mephenytoin 4′-hydroxylation.
[i]Debrisoquine-4-hydroxylation.
[j]Dextromethorphan *O*-demethylation.
[k]Chlorzoxazone-6-hydroxylation.
[l]Caffeine-8-hydroxylation.
[m]Cortisol-6-hydroxylation.
[n]Erythromycin N-demethylation.
[o]Midazolam-1-hydroxylation.
[p]Testosterone 6β-hydroxylation.
[q]Midazolam-4-hydroxylation.
[r]Lauric acid 12-hydroxylation.

Reaction type. Oxidation

$$R{-}N(S, P) \rightarrow \underset{\displaystyle \overset{\downarrow}{O}}{R{-}N(S, Se, P)}$$

Substrates. Compounds containing a heteroatom (N, S, Se, or P).

Cofactors. NADPH and O_2.

TISSUE DISTRIBUTION. Liver, kidney, lung, and skin.

SUBCELLULAR LOCALIZATION. Endoplasmic reticulum.

ISOZYMES. FMO^{I} (fetal liver FMO) and FMO^{II} (major form of FMO in adult liver).

POLYMORPHISM. Unknown.

NOTE: THERMAL INSTABILITY OF FMOS. FMOs are heat labile and can be inactivated in the absence of NADPH by warming microsomes at 50°C for 1 min.

8.2.3. Esterase

METABOLIC IMPLICATIONS. Esters, amides, hydrazides, and carbamates can be hydrolyzed by various esterases. Ester hydrolysis can occur in the plasma mainly by cholinesterase (nonspecific acetylcholine esterases, pseudocholine esterases, and other esterases) or in the liver by specific esterases for particular groups of compounds. Enzymatic hydrolysis of esters and amides can be important in determining the duration of action of certain drugs. Especially, the rate of enzymatic cleavage of ester or amide moiety from ester or amide prodrugs can play a critical role in the onset of pharmacological activity and its duration (Satoh and Hosokawa, 1998).

REACTION TYPE. Hydrolysis:
(a) Hydrolysis of esters: $R_1\text{–CO–OR}_2 \rightarrow R_1\text{–COOH} + R_2\text{–OH}$
(b) Hydrolysis of amides: $R_1\text{–CO–NH-R}_2 \rightarrow R_1\text{–COOH} + R_2\text{–NH}_2$

Hydrolysis of amides can occur by amidases in the liver and in general, enzymatic hydrolysis of amides is slower than that of esters. Amides can be also hydrolyzed by esterases with a much slower rate than the corresponding esters.

SUBSTRATES. Esters and amides.

SUBCELLULAR LOCALIZATION. Endoplasmic reticulum and cytosol.

TISSUE DISTRIBUTION. Ubiquitous, liver (centrilobular region), kidney (proximal tubules), testis, intestine, lung, plasma, and red blood cells.

ISOZYMES.

Main groups	Substrates
A-esterase (arylesterase)	aromatic esters
B-esterase (carboxylesterase)	aliphatic esters
C-esterase (acetylesterase)	acetyl esters
Cholinesterase	choline esters

POLYMORPHISM. Approximately 2% of Caucasians have defective serum cholinesterase activity (Daly *et al.*, 1993).

SPECIES DIFFERENCES. In general, esterase activity is higher in small laboratory animals such as the rat and mouse than in humans.

8.2.4. Alcohol Dehydrogenase (ADH)

METABOLIC IMPLICATIONS. Alcohol dehydrogenase (ADH) is a major enzyme responsible for oxidation of alcohol (ethanol) to aldehyde (acetaldehyde). Other quantitatively less important microsomal and peroxisomal enzymes for ethanol oxidation include CYP2E1 and catalase, respectively.

REACTION TYPE. Oxidation of alcohol to aldehyde:

$$R{-}CH_2{-}OH \rightarrow R{-}CHO$$

SUBSTRATES. Aliphatic or aromatic alcohols.

COFACTORS. NAD^+.

ENZYME STRUCTURE. Zinc-containing dimer of two 40 kDa subunits.

TISSUE DISTRIBUTION. Liver, kidney, lung and gastric mucosa.

SUBCELLULAR LOCATION. Cytosol.

ISOZYMES.

Classes	Substrates
Class I (ADH_1, ADH_3, and ADH_3)	Small alcohols such as ethanol
Class II (ADH_4)	Larger aliphatic and aromatic alcohols
Class III (ADH_5)	Long-chain aliphatic and aromatic alcohols

POLYMORPHISM. Approximately 85% of Asians express the class I isozymes (so-called atypical ADH responsible for rapid conversion of ethanol to acetaldehyde), whereas fewer than 20% of Caucasians express the atypical ADH (Agarwal and Goedde, 1992).

8.2.5. Aldehyde Dehydrogenase (ALDH)

METABOLIC IMPLICATIONS. Aldehyde dehydrogenase (ALDH) is a major enzyme responsible for oxidation of xenobiotic aldehydes to acids. In particular, acetaldehyde formed from ethanol by alcohol dehydrogenase is oxidized to acetic acid by ALDH, which is further oxidized to carbon dioxide and water.

REACTION TYPE. Oxidation of aldehyde to acid:

$$R{-}CH_2{-}CHO \rightarrow R{-}CH_2{-}COOH$$

SUBSTRATES. Aliphatic or aromatic aldehydes.

COFACTORS. NAD^+ or $NADP^+$.

ENZYME STRUCTURE. Tetramer of 54 kDa subunits (ALDH1 and ALDH2) or dimer of 85 kDa subunits ($ALDH_3$).

TISSUE DISTRIBUTION. Liver, kidney, lung, and gastric mucosa.

SUBCELLULAR LOCATION. Cytosol (ALDH1 and ALDH3), mitochondria (ALDH2).

ISOZYMES.

Classes	Substrates
ALDH1	Various xenobiotic aldehydes
ALDH2	Small aldehydes such as acetaldehyde
ALDH3	

POLYMORPHISM. Approximately 50% of Asians have a defective ALDH2 gene causing impaired ALDH2 activity (Goedde and Agarwal, 1992).

8.2.6. Monoamine Oxidase (MAO)

METABOLIC IMPLICATIONS. Monoamine oxidase (MAO) has been seen to be related to the metabolism of exogenous tyramine and the "cheese effect" produced as a result of the ingestion of large amounts of tyramine-containing foods under certain conditions. MAO catalyzes the oxidative deamination of biogenic amines (Benedetti and Tipton, 1998).

REACTION TYPE. Oxidative deamination of amines:

$$RCH_2{-}NR_1R_2 \rightarrow R{-}CHO + NHR_1R_2$$

SUBSTRATES. Primary, secondary, and tertiary amines.

COFACTORS. Oxygen from water, not from molecular oxygen. MAO converts amines into the corresponding imines, which are then further hydrolyzed to aldehydes with oxygen taken from water.

TISSUE DISTRIBUTION. Ubiquitous, liver, intestine, lung, blood platelets, and lymphocytes.

SUBCELLULAR LOCATION. Primarily in mitochondria, although some MAO activity has also been reported in the microsomal fraction.

ISOZYMES. MAO-A and MAO-B.

8.3. PHASE II ENZYMES

8.3.1. Uridine Diphosphate-Glucuronosyltransferase (UDPGT)

METABOLIC IMPLICATIONS. Glucuronidation is quantitatively the most important conjugation reaction of xenobiotics mediated by uridine diphosphate-glucuronosyl-transferase (UDPGT) (Clarke and Burchell, 1994). It is generally considered a low-affinity and high-capacity reaction. More than 95% of the drugs in the market are metabolized by cytochrome P450s, UDPGT, and sulfotransferases.

REACTION TYPE. Glucuronidation (Fig. 8.3):
(a) O-glucuronidation:

R–OH → R–O–glucuronic acid (ether glucuronidation)
R–COOH → R–COO-glucuronic acid (acyl (or ester) glucuronidation)

(b) N, S-glucuronidation:

R–NH_2 (or SH) → R–NH (or S)-glucuronic acid

SUBSTRATES. Glucuronidation can occur on nucleophilic moieties of molecules such as alcohol, acid (O-glucuronidation), amine (N-glucuronidation), and thiol (S-glucuronidation).

ENZYME STRUCTURE. Oligomers of between 1 and 4 subunits wth a molecular weight of between 50 and 60 kDa.

COFACTOR. Uridine-5′-diphospho-α-D-glucuronic acid (UDPGA).

TISSUE DISTRIBUTION. Liver, lung, kidney, stomach, intestine, skin, spleen, thymus, heart, and brain Most tissues have some glucuronidation activity.

SUBCELLULAR LOCATION. Mainly endoplasmic reticulum and some in the nuclear membrane.

ISOZYMES. More than 15 UDPGTs are known in humans, and they can be categorized into two major subfamilies, UDPGT1 and UDPGT2. The UDPGT protein sequences exhibit greater than 60% similarity within the subfamily.

SPECIES DIFFERENCES. Glucuronidation occurs in most mammalian species with the exception of the cat and related felines and the Gunn rat.

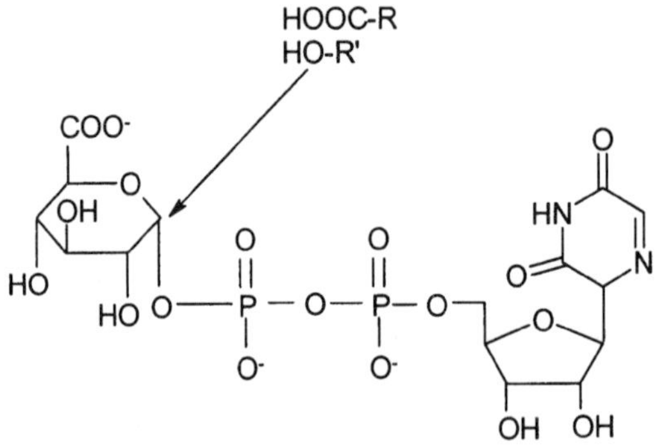

Figure 8.3. Substitution reaction of a nucleophilic substrate (R–OH, R–COOH) on the C_1 carbon atom of uridine diphosphate-glucuronic acid (UDPGA) by uridine diphosphate-glucuronosyltransferase (UDPGT).

INDUCIBILITY. Inducible by phenobarbital, 3-methylcholanthrene (MC), or pregnenolone-16α-carbonitrile (PCN) in rats. In humans, the induction of various UDPGT activities by phenobarbital, phenytoin, and oral contraceptives has been observed.

POLYMORPHISM. Approximately 2–5% of the population have a defective UDPGT1 gene complex causing hyperbilirubinemia (Gilbert's syndrome) (Burchell *et al.*, 1995).

IN VITRO EXPERIMENTAL CONDITIONS. There are important *in vitro* experimental considerations, which can affect the degree of activities and substrate specificity of UDPGTs as a result of the latency of enzyme activity and chemical instability of acyl glucuronides.

8.3.1.1. Latency and Membrane Disruption by Detergents

The active site of the UDPGT lies on the lumenal side of the endoplasmic reticulum (ER), which restricts the access of substrates and UDPGA from cytosol. It has been suggested that UDPGA transport from cytosol onto the lumenal side of the ER across the ER membranes may be the rate-determining step for glucuronidation in the intact microsomes. Owing to this membrane barrier, UDPGTs are latent enzymes and their activities in freshly isolated microsomes cannot be fully revealed without disrupting the membranes to a certain degree. For instance, more than 95% of UDPGT activity can be latent in liver microsomes prepared in 0.25 M sucrose with 5 mM HEPES, pH 7.4. Disruption of microsomal membranes with detergents such as Lubrol PX can remove UDPGT latency and increase enzyme activity by up to 10- to 20-fold under the optimal conditions. Often, the ratio of protein and detergent (0.01–0.5 mg detergent/mg microsomal protein) has to be tested empirically for optimal activation of UDPGT, and preincubation of microsomes with detergent(s) is required before an experiment (Burchell and Coughtrie, 1989; Mulder, 1992).

8.3.1.2. Chemical Instability of Acyl Glucuronides

In buffer or biological matrices at neutral or slightly alkaline pH, acyl glucuronides of many drugs with carboxylic acid moiety can undergo hydrolysis converting back to the parent drugs (futile cycling) and/or rearrangement (intramolecular rearrangement and intermolecular transacylation), whereas ether glucuronides are relatively stable. When acyl glucuronidation is anticipated, it is important to treat biological samples with acetic acid or HCl upon collection from animals, adjusting the pH to below 5, in order to minimize hydrolysis or rearrangement so that the amount of acyl glucuronides can be measured accurately (Kaspersen and Van Boeckel, 1987; Musson *et al.*, 1985; Watt *et al.*, 1991).

8.3.1.3. Acyl Migration

The rearrangement or isomerization reaction of acyl glucuronides involves the nonenzymatic migration of the drug moiety from the biosynthetic C_1 position of the glucuronic acid ring to the neighboring C_2, C_3, or C_4 positions (Fig. 8.4). In particular, intramolecular rearrangement (isomerization via acyl migration) of acyl glucuronide of drugs in plasma or albumin solutions can potentially lead to covalent binding of the drug moiety to proteins (intermolecular transacylation). These protein adducts with the (rearranged) acyl glucuronides of drugs have been proposed as possible causes of the *in vivo* toxicity seen in drugs with acid moieties (Hayball, 1995; Smith *et al.*, 1986).

8.3.2. Sulfotransferase (ST)

METABOLIC IMPLICATIONS. Sulfation is a predominant conjugation reaction of a compound at a low concentration mediated by sulfotransferase (ST), whereas glucuronidation becomes an important conjugation pathway at a higher substrate concentration. Sulfation is considered in general a high-affinity and low-capacity reaction (Weinshilboum and Otterness, 1994).

REACTION TYPE. Sulfation:

(a) O-sulfation:

$$\text{R–OH} \rightarrow \text{R–OSO}_3\text{H}$$

(b) N-sulfation:

$$\text{R–NH-COR}' \rightarrow \underset{\substack{|\\ \text{SO}_3\text{H}}}{\text{R–N–COR}'}, \text{ R: aryl}$$

SUBSTRATES. Sulfation can occur on nucleophilic moieties of such molecules as phenol, alcohol, and arylamine.

ENZYME STRUCTURE. Homodimers *in vivo* with a molecular weight between 32 and 34 kDa.

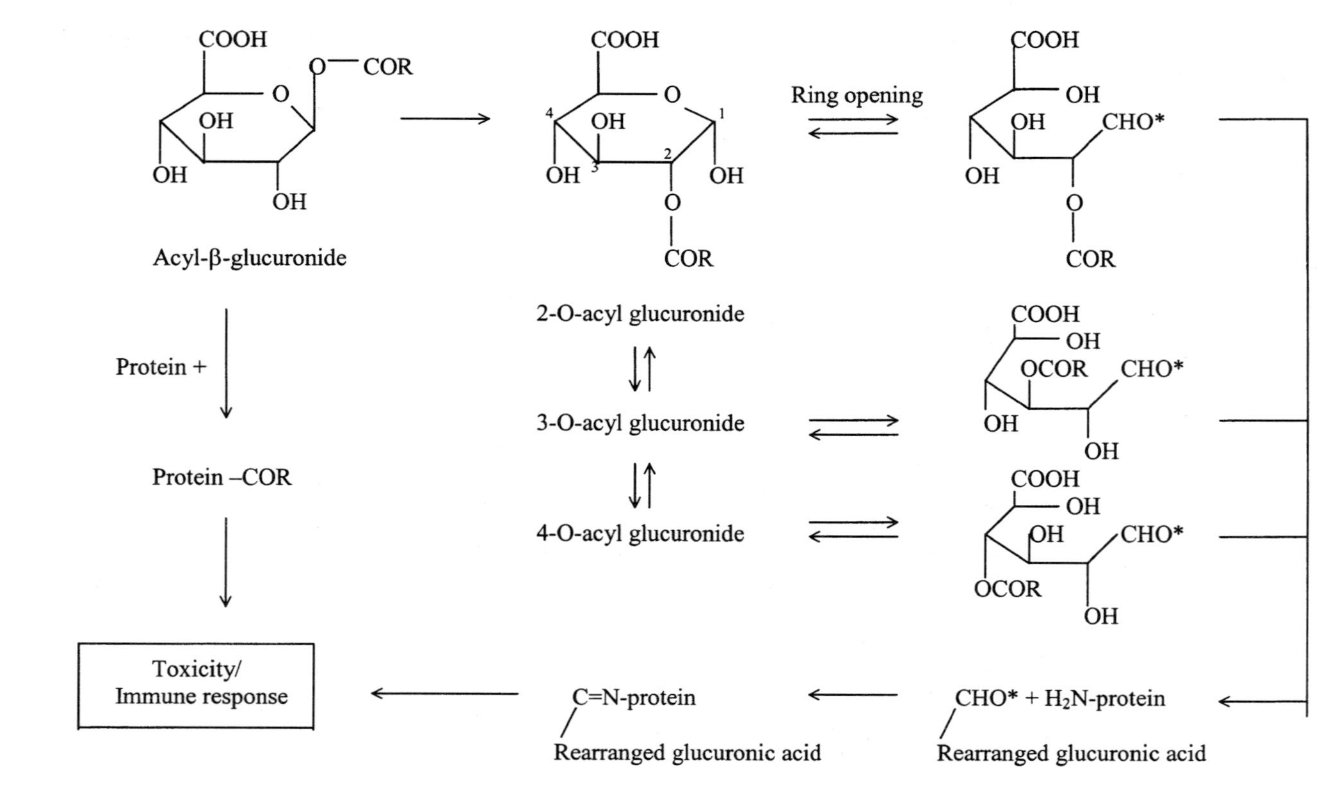

Figure 8.4. Acyl migration of acyl glucuronides. Diagrammatic representation of the migration of acyl group (RCO–) from C_1 to C_2, C_3, or C_4 positions of the glucuronic acid moiety of an acyl glucuronide, and formation of protein adducts potentially related to *in vivo* toxicity of acyl glucuronides.

COFACTOR. 3′-Phosphoadenosine-5′-phosphosulfate (PAPS).

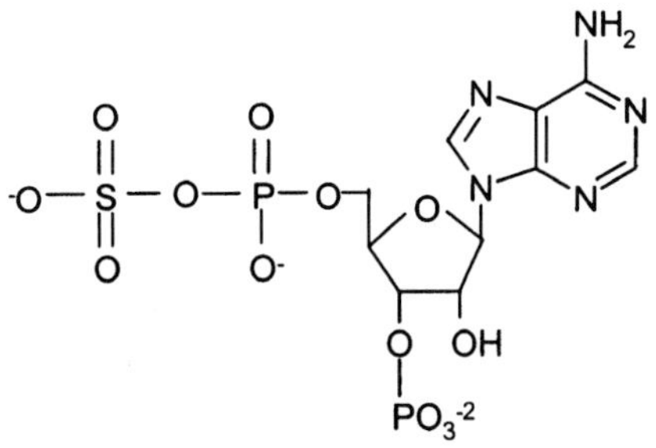

PAPS

TISSUE DISTRIBUTION. Liver, kidney, adrenals, lung, brain, jejunum, and blood platelets, and, to a lesser extent, skin and muscle.

SUBCELLULAR LOCATION. Cytosol.

ISOZYMES. Six different phenol sulfotransferases (PST) and seven different steroid/bile acid sulfotransferases have been characterized in rats. In humans, four subfamilies have been identified: TS ST (thermostable ST or PST), TL ST (thermolabile ST, or monoamine ST), EST (estrogen ST), and DHEA ST (dehydroepiandrosterone ST).

POLYMORPHISM. A low-activity allele has been reported at a frequency of 0.2% for TS ST and 0.08% for TL ST. Bimodal frequency distribution of DHEA ST activity suggests that approximately 75% of the population are poor metabolizers (Weinshilboum and Aksoy, 1994).

SPECIES DIFFERENCES. The pig and opossum are defective in their capability regarding sulfate conjugation of phenolic compounds.

NOTE: RELATIONSHIPS BETWEEN UDPGT AND ST. Often, UDPGT and ST are considered to be complementary to each other for conjugation of the same substrates, except in connection with acyl glucuronidation, which cannot be replaced by sulfation. In general, glucuronidation is considered a low-affinity (K_m) and high-capacity (V_{max}) reaction, whereas sulfation is known as a high-affinity and low-capacity conjugation. Thus, at low substrate concentrations sulfation may be more predominant, but as concentration increases, glucuronidation becomes quantitatively more important. This is sometimes due in part to rapid depletion of the cofactor of ST, i.e., PAPS.

FUTILE CYCLING OF CONJUGATION AND DECONJUGATION: The glucuronide and sulfate conjugates of various compounds in tissues can undergo hydrolysis back to the parent compounds, which can be subsequently reconjugated. This futile cycling of conjugation and deconjugation is further facilitated because the enzymes involved in these processes are localized in the same or adjacent subcellular compartments. For instance, both UDPGT [glucuronidation (conjugation)] and β-glucuronidase [hydrolysis of glucuronide conjugates (deconjugation)] are localized in the endoplasmic reticulum of hepatocytes. Aryl sulfatase (hydrolysis of sulfate conjugates) is located on the cytosolic surface of the endoplasmic reticulum and is readily accessible to sulfate conjugates formed by sulfotransferases in cytosol (Coughtrie *et al.*, 1998). In addition to enzyme activities, biliary excretion, transport of conjugates and free substrates across cellular membranes, and protein binding can also influence the extent and rate of futile cycling. Futile cycling of conjugation and deconjugation may be one of the important mechanisms for regulating the net production of conjugated metabolites of xenobiotics (Kauffman, 1994).

8.3.3. N-Acetyltransferase (NAT)

METABOLIC IMPLICATIONS. The first genetic polymorphism described for an enzyme involved in human drug metabolism was for N-acetyltransferase (NAT). Approximately half of the Caucasian population are poor acetylators.

REACTION TYPE. Acetylation on amine moiety.

$$R{-}NH_2 \rightarrow R{-}NH{-}COCH_3$$

$$R{-}SO_2NH_2 \rightarrow R{-}SO_2NH{-}COCH_3$$

SUBSTRATES. N-acetylation reactions are common for aromatic amines ($R{-}NH_2$), sulfonamides ($R{-}SO_2{-}NH_2$), or hydrazine ($R{-}NH{-}NH_2$) derivatives.

ENZYME STRUCTURE. N-Acetyltransferase has a molecular weight of 26.5 kDa.

COFACTOR. Acetyl-coenzyme A (CoA).

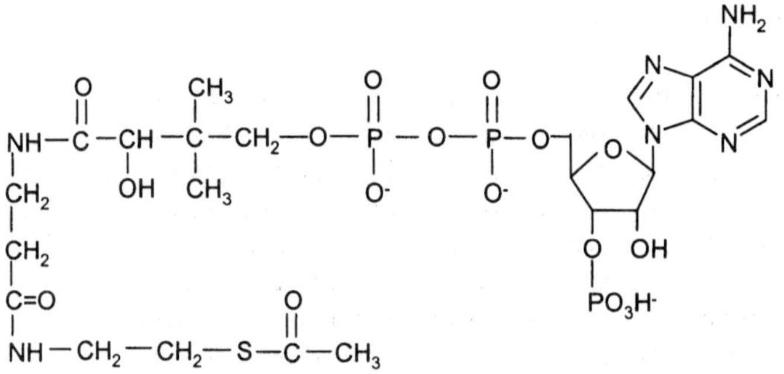

CoA

TISSUE DISTRIBUTION. Liver (in the Kupffer cells, not in the hepatocytes), spleen, lung, and intestine.

SUBCELLULAR LOCATION. Cytosol.

Isozymes. NAT1 and NAT2 enzymes (Vatsis and Weber, 1994).

Polymorphism. The slow acetylator has a genetic defect in the NAT2 gene. Approximately, 40–60% of Caucasians and 10–30% of Asians are slow acetylators.

Species differences. Dogs and guinea pigs have a deficiency in N-acetylation.

Note: Acetylation of sulfonamides leads to a less water-soluble metabolite (acetylsulfonamides) than the parent compound. Precipitation of acetylsulfonamides of some sulfonamide drugs in the kidney is found to be related to renal toxicity.

8.3.4. Glutathione S-Transferase (GST)

Metabolic implications. Glutathione S-transferase (GST) represents an integral part of the phase II detoxification system. GST protects cells from oxidative- and chemical-induced toxicity and stress by catalyzing the glutathione conjugation reaction with an electrophilic moiety of lipophilic and often toxic xenobiotics (van der Aar *et al.*, 1998). In the liver, GST accounts for up to 5% of the total cytosolic proteins.

Reaction type. Glutathione conjugation.

$$\text{R–X} \rightarrow \text{R–S–glutathione} + \text{X}^- \quad \text{(X: halide, sulfate, or phosphate)}$$

$$\text{R–C=C–COR} \rightarrow \text{R–C–}\underset{\substack{|\\ \text{S–glutathione}}}{\text{C}}\text{–COR}$$

Substrates. In general, substrates for GST are lipophilic and have an electrophilic moiety. Glutathione conjugates of xenobiotics formed in the liver are usually excreted in bile and urine, or further metabolized to mercapturic acids in the kidney and excreted in urine. Glutathione conjugation occurs preferably in substrates with reactive or good leaving groups such as epoxide, halide, sulfate, phosphate, or nitro moiety attached to an allylic or a benzylic carbon. The addition of glutathione may be facilitated by electron-withdrawing groups, such as –CHO, –COOR, –COR, or –CN, adjacent to the electrophilic moiety of the compounds.

Cofactor. Glutathione, a tripeptide cofactor (GSH, L-γ-glutamyl-L-cysteinylglycine (Gly-Cys-Glu)), is present in virtually all tissues, often in relatively high (0.1 10 mM) concentrations. Concentration of GSH in the liver is approximately 10 mM.

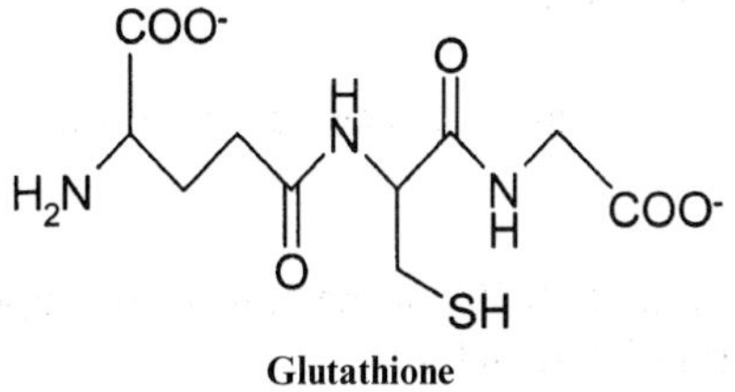

Glutathione

Enzyme structure. Dimer *in vivo* with a molecular weight of 24–28 kDa.

Tissue distribution. Liver, gut, kidney, testis, adrenal, and lung.

Subcellular location. Cytosol (major) and endoplasmic reticulum (minor).

Isozymes. The mammalian GSTs are divided into six classes, i.e., five classes of cytosolic enzymes, α, μ, π, θ, and σ, and one class of microsomal enzyme.

Polymorphism. (a) GSTM1 (the class μ enzyme): 40–50% of individuals from various ethnic groups have a deficiency. (b) GSTT1 (the class θ enzyme): 10–30% of Europeans have a deficiency.

8.3.5. Methyl Transferase

Metabolic implications. Methyl transferases are mainly involved in methylation of endogenous substrates such as histamine, catecholamines, and norepinephrine. Some drugs, however, can be methylated by nonspecific methyl transferases. Methyl transferases are important in the metabolism of chemotherapeutic agents.

Reaction type. O, N, S-methylation:

$$R_2{=}NH \rightarrow R_2{=}N{-}CH_3$$

$$R{-}SH \rightarrow R{-}S{-}CH_3$$

Cofactor. S-adenosylmethionine (SAM).

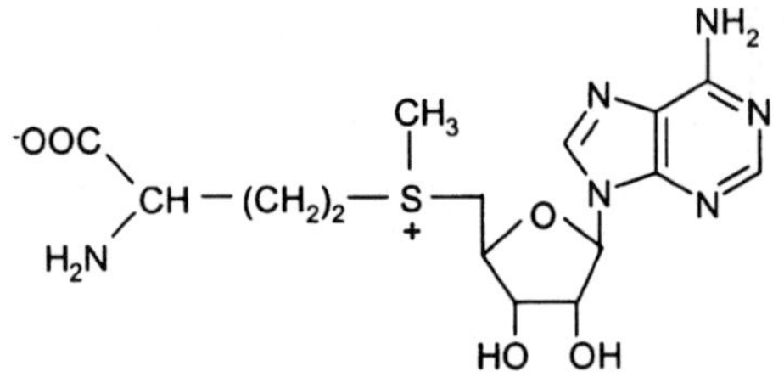

SAM

Tissue distribution. Liver, brain, lung, kidney, adrenals, skin, and erythrocytes.

Subcellular location. Cytosol.

Isozymes. At least four different enzymes can perform S-, N-, or O-methylation reactions.

Polymorphism. Approximately 0.3% of the European population have a deficiency in thiopurine S-methyltransferase activity.

NOTE: In general, methylation of a compound produces a less polar metabolite than the parent compound, and thus, unlike other conjugation reactions, tends to decrease the rate of its excretion.

8.3.6. Amino Acid Conjugation

METABOLIC IMPLICATIONS. Exogenous carboxylic acid, especially acetates, can be activated to coenzyme A derivatives (acyl CoA thioether) *in vivo* by acyl-CoA synthetase and further conjugated with endogenous amines such as amino acids by acyl-CoA:amino acid N-acyltransferase. Amino acid conjugates of xenobiotics are eliminated primarily in urine by tubular active secretion mechanisms.

REACTION TYPE. Two-step amino acid conjugation:

(a) Acylation of substrate:

$$\text{R–COOH} + \text{CoA–SH} \xrightarrow[\text{and ATP}]{\text{acyl-CoA synthetase}} \text{RCO–S–CoA}$$

(b) Conjugation of amino acid:

$$\underset{\text{(amino acids)}}{\text{RCO–S–CoA} + \text{H}_2\text{N–R}'} \xrightarrow[\text{N-acyltransferase}]{\text{acyl-CoA:amino acid}} \text{RCO–NH–R}'$$

COFACTORS. Coenzyme A (CoA-SH) for acyl-CoA synthetase, and amino acids such as glycine, glutamine, ornithine, arginine, and taurine, for acyl-CoA:amino acid N–acyltransferase.

H_2N–CH_2–COO^- Glycine

H_2N–CH_2CH_2–SO_3^- Taurine

H_2N–CH(COO^-)–CH_2CH_2–C(=O)–NH_2 Glutamine

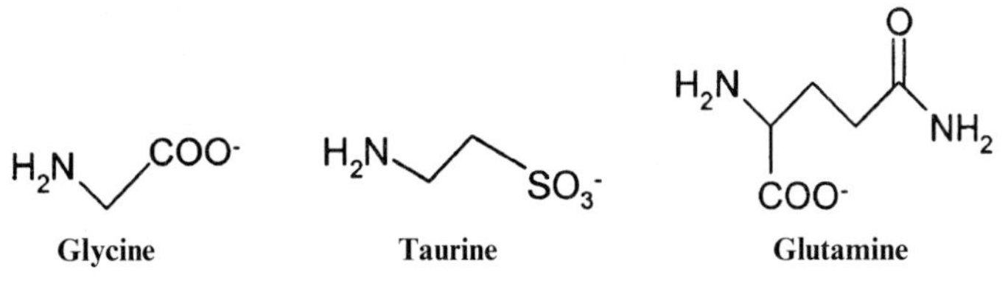

TISSUE DISTRIBUTION. Liver and kidney.

SUBCELLULAR LOCATION. Mitochondria and endoplasmic reticulum for acyl-CoA synthetase, and cytosol and mitochondria for acyl-CoA:amino acid N-acyltransferase.

SPECIES DIFFERENCES. The amino acid used for conjugation is both species- and compound-dependent. For instance, amino acid conjugation of bile acids occurs with both glycine and taurine in most species, whereas in cats and dogs, conjugation of bile acids occurs only with taurine.

Table 8.4. Summary of Characteristics of Important Metabolizing Enzymes in Humans

	Reaction	Cofactor	Tissue distribution	Subcellular location	Isozymes	Polymorphism	Route of elimination	Inducibility
Cytochrome P450 (P450 or CYP)	Mainly oxidation	NADPH	Most tissues (liver, intestine, etc)	Endoplasmic reticulum (ER)	~20	2C18, 2C19, 2D6, 3A5	Urine, bile	Yes
Flavin-mono-oxygenase	Oxidation	NADPH	Liver, kidney, lung, skin	ER	FMOI and FMOII	—	—	—
Esterase	Hydrolysis	—	Liver, blood, intestine	ER, cytosol	A, B, C-Esterase and choline esterase	Rare in cholinesterase	—	Yes
Uridine diphosphate-glucuronosyl-transferase (UDPGT)	Glucuronylation on –OH, –COOH, =NH, $-NH_2$, –SH	UDPGA	Most tissues (liver, intestine, etc)	ER	>15 in two subfamilies (UDPGT1 and UDPGT2)	UDPGT1	Mainly bile (MW > 350)	Yes
Sulfotransferase (ST)	Sulfation on –OH, $-NH_2$, –NHOH	PAPS	Liver, adrenals lung, brain, jejenum, blood platelets	Cytosol	TS ST, TL ST, EST, DHEA ST	Rare in TS ST, TL ST	Mainly urine	—
N-acetyltransferase (NAT)	Acetylation on $-NH_2$	Acetyl-CoA	Liver, spleen, lung, intestine	Cytosol	NAT1, NAT2	NAT2	—	—
Glutathione S-transferase (GST)	Glutathione conjugation on electrophilic moieties with good leaving group	GSH	Liver, kidney, intestine	Cytosol, ER	6 classes	GSTM1, GSTT1	Mainly bile	—
Amino acid conjugation	Amino acid conjugation on acyl-CoA of substrates	CoA, glycine, taurine, etc	Liver, kidney	Mitochondria,[a] ER,[a] cytosol,[b] mitochondria[b]	—	—	Mainly urine	—

[a]Subcellular locations for acyl-CoA synthetase.
[b]Subcellular locations for acyl-CoA:amino acid N-acyltransferase.

NOTE: Amino acid conjugation of carboxylic acid–containing xenobiotics is an alternative metabolism pathway to glucuronidation. Acyl glucuronides of xenobiotics can be potentially toxic, whereas conjugation with amino acids is a detoxification reaction.

8.4. EXTRAHEPATIC METABOLISM

Most tissues have some metabolic activity; however, quantitatively the liver is by far the most important organ for drug metabolism. Important organs for extrahepatic metabolism include the intestine (enterocytes and intestinal microflora), kidney, lung, plasma, blood cells, placenta, skin, and brain. In general, the extent of metabolism in the major extrahepatic drug-metabolizing organs such as the small intestine, kidney, and lung is approximately 10–20% of the hepatic metabolism. Less than 5% of extrahepatic metabolism compared to hepatic metabolism can be considered low with negligible pharmacokinetic implications (Connelly and Bridges, 1980; deWaziers *et al.*, 1990; Krishna and Klotz, 1994; Ravindranath and Boyd, 1995).

8.4.1. Intestinal Metabolism

Because most drugs are administered orally, there has been much emphasis in presystemic metabolism on the effects of gastrointestinal metabolism on the bioavailability of drugs. Recent studies have indicated that P450 isoforms such as CYP2C19 and 3A4 in enterocytes might play an important role in the presystemic intestinal metabolism of drugs and the large interindividual variability in systemic exposure after oral administration (Ilett *et al.*, 1990; Kaminsky and Fasco, 1992; Schwenk, 1988; Zhang *et al.*, 1996). The cytochrome P450 content of the intestine is about 35% of the hepatic content in the rabbit, but accounts for only 4% of the hepatic content in the mouse. Cytochrome P450 levels and activities are highest in the duodenum near the pyrolus, and then decrease toward the colon. A similar trend in regional activity levels along the intestine has been observed for glucuronide, sulfate, and glutathione conjugating enzymes. The rate and extent of first-pass intestinal metabolism of a drug after oral administration are dependent on various physiological factors such as:

1. Site of absorption: If the absorption site in the intestine is different from the metabolic site, first-pass intestinal metabolism of a drug may not be significant.
2. Intracellular residence time of drug molecules in enterocytes: The longer the drug molecules stay in the enterocytes prior to entering the mesenteric vein, the more extensive the metabolism.
3. Diffusional barrier between splanchnic bed and enterocytes: The lower the diffusibility of a drug from the enterocytes to the mesenteric vein, the longer its residence time.
4. Mucosal blood flow: Blood in the splanchnic bed can act as a sink to carry drug molecules away from the enterocytes, which reduces intracellular residence time of drug in the enterocytes.

Treatment with certain drugs such as methylcholanthren (MC) and phenobarbital can increase metabolizing enzyme levels in the intestine. Enzyme induction requires 2–4 days and is more extensive when the inducer is administered orally as opposed to parenterally.

8.4.2. Renal Metabolism

In addition to physiological functions of homeostasis in water and electrolytes and the excretion of endogenous and exogenous compounds from the body, the kidneys are the site of significant biotransformation activities for both phase I and phase II metabolism. The renal cortex, outer medulla, and inner medulla exhibit different profiles of drug metabolism, which appears to be due to heterogeneous distribution of metabolizing enzymes along the nephron. Most metabolizing enzymes are localized mainly in the proximal tubules, although various enzymes are distributed in all segments of the nephron (Guder and Ross, 1984; Lohr *et al.*, 1998). The pattern of renal blood flow, pH of the urine, and the urinary concentrating mechanism can provide an environment that facilitates the precipitation of certain compounds, including metabolites formed within the kidneys. The high concentration or crystallization of xenobiotics and/or their metabolites can potentially cause significant renal impairment in specific regions of the kidneys.

8.4.3. Metabolism in Blood

Blood contains various proteins and enzymes. As metabolizing enzymes, esterases, including cholinesterase, arylesterase, and carboxylesterase, have the most significant effects on hydrolysis of compounds with ester, carbamate, or phosphate bonds in blood (Williams, 1987). Esterase activity can be found mainly in plasma, with less activity in red blood cells. Plasma albumin itself may also act as an esterase under certain conditions. For instance, albumin contributes about 20% of the total hydrolysis of aspirin to salicylic acid in human plasma. The esterase activity in blood seems to be more extensive in small animals such as rats than in large animals and humans. Limited, yet significant monoamine oxidase activities can be also found in blood.

8.5. VARIOUS EXPERIMENTS FOR DRUG METABOLISM

In this section, various *in vitro*, *in situ*, and *in vivo* experiments for drug metabolism and important considerations in regard to *in vitro* drug metabolism are discussed.

8.5.1. Examining Metabolic Profiles of Drugs

Suitability of a particular experimental system for the metabolism of a compound of interest can be significantly affected by experimental conditions such as compound availability and assay sensitivity (Rodrigues, 1994). The advantages and limitations of various *in vitro*, *in situ*, and *in vivo* experimental systems are summarized in Table 8.5. A schematic description of the preparation of liver S9 and microsomes is shown in Fig. 8.5.

Table 8.5. Advantages and Limitations of Various Experimental Methods for Investigating Drug Metabolism in the Liver[a]

Method	Advantages	Limitations
In vitro system		
S9 fraction[b]	Useful for both phase I and phase II metabolism	Difficult to assess the effects of membrane transport on metabolism
Microsomes[c]	Useful for cytochrome P450 metabolism Can be used for UDPGT activity Hight throughput screening for metabolic stability of compounds is possible Easy to maintain and stable for a long-term storage at −80°C Freezing and thawing (up to 10 cycles) with little loss of cytochrome P450 activity Simple sample preparation for assay	Difficult to asses the effects of membrane transport on metabolism In case of UDPGT activity measurement, proper extrapolation of UDPGT activity to *in vivo* conditions may be difficult owing to the usage of detergent to enhance UDPGT activity Production of metabolites potentially different from those under *in vivo* conditions owing to the closed experimental system of microsome studies (no further elimination of metabolites formed in microsomes) Limited integrated drug metabolism for phase I and II simultaneously
Purified enzumes	Useful for studying metabolism by specific metabolizing enzymes	Cross-contamination of enzymes during isolation
Recombinant enzymes	Useful for identifying specific isozyme(s) of the metabolizing enzyme responsible for metabolism of substrate of interest Potential replacement of enzymes obtained from tissues (large-scale preparation is possible) Easy to handle; unlike enzymes prepared from animal tissues, recombinant enzymes are not hazardous	Technical difficulties in preparing recombinant enzymes Difficult to fully characterize enzyme activities Apparent discrepancy in enzyme affinities (K_m or K_i) between some recombinant enzymes and those in other *in vitro* systems for certain compounds
Hepatocytes[d]	Useful for both phase I and phase II metabolism Can examine the effects of membrane transport in metabolism	Only freshly isolated hepatocytes are suitable for metabolism studies, since metabolic activities diminishes within a few hours, especially in rat hepatocytes, after isolated from the

Table 8.5. Continued

Method	Advantages	Limitations
Hepatocytes	Primary cultures of hepatocytes can be used for studying the inducibility of metabolizing enzymes such as P450 under certain incubation conditions Cryopreservation of hepatocytes is possible for long-term storage	liver (this limitation has been significantly improved by recent progress in cryopreservation techniques for hepatocytes) Not suitable for high throughput screening Need to assay not only buffer but also hepatocytes for metabolite identification and extent of metabolism Difficult to prepare mixture of hepatocytes, which can reflect average population for metabolic profiles of compounds tested
Liver slices[e]	Integration of all cell types in the liver for metabolism study Tissue architecture and cell-to-cell communication maintained Useful for both phase I and phase II metabolism Can examine the effects of membrane transport on metabolism	Only freshly obtained liver slices are suitable for metabolism studies (limited applicability of cryopreservation to liver slices for metabolism studies) Release of cytosolic enzymes from damaged cells at the slice surface and poor oxygen supply to the center of the slice can cause impaired metabolic activities Not suitable for high throughput screening Need to assay both buffer and liver slices for metabolite identification and extent of metabolism Difficult to prepare mixture of liver slices representing population Lower estimate of intrinsic clearance as compared to that from studies with hepatocytes
In situ system		
Liver perfusion	Useful for both phase I and phase II metabolism Sequential metabolism can be studied Useful for examining biliary excretion Can study the effects of blood flow, protein binding, etc, on metabolism Closer system to *in vivo* metabolism and organ clearance than *in vitro* systems (maintaining the spatial heterogeneity and architecture of the liver)	Requires more sophisticated equipment than other *in vitro* methods Limited supply of the fresh liver (especially fresh human liver for perfusion study) Slow throughout Similar limitations to those of liver slices

Table 8.5. Continued

Method	Advantages	Limitations
In vivo system		
Metabolism study in bile-cannulated animals after intravenous administration of drug	The most comprehensive model for investigating drug metabolic profiles By measuring the amount of unchanged drug excreted in the urine and bile, the quantitative contributions of metabolism to the overall clearance can be estimated	Slow throughput Labor intensive

[a] Data taken from deKanter *et al.* (1998), Elkins (1996), Hawksworth (1994), Pearce *et al.* (1996), Price *et al.* (1998), Remmel and Burchell (1993), Rodrigues (1994), Silvo *et al.* (1998), Shett (1994), Thummel (1994), Vickers (1997), and Wrighton *et al.* (1993).

[b] S9 subcellular fraction of the liver can be obtained as a supernatant after centrifugation of liver homogenate at 9000 g, which removes nuclei and mitochondria pellets. S9 contains both cytosolic (soluble) and microsomal metabolizing enzymes and is capable of performing both phase I and phase II metabolism.

[c] A microsomal fraction of the liver can be prepared as a pellet after centrifuging S9 fraction at 100,000 g. Microsomes consist mainly of the endoplasmic reticulum, where important phase I metabolizing enzymes such as cytochrome P450s are located. Microsomes are not present in phase II metabolizing enzymes, except a few transferases including UDPGT.

[d] Primary hepatocyte suspension can be obtained by perfusing the liver with a collagenase-containing perfusate. In terms of drug metabolism, the one important fact that limits the use of isolated hepatocytes is the rapid loss of activities of metabolizing enzymes including cytochrome P450s, when the cells are kept for any extended period, sometimes, within a matter of several hours. Recent progress in cryopreservation techniques, however, greatly improved utilities of hepatocytes for metabolism studies. In general, cultured hepatocytes such as HepG2 cells are consdered to be not suitable for metabolism studies due to the lower levels of metabolizing enzymes including cytochrome P450s compared to primary hepatocytes.

[e] A commercially available tissue slicer can produce liver slices with a thickness of approximately 200 μm, containing five to six layers of hepatocytes from cylindrical cores of the liver. Liver slices are capable of both phase I and II metabolism. Owing to limitations of drug penetration through the multilayers of hepatocytes in liver slices, the metabolism of drugs is often limited to hepatocytes located in an outer layer of the slices, resulting in an underestimate of the extent of the metabolism.

8.5.2. Phenotyping of Cytochrome P450 Isoforms

Despite the fact that many of P450 isoforms exhibit partially overlapping substrate specificity, it has become apparent that in most cases a single P450 isoform may be exclusively or primarily responsible for metabolism of a particular drug at therapeutic concentrations *in vivo* (Parkinson, 1996*b*; Guengerich, 1996). Information on P450 isoform(s) responsible for the metabolism of the drug of interest is important for an understanding of two critical aspects of drug metabolism in humans. (a) drug–drug interaction in metabolism between coadministered drugs, and (b) metabolic polymorphism of a drug. Four *in vitro* methods have been used for phenotyping P450 isoforms, and a combination of at least two different approaches is generally necessary to identify the P450 isoform(s) responsible for the metabolism of the substrate of interest.

8.5.2.1. Correlation between Metabolism Rates and P450 Isoform Activities

The particular P450 isoform responsible for the metabolism of a particular compound can be identified by examining the relationship between the initial rate

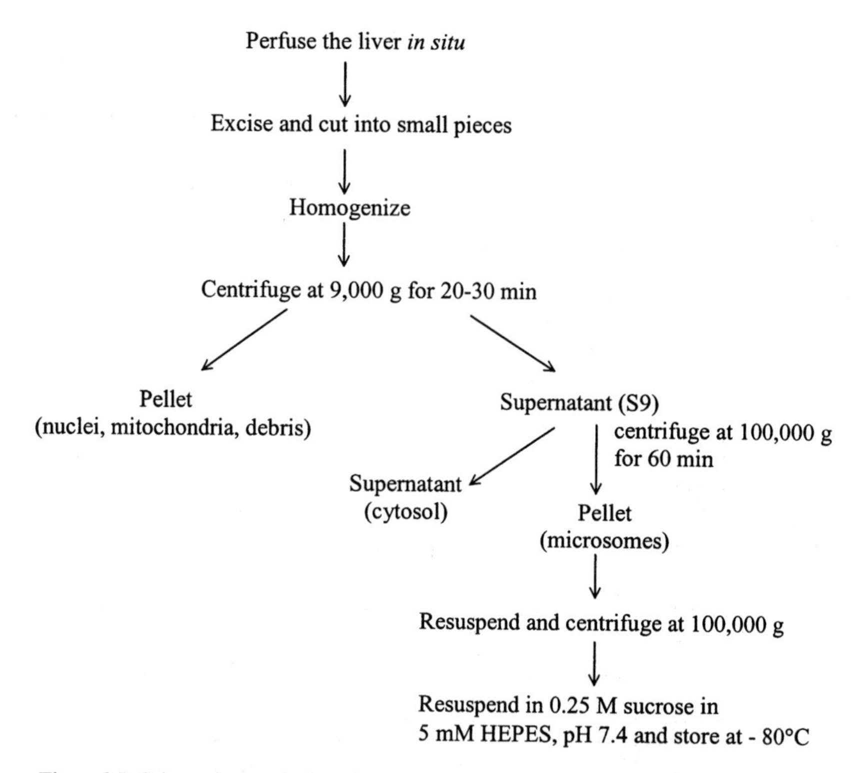

Figure 8.5. Schematic description of preparation processes for liver S9 and microsomes.

of metabolism of the compound in several different human microsomes and the level of activity of the individual P450 isoforms in the same microsomes. Information on activities of the individual P450 isoforms can be obtained by measuring the disappearance rates of known substrates for those P450 isoforms in the microsomes (Beaune *et al.*, 1986). If the rates of metabolism of the two different reactions show a linear correlation, both metabolic reactions can be considered to be mediated predominantly by the same P450 isoform. This approach is possible because the levels of the individual P450 isoforms in liver microsomes vary significantly among different subjects. Let us assume that the initial rates of disappearance of compound A and known substrates for different cytochrome P450 isoforms in a panel of human liver microsomes (H1-H10) are as shown in Table 8.6. Disappearance rates of compound A exhibit a good linear correlation only with those of a known substrate for CYP2D6 among the three isoforms examined in the same human microsome samples, suggesting that a major cytochrome P450 responsible for metabolism of compound A is CYP2D6.

8.5.2.2. Competitive Inhibition

If metabolism (disappearance rate) of the compound of interest in human liver microsomes is markedly inhibited by a known inhibitor against a particular P450

Table 8.6. Initial Rates of Disappearance of Compound A and Known Substrates for Different Cytochrome P450 Isoforms in a Panel of Human Liver Microsomes

Human liver microsome ID	Initial disappearance rates in microsomes			
		Known substrates for		
	Compound A	CYP2C9	CYP2D6	CYP3A4
H1	1	3	2	57
H2	2	6	4	35
H3	3	5	6	27
H4	4	9	8	3
H5	5	10	10	48
H6	6	3	12	5
H7	7	6	14	23
H8	8	9	16	39
H9	9	14	18	42
H10	10	7	20	7

isoform, then that P450 isoform may be involved in the metabolism of the compound to a significant extent. Results from these studies must be interpreted with caution, because most of the known inhibitors for cytochrome P450s can act on more than one isoform. These studies can be done using either liver microsomes or recombinant P450 enzymes.

8.5.2.3. Antibody Specific for Particular P450 Isoforms

The inhibitory effects of specific antibodies against selected cytochrome P450 isoforms on the metabolism of a compound in human liver microsomes are evaluated. Owing to the ability of an antibody to selectively inhibit a specific cytochrome P450 isoform, information from these studies is sufficient to establish which P450 isoform is responsible for metabolism of the compound (Gelbonin, 1993).

8.5.2.4. Metabolism with Purified or Recombinant P450 Isozymes

The initial rates of metabolism of a compound of interest by purified or recombinant P450 isozymes are measured. If a particular P450 isoform causes a faster initial disappearance rate compared to the other P450 isoforms tested, the metabolism of the compound may be mediated predominantly by that isoform (Guengerich *et al.*, 1996). The information from this type of study does not address the quantitative contribution of the particular isoform to the overall metabolism of the compound by cytochrome P450s.

8.5.3. Important Factors in Drug Metabolism Experiments

There are several important factors to be considered for *in vitro* and *in situ* drug metabolism studies: concentrations of enzymes, e.g., microsomal protein concentrations or the number of hepatocytes, concentrations of substrates and cosubstrate(s), effects of organic solvents for compounds with low aqueous solubility, and assay sensitivity, among others.

8.5.3.1. Substrate Concentrations for Metabolism Studies

In order to make a meaningful extrapolation of *in vitro* findings to *in vivo* situations, *in vitro* metabolism experiments should be conducted at physiologically or toxicologically relevant drug concentrations. Important differences between *in vivo* and *in vitro* metabolism in terms of drug concentrations are summarized in Table 8.7.

The most physiologically relevant concentrations of substrates and cosubstrates in an incubation buffer for *in vitro* metabolism studies would be those that can produce unbound drug concentrations adjacent to metabolizing enzymes similar to those within cells *in vivo*. In practice, it is almost impossible to accurately measure

Table 8.7. Differences between *In Vivo* and *In Vitro* Metabolism Experiments in Terms of Drug Concentrations

Factor	*In vivo*	*In vitro*
Drug concentrations[a]	0.01–10 μM (a typical range of total therapeutic drug concentration in plasma)	1–1000 μM
Protein binding	Can be extensive	Moderate at a typical microsomal protein concentration (⩽1 mg/ml) used for *in vitro* experiments
Unbound drug concentration	Can be substantially lower than total drug concentration in plasma	Lower than total concentration, but not to the extent observed *in vivo*
Concentration *vs.* time profile	Continuously changing	Fixed initial concentrations
Duration of drug exposure	Long	Short
Responsible metabolizing enzymes[b]	A few enzyme systems with high affinity and low capacity	Potentially more enzyme systems involved depending on experimental conditions used due to higher substrate concentrations (in a closed system)
Effects of other elimination pathways on drug concentrations	Effects of biliary excretion, renal elimination, and intestinal secretion	Cannot be examined

[a]In most cases, drug concentrations for *in vitro* metabolism studies are higher than *in vivo* unbound (or even total) therapeutic drug concentrations, mainly owing to assay limitations associated with smaller sample volumes obtained from *in vitro* studies. This can cause physiologically irrelevant extrapolation of *in vitro* data to *in vivo* in the rate and route of metabolism.

[b]It is more likely that a few (or even single) metabolizing enzyme(s) with high affinity and low capacity would be mainly responsible for biotransformation of a drug over a low therapeutic concentration range *in vivo*. At higher drug concentrations used in *in vitro* studies enzyme systems, which may not be important *in vivo*, can have significant effects on drug metabolism. Besides, further elimination of primary metabolites produced in a closed system such as microsomes, in which many phase I and II metabolizing enzyme acivities are absent, can be hindered.

the unbound substrate concentrations available to the metabolizing enzymes in endoplasmic reticulum or cytosol in hepatocytes *in vivo*. Several *in vitro* techniques, such as digitonin treatment of hepatocytes, for estimating unbound drug concentrations within hepatocytes have been reported in the literature; however, none of them appears to be reliable because of experimental difficulties and/or unrealistic assumptions. In addition, various active transport systems located in the sinusoidal membrane of hepatocytes can make predicting an intracellular drug concentration even more complicated.

For metabolism studies with primary hepatocytes or liver slices, unbound drug concentrations in sinusoidal blood (blood within sinusoids of the liver) available to hepatocytes *in vivo* would be the most desirable. It is also difficult to measure or make any reliable estimate of sinusoidal unbound drug concentrations from systemic plasma drug concentrations, because the former can be significantly different from the latter owing to metabolic activities in the liver. Therefore, the assumption that unbound drug concentrations in plasma are equal to those in the sinusoidal blood and thus to those adjacent to the metabolizing enzymes within hepatocytes may not be valid.

As a result of these difficulties in determining approprate drug concentrations, it is desirable to conduct *in vitro* metabolism studies over a wide range of substrate concentrations, e.g., 0.01–100 μM, especially for investigational drugs, for which little or no clinical pharmacology or safety data are available. This range of drug concentrations is likely to cover most of the therapeutic exposure levels of the drug *in vivo*. It is equally important to realize that in many cases the selection of drug concentrations used in *in vitro* studies is also affected by the experimental conditions, such as drug availability, aqueous solubility of the drug in an incubation buffer, and assay sensitivity (Rodrigues, 1994).

8.5.3.2. Effects of Organic Solvents on In Vitro P450 Metabolism Studies

For compounds with poor aqueous solubility, water-miscible organic solvents such as dimethyl sulfoxide are often used to enhance solubility of compounds in aqueous incubation media for *in vitro* metabolism studies. In general, compounds are dissolved into those organic solvents as stock solutions at high concentrations and then diluted to proper concentrations for the studies in aqueous incubation media. It has been found that the organic solvents used for solubilizing lipophilic compounds can have significant inhibitory effects on the activity of metabolizing enzymes. The inhibitory effects of three widely used organic solvents, i.e., dimethyl sulfoxide, methanol, and acetonitrile, on the activity of cytochrome P450s in isolated human liver microsomes are summarized below (Chauret *et al.*, 1998; Kawalek and Andrews, 1980).

(a) Dimethyl Sulfoxide. Although dimethyl sulfoxide (DMSO) is considered to be a good universal organic solvent for solubilizing lipophilic compounds, it appears not to be an optimal solvent for *in vitro* cytochrome P450-mediated metabolism studies using human liver microsomes. It has been reported that DMSO showed significant inhibitory effects (10–60%) on the activities of several P450 isoforms (CYP2C8/9, 2C19, 2D6, 2E1, and 3A4) even at low levels (0.2% v/v).

(b) Methanol. Methanol exhibited no measurable inhibitory effects on the catalytic activities of CYP1A2, 2A6, 2C19, 2D6, and 3A4 at 0.5–1%. However, significant inhibition has been reported on the activities of CYP2C9 and 2E1 at the same concentration range.

(c) Acetonitrile. Acetonitrile (ACN) appears to be the most suitable organ solvent among the three, as long as its concentration is kept at a relatively low level. At up to 1% ACN, no significant inhibition was noted on the activities of CYP1A2, 2A6, 2C8/9, 2C19, 2D6, 2E1, and 3A4. It is important to note that the effects of an organic solvent described above can vary with the experimental conditions, such as the types and concentrations of substrates, the integrity of microsomes used, and protein concentrations of microsomes, among others. In general, less than 0.2%, 0.5%, and 1% (v/v) of DMSO, methanol, and acetonitrile, respectively, are recommended for solubilization of lipophilic substrates for *in vitro* microsome studies, in order to minimize their inhibitory effects on the activities of cytochrome P450s.

8.6. PHYSIOLOGICAL AND ENVIRONMENTAL FACTORS AFFECTING DRUG METABOLISM

Physiological factors affecting the rate and pathway of drug metabolism include, e.g., species, genetics, gender, age, hormone, disease, and pregnancy. Environmental factors include, among others, diet, smoking, heavy metals, pollutants, and insecticides.

8.6.1. Physiological Factors

8.6.1.1. Species-Related Differences in Metabolism

There are species-related differences in both quantitative (the same metabolites with different rates) and qualitative (different metabolites via different metabolic pathways) aspects of drug metabolism. Information on species-related differences in the metabolism of investigational drugs is especially critical for drug safety evaluation in both animals and humans, because metabolic profiles in animals can be substantially different from those in humans, and the toxicity caused by the metabolites seen in the one may not be observed in the other and vice versa.

Common laboratory animals, especially rats, metabolize drugs considerably faster than humans; however, variability in drug-metabolizing enzymes is generally considered to be greater in humans than in animals (Nedelcheva and Gut, 1994; Smith, 1991; Soucek and Gut, 1992). Species differences can occur in both phase I and II metabolism, and are considered to be due mainly to evolutionary divergence among various species. For instance, there are significant differences in expression and the extent of activities of cytochrome P450s among different species (Table 8.8). The activity of β-glucuronidase in the intestine is much higher in rats than in humans. The activity of glutathione-S-transferase is substantially higher in mice and

Table 8.8. Differences in Important Cytochrome P450 Isoforms between Humans and Rats

Species	Cytochrome P450 isoforms			
Human	1A1,[a] 1A2[a]	2A6 2B6 2C8, 2C9, 2C10, 2C18, 2C19 2D6 2E1[a]	3A3, 3A4, 3A5, 3A7	4A9, 4A11
Rat[b]	1A1, 1A2	2A1, 2A2, 2A3 2B1, 2B2 2C11, 2C12, 2C13 2D1, 2D2, 2D3, 2D4, 2D5 2E1	3A1, 3A2	4A1, 4A2, 4A3

[a]There are three species-independent cytochrome P450s (the same name across all mammalian species), i.e., CYP1A1, 1A2, and 2E1, of which the genetic regulation is highly conserved among different species.
[b]Rats have approximately 40 different cytochrome P450 isoforms.

hamsters than in rats and humans. Glucuronidation activities are usually higher in rabbits. In general, the metabolism of the old world monkey, notably the rhesus monkey, is the one that resembles human metabolism most closely. In some animals, particular metabolic pathways are deficient (Table 8.9).

8.6.1.2. Genetics-Related Differences in Metabolism

Distinctive population subgroups may exhibit differences in their ability to metabolize certain drugs as compared to the general population. These differences among individuals in the extent of drug metabolism often follow a bimodal distribution pattern as shown in Fig. 8.6, which is indicative of genetic polymorphism in metabolism (pharmacogenetics). Genetic polymorphism (or simply polymorphism) is defined as a Mendelian or monogenic trait that is found in at least two phenotypes (and presumably at least two genotypes) in more than 1–2% of the population. If the frequency is lower than 1–2%, it is called a rare trait.

Table 8.9. Species Defective in Particular Xenobiotic Metabolism

Reactions or enzymes	Defective species
N-hydroxylation of aliphatic amines	Rat, marmoset
N-hydroxylation of arylacetamide	Guinea pig
N-acetylation of primary amines	Dog, guinea pig
Glucuronidation	Cat
Sulfation	Pig, opossum
Mercapturic acid formation	Guinea pig
Epoxide hydrolase	Mouse

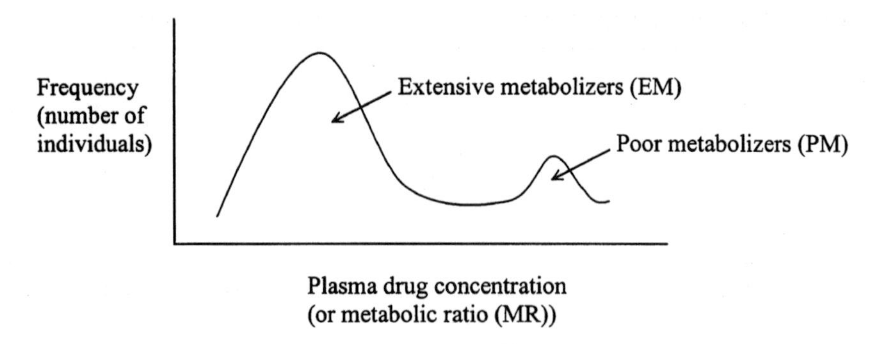

Figure 8.6. Metabolic polymorphism (a bimodal distribution of phenotypes in populations).

(a) Extensive Metabolizer and Poor Metabolizer. The poor metabolizer (PM) is an individual with deficient metabolic ability for a certain drug owing to genetic defects in a particular metabolizing enzyme(s), usually resulting in a higher exposure of the drug as compared to the normal population [extensive metabolizers (EM)].

Several metabolizing enzymes are known to exhibit polymorphism in humans. Table 8.10 summarizes the percentage of poor metabolizers in different populations who do not express the corresponding CYP isoforms in the liver. For instance, plasma exposure levels of certain drugs such as debriosquine metabolized by CYP2D6 are substantially higher in approximately 5–10% of Caucasians, as compared to the rest of population, which corresponds to the extent of polymorphic distribution of the poor CYP2D6 metabolizers in Caucasians. Genetic polymorphism becomes important in therapeutic monitoring, especially when drug elimination occurs mainly via a single metabolic pathway subject to polymorphism. Careful dosage monitoring should be implemented to avoid any adverse effects (idiosyncratic responses) of the drug at the high exposure levels produced in poor metabolizers (Daly *et al.*, 1993; Daly, 1995; Smith *et al.*, 1994).

(b) Phenotyping and Genotyping. For polymorphic phenotyping to identify the P450 isoform or N-acetyltransferase responsible for metabolism of a drug, a metabolic ratio (MR) between the amount of parent drug and certain metabolite(s)

Table 8.10. Important Metabolizing Enzymes Exhibiting Polymorphism in Humans

Enzymes	% Population as poor metabolizers	Known substrates
CYP2C18	2–3% of Caucasians	S-mephenytoin
CYP2C19	2–3% of Caucasians and 20% of Asians	—
CYP2D6	5–10% of Caucasians and 1–2% of Asians	Debriosquine, spartein, dextrometorphan
CYP3A5	80% of Caucasians	Midazolam
N-acetyl-transferase	50% of Caucasians and <25% of Asians	Isoniazid, sulfametazine

excreted in the urine produced by particular polymorphic enzyme(s) of interest can be measured (Bertilsson, 1995). Estimates of MR are significantly higher in poor metabolizers as compared to those in extensive metabolizers:

$$\text{MR} = \frac{\text{Amount of parent drug excreted in urine}}{\text{Amount of metabolite excreted in urine}} \tag{8.1}$$

Polymorphic genotype screening can be done using a polymerase chain reaction (PCR) followed by specific probes for many of the mutations in genes responsible for defective or absent metabolizing enzymes.

8.6.1.3. Gender-Related Differences in Metabolism

There are significant gender-related differences in cytochrome P450 expression in rats (male rats have a higher metabolic activity than female rats), due mainly to the different patterns of growth hormone secretion (pulsatile in male rats and continuous in female rats, Table 8.11) (Agrawal and Shapiro, 1997). Gender differences in hepatic metabolism have also been seen for many steroid hormones such as androgens, estrogens, and corticosteroids (Skett, 1988). Gender-related differences in metabolism are, in general, not apparent in mice and dogs. Although data are limited and not in complete agreement, CYP3A4 activity appears to be higher (approximately 1.4-fold) in women than in men, whereas the activities of many other metabolizing enzymes (e.g., CYP1A2, 2C19, UDPGT) may be higher in men than in women (Gleiter and Gundert-Remy, 1996; Harris *et al.*, 1995; Mugford and Kedderis, 1998).

8.6.1.4. Effects of Aging on Metabolism

It has been found that in rats the activity of cytochrome P450 decreases with age, but there is no change in uridine diphosphate-glucuronosyltransferase activity. In humans, cytochrome P450 activities decrease with age.

8.6.1.5. Effects of Disease on Metabolism

The level of cytochrome P450 isozymes in rats can be increased or decreased by the physiological status of the animals, including starvation (2B1/2, 2E1, 3A2,

Table 8.11. Gender-Related Differences in Cytochrome P450 Expression in Rats

Male specific isoforms	2C11, 2C13, 2C22, 2A2, 3A2
Female specific isoforms	2C12
Male dominant isoforms	2B1, 2B2, 3A1
Female dominant isoforms	1A2, 2A1, 2C7, 2E1
No gender difference	1A1, 2C6

4A1-3), diabetes (2A1, 2C7/11/12/13, 2E1, 3A2), or high blood pressure (2A1, 2C11, 3A1) (Shimojo, 1994). In humans, elevated hepatic cytochrome P450 activity has been reported in diabetic patients. Liver diseases such as cirrhosis and hepatoma can impair hepatic metabolism (George *et al.*, 1995).

8.6.2. Environmental Factors

Important environmental factors affecting drug metabolism in humans include diet, smoking, and pollutants (Baijal and Fitzpatrick, 1996; Guengerich, 1995; O'Mahony and Woodhouse, 1994; Williams *et al.*, 1996). Some of the important relevant chemical effects include:

- Butylated hydroxytoluene (BHT, food additive) inhibits lipid peroxidation.
- Caffeine induces or inhibits oxidative metabolism of some drugs.
- Charcoal-broiled meat [polycyclic aromatic hydrocarbons (PAH)] or cigarette smoking (PAH) induces CYP1A1/2.
- Cruciferous vegetables (cabbage, cauliflower, or brussels sprouts) induces CYP1A1/2.
- Grapefruit juice inhibits CYP3A4.
- Vitamin C induces oxidative metabolism in elderly patients with vitamin C deficiency.

8.7. METABOLITE KINETICS

A thorough understanding of *in vivo* disposition profiles of metabolite(s) is important in assessing a drug's pharmacological and/or toxicological effects. After administration, a drug is excreted in urine or bile, and/or is converted to metabolites, which will be eliminated. The rates of change in the amount of drug metabolites in the body are affected by how fast they are generated (formation rate of the metabolites) as well as how rapidly they are eliminated (elimination rate of the metabolites). To elucidate the pharmacokinetic relationship between these two processes, let us assume simple one-compartment models for both the parent drug and its metabolite. A drug plasma concentration–time profile after intravenous bolus injection is assumed to follow a first-order decline, and the exposure profile of the metabolite, if the metabolite itself is administered intravenously, is also assumed to follow a first-order elimination (Fig. 8.7).

The rate of change in the amount of drug can be described as first-order kinetics of the amount present in the body after intravenous injection:

$$dA(t)/dt = -(k_m + k_{other}) \cdot A(t) = -k \cdot A(t) \tag{8.2}$$

A(t) is the amount of drug in the body at time t after intravenous bolus injection. k, k_m, and k_{other} are the rate constants representing systemic elimination of the drug, metabolism of the drug to the metabolite, and elimination processes of the drug other than metabolism, respectively. From Eq. (8.2), the concentration of a drug in

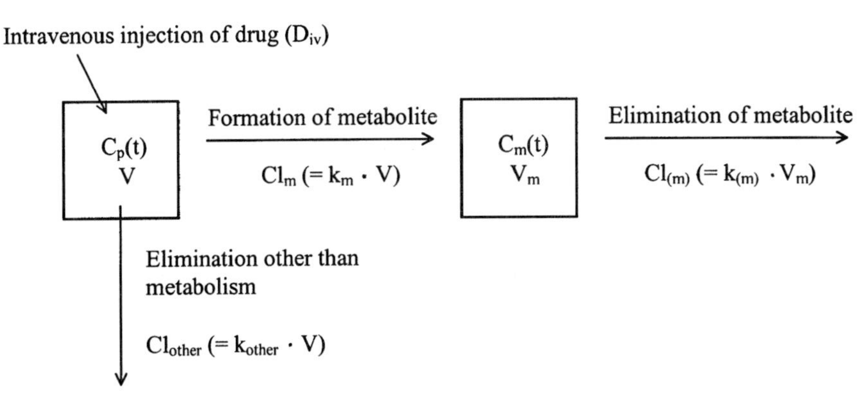

Figure 8.7. Processes affecting concentrations of the parent drug and its metabolite under a linear condition after intravenous administration, assuming one-compartment models for both the drug and its metabolite. $C_p(t)$, $C_m(t)$: concentrations of the drug and its metabolite in plasma at time t, respectively; Cl_m and Cl_{other}; metabolic clearance of the drug to produce the metabolite and drug clearance other than metabolism, respectively; $Cl_{(m)}$: systemic clearance of the metabolite, which is the same as the systemic clearance measured after intravenous bolus injection of the metabolite, itself; D_{iv}: intravenous dose of drug; k_m and k_{other}: rate constants representing the metabolism of the drug to the metabolite and elimination processes of the drug other than metabolism, respectively; $k_{(m)}$: rate constant representing systemic elimination processes of the metabolite, when the metabolite itself is administered intravenously; V and V_m: volumes of distribution of the drug and the metabolite, respectively.

plasma at time t [Cp(t)] can be derived as

$$C_p(t) = (D_{iv}/V) \cdot e^{-k \cdot t} \tag{8.3}$$

where D_{iv} is an intravenous dose, and V is the volume of distribution of the drug. The rate of change in the amount of the metabolite depends on the difference between the rate of formation of the metabolite from the drug and the rate of elimination of the metabolite itself:

Rate of change of amount of metabolite in the body ↓ *Rate of formation of metabolite from the drug* ↓ *Rate of elimination of metabolite from the body* ↓

$$dA_m(t)/dt = k_m \cdot A(t) - k_{(m)} \cdot A_m(t) \tag{8.4}$$

where $A_m(t)$ is the amount of the metabolite in the body at time t after intravenous administration of the drug, and $k_{(m)}$ is the rate constant for the systemic elimination processes of the metabolite. From Eqs. (8.3) and (8.4), the concentration of the metabolite in plasma at time t [$C_m(t)$] can be described as follows:

$$C_m(t) = \frac{k_m \cdot D_{iv}}{V_m \cdot (k_{(m)} - k)} \cdot (e^{-k \cdot t} - e^{-k_{(m)} \cdot t}) \tag{8.5}$$

where V_m is the volume of distribution of the metabolite. Equation (8.5) can be simplified at later time points under two different conditions, depending on the magnitude of k and $k_{(m)}$.

8.7.1. "Formation-Rate-Limited" Metabolite Kinetics

If the elimination rate constant of the drug is much smaller than that of the metabolite, i.e., $k \ll k_{(m)}$, then, the semilogarithmic plasma concentration *vs.* time curve of the metabolite declines in parallel with that of the drug during terminal phases with similar slopes and half-lives (Fig. 8.8A). In this case, the formation of the metabolite from the drug is much slower than the elimination of the metabolite, and becomes rate-determining in overall changes in metabolite concentrations in the body. According to Eq. (8.5), when $k \ll k_{(m)}$, $e^{-k \cdot t} - e^{-k_{(m)} \cdot t}$ approaches $e^{-k \cdot t}$ during later time points, since $e^{-k_{(m)} \cdot t}$ is negligible compared to $e^{-k \cdot t}$. Equation (8.6) describes $C_m(t)$ during the later phase:

$$C_m(t) = \frac{k_m \cdot D_{iv}}{V_m \cdot k_{(m)}} \cdot e^{-k \cdot t} \tag{8.6}$$

The exponential term of $C_m(t)$ during the later time points, $-k \cdot t$ is equal to that of $C_p(t)$ [Eq. (8.3)], and thus terminal half-lives ($t_{1/2}$) of the drug and the metabolite become similar. If the metabolite itself is administered intravenously, its $t_{1/2}$ becomes shorter than that of the parent drug after intravenous injection. Rate constant terms in Eq. (8.6) can be replaced with clearance (Cl) and volume of distribution (V) terms as follows:

$$k_m = Cl_m/V \qquad k_{(m)} = Cl_{(m)}/V_m$$

and

$$C_m(t) = \frac{(Cl_m/V) \cdot D_{iv}}{\cancel{V_m} \cdot Cl_{(m)}/\cancel{V_m}} \cdot e^{-k \cdot t}$$

$$= \frac{Cl_m}{Cl_{(m)}} \cdot \underbrace{\frac{D_{iv}}{V} \cdot e^{-k \cdot t}}_{C_p(t)} \quad \text{during the terminal phase} \tag{8.7}$$

Cl_m is the metabolic clearance of the drug to produce the metabolite, and $Cl_{(m)}$ is the systemic clearance of the metabolite, which can be determined following intravenous injection of the metabolite itself. According to Eq. (8.7), the relative extent of plasma concentration levels of the drug [$C_p(t)$] and its metabolite [$C_m(t)$] during the terminal phase depends on the ratio between Cl_m and $Cl_{(m)}$ under the formation-rate-limited metabolite kinetic conditions.

Summary of formation-rate-limited metabolite kinetics:

1. The terminal half-life of the metabolite formed from the drug is similar to that of the drug itself.
2. If the metabolite itself is administered intravenously, its terminal half-life is shorter than that of the parent drug after intravenous injection.

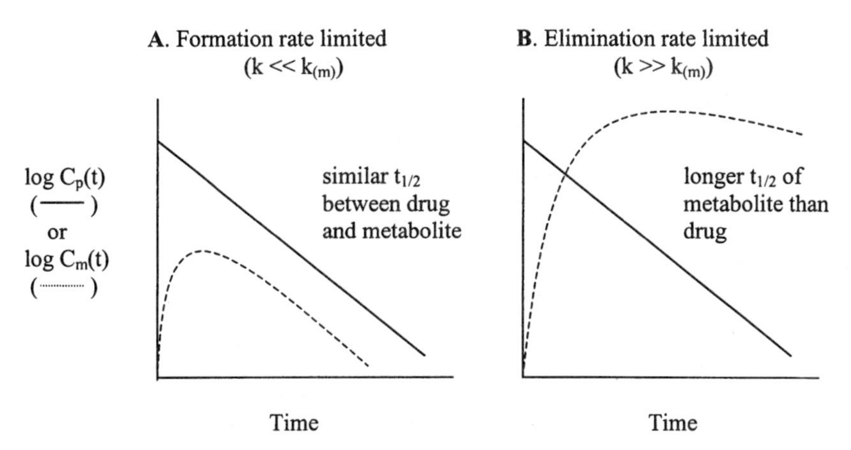

Figure 8.8. Plasma concentration *vs.* time profile of a drug [$C_p(t)$, —] and its metabolite [$C_m(t)$, ---], following intravenous drug administration, based on one-compartment models for both drug and metabolite on a semilog scale. A and B represent the formation-rate-limited and the elimination-rate-limited metabolite kinetics, respectively.

3. Relative plasma concentration levels of the drug and its metabolite during the terminal phase are dependent on the ratio between the metabolic clearance of the drug for producing the metabolite and the systemic clearance of the metabolite.

8.7.2. "Elimination-Rate-Limited" Metabolite Kinetics

The other extreme case is when k is much greater than $k_{(m)}$. In this case, the semilogarithmic plasma concentration *vs.* time curve of the metabolite will show a shallower slope with a longer half-life than the drug during the terminal phase (Fig. 8.8B). This is because when $k \gg k_{(m)}$, $e^{-k \cdot t}$ becomes much smaller than $e^{-k_{(m)} \cdot t}$ at later time points:

$$C_m(t) = \frac{k_m \cdot D_{iv}}{V_m \cdot k} \cdot e^{-k_{(m)} \cdot t} \quad \text{during the terminal phase} \tag{8.8}$$

As indicated in Eq. (8.8), the exponential term of $C_m(t)$ during the terminal phase, $-k_{(m)} \cdot t$, becomes greater than that of $C_p(t)$, $-k \cdot t$ [Eq. (8.3)], and thus $t_{1/2}$ of the metabolite is longer than that of the drug. In this case, if the metabolite itself is administered intravenously, its $t_{1/2}$ is also longer than that of its parent drug after intravenous injection. Rate constants in Eq. (8.8) can be replaced with Cl and V terms as follows:

$$C_m(t) = \frac{(Cl_m/V) \cdot D_{iv}}{V_m \cdot (Cl/V)} \cdot e^{-k_{(m)} \cdot t}$$

$$= \frac{Cl_m}{Cl} \cdot \frac{D_{iv}}{V_m} \cdot e^{-k_{(m)} \cdot t} \quad \text{during the terminal phase} \tag{8.9}$$

Summary of elimination-rate-limited metabolite kinetics:

1. The terminal half-life of the metabolite formed from the drug is longer than that of the drug itself.
2. If the metabolite itself is administered intravenously, $t_{1/2}$ of the metabolite is longer than that of drug following intravenous administration.

Exception: As shown above, $t_{1/2}$ of the metabolite produced after administration of the parent drug cannot be shorter than that of the drug, regardless of the route of administration under linear conditions. It is, however, possible to have $t_{1/2}$ of the metabolite shorter than that of the drug during the later time points. This can be due either to product inhibition (Perrier *et al.*, 1973), i.e., the metabolite inhibits metabolism of its parent drug or to rapid depletion of the cosubstrate(s) required for conversion of the drug to the metabolite during the early phase.

8.7.3. Pharmacokinetic Properties of Metabolites

In general, both $Cl_{(m)}$ and V_m of metabolites tend to be smaller than those of the parent drugs, which appears to be due to the increased hydrophilicity of the metabolites as compared with the parent drugs. The following summarizes the general pharmacokinetic properties of metabolites:

1. Metabolites are more polar and hydrophilic than their parent drugs, and thus more readily excreted in the urine.
2. Metabolites are often acidic (by oxidation and/or glucuronide or sulfate conjugation of the neutral or basic parent drugs).
3. The volume of distribution of the acidic metabolite is usually smaller than that of the neutral or basic parent drug. This is because acidic metabolites tend to be highly albumin-bound, which may restrict their distribution from plasma to other tissues or organs, and the extent of tissue binding of acidic metabolites tends to be less than for their neutral or basic parent drugs.

8.7.4. Estimating Systemic Clearance of Metabolites

One of important concepts in metabolite kinetics is mass balance between the total amount of drug converted to the metabolite and the total amount of metabolite eliminated. Equation (8.4) can be expressed in clearance and concentration terms of a drug and its metabolite:

$$dA_m(t)/dt = Cl_m \cdot C_p(t) - Cl_{(m)} \cdot C_m(t) \tag{8.10}$$

Integration of Eq. (8.10) from 0 to ∞ yields

$$\underbrace{\int_0^\infty \frac{dA_m(t)}{dt}\, dt}_{\uparrow} = Cl_m \cdot \int_0^\infty C_p(t)\, dt - Cl_{(m)} \cdot \int_0^\infty C_m(t)\, dt$$

$$0 = Cl_m \cdot AUC_{0-\infty} - Cl_{(m)} \cdot AUC_{m,0-\infty} \tag{8.11}$$

where $AUC_{0-\infty}$ and $AUC_{m,0-\infty}$ are the AUCs of a drug and its metabolite following drug administration from 0 to ∞, respectively. There is no metabolite in the body at time 0 and ∞; therefore integration of the rate of change in the amount of the metabolite from 0 to ∞ becomes 0. Equation (8.12) represents a mass balance between the amount of drug converted to metabolite and the amount of metabolite eliminated from the body from time 0 to ∞:

MASS BALANCE OF METABOLISM:

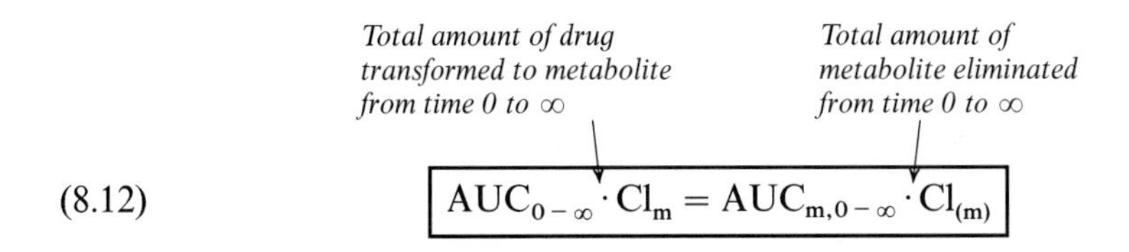

(8.12) $$AUC_{0-\infty} \cdot Cl_m = AUC_{m,0-\infty} \cdot Cl_{(m)}$$

This relationship is true regardless of the route of drug administration. Important assumptions include linear kinetics, i.e., drug or metabolite concentration-independent Cl_m and $Cl_{(m)}$, and no effects of the metabolite on the drug elimination mechanism(s). When the drug is also eliminated other than via metabolism (Cl_{other}), Cl_m can be expressed as the difference between the systemic clearance of the drug (Cl_s) and Cl_{other}:

(8.13) $$Cl_m = Cl_s - Cl_{other}$$

It can be difficult to measure Cl_m experimentally for a particular metabolite when there several different metabolites are generated from the drug. $Cl_{(m)}$ of the particular metabolite can be determined following intravenous administration of the metabolite itself.

8.8. INDUCTION OF METABOLISM

Administration of certain xenobiotics sometimes results in a selective increase in the concentration of metabolizing enzymes in both phase I and II metabolism, and thereby in their activities (Barry and Feely, 1990; Okey, 1990; Park, 1987). Enzyme induction becomes important especially when polypharmacy involves drugs with narrow therapeutic windows, since the induced drug metabolism could result in a significant decrease in its exposure and therapeutic effects. In addition, enzyme induction may cause toxicity, associated with increased production of toxic metabolites.

8.8.1. Mechanisms of Induction

- Stimulation of transcription of genes and/or translation of proteins, and/or stabilization of mRNA and/or enzymes by inducers, resulting in elevated enzyme levels.

- Stimulation of preexisting enzymes resulting in apparent enzyme induction without an increase in enzyme synthesis (this is more common *in vitro* than *in vivo*).
- In many cases, the details of the induction mechanisms are unknown. The following two receptors have been identified for CYP1A1/2 and CYP4A1/2 induction: (a) *Ah* (aromatic hydrocarbon) receptor in cytosol, which regulates enzyme (CYP1A1 and 1A2) induction by polycyclic aromatic hydrocarbon (PAH)-type inducers; and (b) peroxisome proliferator activated receptor (PPAR), where hypolipidemic agents cause peroxisome proliferation in rats (CYP4A1 and 4A2); humans have low PPAR and show no effects from hypolipidemic agents.

8.8.2. Characteristics of Induction

- Induction is a function of intact cells and cannot be achieved by treating isolated cell fractions such as microsomes with inducers. Evaluation of enzyme induction is usually conducted in *ex vivo* experiments, i.e., treating animals *in vivo* with potential inducers and measuring enzyme activities *in vitro* or in cell-based *in vitro* preparations such as hepatocytes, liver slices, or cell lines. Recent studies have demonstrated that primary cultures of hepatocytes can be used for studying the inducibility of metabolizing enzymes such as P450 under certain incubation conditions (Silva, 1998).
- Enzyme induction is usually inducer-concentration–dependent. The extent of induction increases as the inducer concentration increases; however, above certain values, induction starts to decline.
- In general, inducers increase the content of endoplasmic reticulum within hepatocytes as well as liver weight.
- In some cases, an inducer induces enzymes responsible for its own metabolism (so-called "autoinduction").

8.8.3. Inducing Agents

In general, enzyme inducers are lipophilic at physiological pH and exhibit relatively long $t_{1/2}$ with high accumulation in the liver. There are several different classes of enzyme inducers.

1. Barbiturates: phenobarbitone, phenobarbital.
2. Polycyclic aromatic hydrocarbons (PAH): 3-methylcholanthrene (3-MC), 2,3,7,8,-tetrachlorodibenzo-*p*-dioxin (TCDD), β-naphthoflavone (β-NF).
3. Steroids: pregnenalone 16-α-carbonitrile (PCN), dexamethasone.
4. Simple hydrocarbons with aliphatic chains: ethanol (chronic), acetone, isoniazid.
5. Hypolipidemic agents: clofibrate, lauric acids.
6. Macrolide antibiotics: triacetyloleandomycin (TAO).
7. A wide variety of structurally unrelated compounds: e.g., antipyrine, carbamazepine, phenytoin, and rifampicin.

Table 8.12 summarizes cytochrome P450 isoforms subject to induction and their corresponding inducing agents in humans.

Table 8.12. A Summary of Inducible Cytochrome P450 Enzymes in Human Liver and Known Inducers[a]

CYP	Inducers
1A1/2	Lansoprazole, omeprazole
2A6	Anticonvulsants, barbiturates, dexamethasone, rifampicin
2B6	Anticonvulsants, rifampicin
2C8, 9, 19	Anticonvulsants, barbiturates, rifampicin
2E1	Ethanol, isoniazid
3A4	Anticonvulsants, barbiturates, dexamethasone, rifampicin

[a]Among human cytochrome P450 isoforms, CYP2D6, 3A5, 4A9, and 4A11 are not inducible.

8.8.4. Time- and Dose-Dependence of Induction

The time course of induction may vary with different inducing agents. Increased transcription of P450 mRNA has been detected in a nucleus as early as 1 hr after administration of phenobarbiton in rats, although the maximum induction may take 2–3 days. Induction is usually reversible, i.e., when the inducer is removed, enzyme levels return to normal. In general, the degree of induction increases with the inducer dose; however, after certain dose levels, induction starts to decline.

8.8.5. Species Differences in Induction

There are significant differences in the inducibility of inducers among different species. Table 8.13 lists some of the important inducers with species-dependent P450 induction.

Table 8.13. Differences in Induction between Human and Laboratory Animals for Cytochrome P450 Enzymes

CYP	Species	Inducibility
CYP1A	Human and dog	Inducible by omeprazole
	Rabbit and mouse	Noninducible by omeprazole
CYP3A	Human and rabbit	Inducible by rifampicin, but not by PCN[a]
	Rat	Inducible by PCN, but not by rifampicin

[a]PCN: pregnenalone 16-α-carbonitrile.

REFERENCES

Agarwal D. P. and Goedde H. W., Pharmacogenetics of alcohol dehydrogenase, in W. Kalow (ed.), *Pharmacogenetics of Drug Metabolism*, Pergamon, New York, 1992, pp. 263–280.

Agrawal A. K. and Shapiro B. H., Gender, age and dose effects of neonatally administered aspartate on the sexually dimorphic plasma growth hormone profiles regulating expression of the rat sex-dependent hepatic CYP isoforms, *Drug Metab. Dispos.* **25**: 1249–1256, 1997.

Baijal P. K. and Fitzpatrick D. W., Effect of dietary protein on hepatic and extrahepatic phaseI and phase II drug metabolizing enzymes, *Toxicol. Lett.* **89**: 99–106, 1996.
Barry M. and Feely J., Enzyme induction and inhibition, *Pharmacol. Ther.* **48**: 71–94, 1990.
Beaune P. *et al.*, Comparison of monooxygenase activities and cytochrome P450 isozyme concentrations in human liver microsomes, *Drug Metab. Dispos.* **14**: 437–442 1986.
Benedetti M. S. and Tipton K. F., Monoamine oxidases and related amine oxidases as phase I enzymes in the metabolism of xenobiotics, *J. Neural. Transm.* **52**: 149–171, 1998.
Bertilsson L., Geographical/interracial differences in polymorphic drug oxidation, current state of knowledge of cytochrome P450 (CYP) 2D6 and 2C19, *Clin. Pharmacokinet.* **29**: 192–209, 1995.
Birkett D. J. *et al.*, *In vitro* approaches can predict human drug metabolism, *Trends Pharmacol. Sci.* **14**: 292–294, 1993.
Bourrié M. *et al.*, Cytochrome P450 isoform inhibitors as a tool for the investigation of metabolic reactions catalyzed by human liver microsomes, *J. Pharmacol. Exp. Ther.* **277**: 321–332, 1996.
Burchell B. and Coughtrie M. W. H., UDP-glucuronosyltransferases, *Pharmacol. Ther.* **43**: 261–289, 1989.
Burchell B. *et al.*, Specificity of human UDP-glucuronosyl transferases and xenobiotic glucuronidation, *Life Sci.* **57**: 1819–1831, 1995.
Caldwell J. *et al.*, An introduction to drug disposition: the basic principles of absorption, distribution, metabolism and excretion, *Toxicol. Pathol.* **23**: 102–114, 1995.
Cashman J. R., Structural and catalytic properties of the mammalian flavin-containing monooxygenase, *Chem. Res. Toxicol.* **8**: 166–181, 1995.
Chauret N. *et al.*, Effect of common organic solvents on *in vitro* cytochrome P450-mediated metabolic activities in human liver microsomes, *Drug Metab. Dispos.* **26**: 1–4, 1998.
Clarke D. J. and Burchell B., The uridine diphosphate glucuronosyltransferase multigene family: function and regulation, in F. C. Kauffman (ed.), *Conjugation-Deconjugation Reactions in Drug Metabolism and Toxicity*, Springer-Verlag, New York, 1994, pp. 3–43.
Connelly J. C. and Bridges J. W., The distribution and role of cytochrome P450 in extrahepatic organs, in J. W. Bridges and L. F. Chasseaud (eds.), *Progress in Drug Metabolism, Vol. 5*, John Wiley & Sons, Chichester, 1980, pp. 1–112.
Coughtrie M. W. H. *et al.*, Biology and function of the reversible sulfation pathway catalyzed by human sulfotransferases and sulfatases, *Chem. Biol. Interact.* **109**: 3–27, 1998.
Daly A. K. *et al.*, Metabolic polymorphisms, *Pharmac. Ther.* **57**: 129–160, 1993.
Daly A. K., Molecular basis of polymorphic drug metabolism, *J. Mol. Med.* **73**: 539–553, 1995.
DeGroot M. J. and Vermeulen N. P. E., Modeling the active site of cytochrome P450s and glutathione *S*-transferases, two of the most important biotransformation enzymes, *Drug Metab. Rev.* **29**: 747–799, 1997.
deKanter R. *et al.*, A rapid and simple method for cryopreservation of human liver slices, *Xenobiotics* **28**: 225–234, 1998.
deWaziers L. *et al.*, Cytochrome P450 isoenzymes, epoxide hydrolase and glutathione transferases in rat and human hepatic and extrahepatic tissues, *J. Pharmacol. Exp. Ther.* **253**: 387–394, 1990.
Ekins S., Past, present, and future applications of precision-cut liver slices for *in vitro* xenobiotic metabolism, *Drug Metab. Rev.* **28**: 591–623, 1996.
Gelbonin H. V., Cytochrome P450 and monoclonal antibodies, *Pharmacol. Rev.* **45**: 413–453, 1993.
George J. *et al.*, Differential alterations of cytochrome P450 proteins in livers from patients with severe chronic liver diseases, *Hepatology* **21**: 120–128, 1995.
Gleiter C. H. and Gundert-Remy U., Gender differences in pharmacokinetics, *E. J. Drug Metab. Pharmacokinet.* **21**: 123–128, 1996.
Goedde H. W. and Agarwal D. P., Pharmacogenetics of aldehyde dehydrogenase, in W. Kalow (ed.), *Pharmacogenetics of Drug Metabolism*, Pergamon, New York, 1992, pp. 281–311.
Gonzalez F. J., Human cytochrome P450: problems and prospects, *Trends Pharmacol. Sci.* **13**: 346–352, 1992.
Guder W. G. and Ross B. D., Enzyme distribution along the nephron, *Kidney Int.* **26**: 101–111, 1984.
Guengerich F. P., Influence of nutrients and other dietary materials on cytochrome P-450 enzymes, *Am. J. Clin. Nutr.* **61**(Suppl): 651S–658S, 1995.
Guengerich F. P., *In vitro* techniques for studying drug metabolism, *J. Pharmacokinet. Biopharm.* **24**: 521–533, 1996.
Guengerich F. P. *et al.*, New applications of bacterial systems to problems in toxicology, *Crit. Rev. Toxicol.* **26**: 551–583, 1996.

Halpert J. R. *et al.*, Contemporary issues in toxicology: selective inhibitors of cytochromes P450, *Toxicol. Appl. Pharmacol.* **125**: 163–175, 1994.

Harris R. Z. *et al.*, Gender effects in pharmacokinetics and pharmacodynamics, *Drugs* **50**: 222–239, 1995.

Hawksworth G. M., Advantages and disadvantages of using human cells for pharmacological and toxicological studies, *Human Exp. Toxicol.* **13**: 568–573, 1994.

Hayball P. J., Formation and reactivity of acyl glucuronides: the influence of chirality, *Chirality* **7**: 1–9, 1995.

Ilett K. F. *et al.*, Metabolism of drugs and other xenobiotics in the gut lumen and wall, *Pharmacol. Ther.* **46**: 67–93, 1990.

Kaminsky L. S. and Fasco M. J., Small intestinal cytochromes P450, *Crit. Rev. Toxicol.* **21**: 407–422, 1992.

Kaspersen F. M. and Van Boeckel C. A. A., A review of the methods of chemical synthesis of sulphate and glucuronide conjugates, *Xenobiotica* **17**: 1451–1471, 1987.

Kauffman F. C., Regulation of drug conjugate production by futile cycling in intact cells, in F. C. Kauffman (ed.), *Conjugation-Deconjugation Reactions in Drug Metabolism and Toxicity*, Springer-Verlag, New York, 1994, pp. 247–255.

Kawalek J. C. and Andrews A. W., The effect of solvents on drug metabolism *in vitro*, *Drug Metab. Dispos.* **8**: 380–384, 1980.

Krishna D. R. and Klotz U., Extrahepatic metabolism of drugs in humans, *Clin. Pharmacokinet.* **26**: 144–160, 1994.

Lohr J. W. *et al.*, Renal drug metabolism, *Pharmacol. Rev.* **50**: 107–141, 1998.

Meyer U. A., Overview of enzymes of drug metabolism, *J. Pharmacokinet. Biopharm.* **24**(5): 449–459, 1996.

Mugford C. A. and Kedderis G. L., Sex-dependent metabolism of xenobiotics, *Drug Metab. Rev.* **30**: 441–498, 1998.

Mugford C. A. *et al.*, 1-Aminobenzotirazole-induced destruction of hepatic and renal cytochromes P450 in male Sprague-Dawley rats, *Fund. Appl. Toxicol.* **19**: 43–49, 1992.

Mulder G. J., Glucuronidation and its role in regulation of biological activity of drugs, *Ann. Rev. Pharmacol. Toxicol.* **32**: 25–49, 1992.

Musson D. G. *et al.*, Assay methodology for quantitation of the ester and ether glucuronide conjugates of difflunisal in human urine, *J. Chromatogr.* **337**: 363–378, 1985.

Nedelcheva V. and Gut I., P450 in the rat and man: methods of investigation, substrate specificities and relevance to cancer, *Xenobiotica* **24**: 1151–1175, 1994.

Nelson D. R., P450 superfamily: update on new sequences, gene mapping, accession numbers and nomenclature, *Pharmacogenetics* **6**: 1–42, 1996.

Newton D. J. *et al.*, Cytochrome P450 inhibitors: evaluation of specificities in the *in vitro* metabolism of therapeutic agents by human liver microsomes, *Drug Metab. Dispos.* **23**: 154–158, 1995.

Okey A. B., Enzyme induction in the cytochrome P-450 system, *Pharmacol. Ther.* **45**: 241–298, 1990.

O'Mahony M. S. and Woodhouse K. W., Age, environmental factors and drug metabolism, *Pharmacol. Ther.* **61**: 279–287, 1994.

Park B. K., *In vivo* methods to study enzyme induction and enzyme inhibition, *Pharmacol. Ther.* **33**: 109–113, 1987.

Parkinson A., Biotransformation of xenobiotics, in C. D. Klaassen (ed.), *Casarett & Doull's Toxicology: The Basic Science of Poisons, 5th Ed.*, McGraw-Hill, New York, 1996*a*, pp. 113–186.

Parkinson A., An overview of current cytochrome P450 technology for assessing the safety and efficacy of new materials, *Toxicol. Pathol.* **24**: 45–57, 1996*b*.

Pearce R. E. *et al.*, Effects of freezing, thawing, and storing human liver microsomes on cytochrome P450 activity, *Arch. Biochem. Biophys.* **331**: 145–169, 1996

Perrier D. *et al.*, Effect of product inhibition on kinetics of drug elimination, *J. Pharmacokinet. Biopharm.* **1**: 231–242, 1973.

Price R. J. *et al.*, Influence of slice thickness and culture conditions on the metabolism of 7-ethoxycoumarin in precision-cut rat liver slices, *ATLA* **26**: 541–548, 1998.

Ravindranath V. and Boyd M. R., Xenobiotic metabolism in brain, *Drug Metab. Rev.* **27**: 419–448, 1995.

Remmel R. P. and Burchell B., Validation and use of cloned, expressed human drug-metabolizing enzymes in heterologous cells for analysis of drug metabolism and drug–drug interactions, *Biochem. Pharmacol.* **46**: 559–566, 1993.

Rendic S. and DiCarlo F. J., Human cytochrome P450 enzymes: a status report summarizing their reactions, substrates, inducers, and inhibitors, *Drug Metab. Rev.* **29**: 413–580, 1997.

Rodrigues A. D., Use of *in vitro* human metabolism studies in drug development: an industrial perspective, *Biochem. Pharmacol.* **48**: 2147–2156, 1994.

Satoh T. and Hosokawa M., The mammalian carboxyesterases: from molecules to functions, *Ann. Rev. Pharmacol. Toxicol.* **38**: 257–288, 1998.

Schwenk M., Mucosal biotransformation, *Toxicol. Pathol.* **16**: 138–146, 1988.

Shimada T. *et al.*, Interindividual variations in human liver cytochrome P-450 enzymes involved in the oxidation of drugs, carcinogens and toxic chemicals: studies with liver microsomes of 30 Japanese and 30 Caucasians, *J. Pharmacol. Exp. Ther.* **270**: 414–422, 1994.

Shimojo N., Cytochrome P450 changes in rats with streptozocin-induced diabetes, *Int. J. Biochem.* **26**: 1261–1268, 1994.

Silva J. M. *et al.*, Refinement of an *in vitro* cell model for cytochrome P450 induction, *Drug Metab. Dispos.* **26**: 490–496, 1998.

Skett P., Biochemical basis of sex differences in drug metabolism, *Pharmacol. Ther.* **38**: 269–304, 1988.

Skett P., Problems in using isolated and cultured hepatocytes for xenobiotic metabolism/metabolism-based toxicity testing-solutions, *In Vitro Toxicol.* **8**: 491–504, 1994.

Smith C. A. D. *et al.*, Genetic polymorphisms in xenobiotic metabolism, *Eur. J. Cancer* **30A**: 1921–1935, 1994.

Smith D. A., Species differences in metabolism and pharmacokinetics: are we close to an understanding? *Drug Metab. Rev.* **23**: 355–373, 1991.

Smith D. A. and Jones B. C., Speculations on the substrate structural–activity relationships (SSAR) of cytochrome P450 enzymes, *Biochem. Pharmacol.* **44**: 2089–2098, 1992.

Smith D. A., Chemistry and enzymology: their use in the prediction of human drug metabolism, *Eur. J. Pharm. Sci.* **2**: 69–71, 1994*a*.

Smith D. A., Design of drugs through a consideration of drug metabolism and pharmacokinetics, *Eur. J. Drug Metab. Pharmacokinet.* **3**: 193–199, 1994*b*.

Smith P. C. *et al.*, Irreversible binding of Zomepirac to plasma protein *in vitro* and *in vivo*, *J. Clin. Invest.* **77**: 934–939, 1986.

Soucek P. and Gut I., Cytochromes P450 in rats: structure, functions, properties and relevant human forms, *Xenobiotica* **22**: 83–103, 1992.

Thummel K. E., Use of midazolam as a human cytochrome P450 3A probe: II. characterization of inter- and intraindividual hepatic CYP3A variability after liver transplantation, *J. Pharmacol. Exp. Ther.* **271**: 557–566, 1994.

van der Aar E. M. *et al.*, Strategies to characterize the mechanisms of action and the active sites of glutathione *S*-transferases: a review, *Drug Metab. Rev.* **30**: 569–643, 1998.

Vatsis K. P. and Weber W. W., Human *N*-acetyltransferases, in F. C. Kauffman (ed.), *Conjugation-Deconjugation Reactions in Drug Metabolism and Toxicity*, Springer-Verlag, New York, 1994, pp. 109–130.

Vickers A. E. M., Liver slices: an *in vitro* tool to predict drug biotransformation and to support risk assessment, *In Vitro Toxicol.* **10**: 71–80, 1997.

Watt J. A. *et al.*, Contrasting systemic stabilities of the acyl and phenolic glucuronides of diflunisal in the rat, *Xenobiotica* **2**: 403–415, 1991.

Weinshilboum S. and Aksoy I., Sulfation pharmacogenetics in humans, *Chem. Biol. Interact.* **92**: 233–246, 1994.

Weinshilboum S. and Otterness D., Sulfotransferase enzymes, in F. C. Kauffman (ed.), *Conjugation-Deconjugation Reactions in Drug Metabolism and Toxicity*, Springer-Verlag, New York, 1994, pp. 45–78.

Williams F. M., Serum enzymes of drug metabolism, *Pharmacol. Ther.* **34**: 99–109, 1987.

Williams L. *et al.*, The influence of food on the absorption and metabolism of drugs: an update, *E. J. Drug Metab. Pharmacokinet.* **21**: 201–211, 1996.

Wrighton S. A. and Stevens J. C., The human hepatic cytochrome P450 involved in drug metabolism, *Crit. Rev. Toxicol.* **22**: 1–21, 1992.

Wrighton S. A. *et al.*, *In vitro* methods for assessing human hepatic drug metabolism: their use in drug development, *Drug Metab. Rev.* **25**: 453–484, 1993.

Zhang Q-Y. *et al.*, Characterization of rat small intestinal cytochrome P450 composition and inducibility, *Drug Metab. Dispos.* **24**: 322–328, 1996.

9

Biliary Excretion

Biliary excretion can be an important hepatic elimination pathway for many drugs (Yamazaki *et al.*, 1996).

9.1. RELATIONSHIP BETWEEN HEPATIC AND BILIARY CLEARANCES

Hepatic clearance (Cl_h) of a drug is the sum of its hepatic metabolic clearance (Cl_{hm}) and its biliary clearance (Cl_{bl}):

(9.1) $$\boxed{Cl_h = Cl_{hm} + Cl_{bl}}$$

Cl_{bl} of a drug can be experimentally determined by measuring the amount of unchanged drug excreted in bile over an extended period of time:

(9.2) $$Cl_{bl} = \frac{\text{Amount of unchanged drug excreted in the bile from time 0 to t}}{AUC_{0-t,iv}}$$

$AUC_{0-t,iv}$ is the AUC in blood from time 0 to t and t is usually more than 24 hr for sufficient bile collection after intravenous administration of a drug.

9.2. SPECIES DIFFERENCES IN BILIARY EXCRETION

Rats and dogs are perhaps the most efficient biliary excretors, while guinea pigs and monkeys are very inefficient. The limited evidence available also suggests that humans are not efficient excretors for compounds with intermediate molecular weights.

Molecular-weight-threshold-theory: Important physicochemical properties of compounds governing the extent of biliary excretion include molecular weight, charge, lipophilicity (log P), and molar refractivity (MR). There have been attempts to establish a correlation between the molecular weight of a compound and the

Table 9.1. Molecular Weight Threshold for Biliary Excretion of Organic Anions and Cations in Different Species

Compounds	Species			
	Rat	Guinea pig	Rabbit	Human
Organic anions	325	400	475	500
Organic cations				
Monovalent	200 with little or no species variation			
Bivalent	500 with little or no species variation			

extent of its biliary excretion in animals (Hirom *et al.*, 1974). The findings from these studies have indicated that there might be a threshold for the molecular weight of a compound subject to relatively extensive biliary excretion. Rough guidelines for the molecular-weight threshold for biliary excretion of organic anions and cations among different species are summarized in Table 9.1. For a compound with a molecular weight well below the threshold, there will be more extensive renal elimination than biliary excretion. A compound of intermediate molecular weight (325–465) tends to be excreted in both the urine and the bile.

9.3. ACTIVE TRANSPORTERS FOR BILIARY EXCRETION

Several active transport systems for endogenous and exogenous substrates have been identified in the canalicular membrane of hepatocytes (Müller and Jansen, 1997; Smit *et al.*, 1995). These active transporters may have a significant effect on biliary excretion of certain organic compounds. In particular, recent studies with gene knock-out animals have revealed the physiological and toxicological functions of some of the transporter systems. A brief description of various transporters and their substrates is provided in Table 9.2.

Table 9.2. Various Active Transport Systems Located in Canalicular Membranes of Hepatocytes and Their Substrate Specificity

Transporters[a]	Substrate specificity	Known substrates
cBAT	Organic anions	Monovalent bile acids
MDR1	Organic cations	Daunomycin
MDR3	Organic cations (?)	Phosphatidyl choline
MRP1	Organic anions Organic cations (?)	Glutathione, glucuronide, and sulfate conjugates: dinitrophenol S-glutathione, leukotriene C_4 (CTC_4), oxidized glutathione, calcein
MRP2 (cMOAT)	Organic anions Organic cations (?)	Glutathione, glucuronide, and sulfate conjugates

[a]cBAT: canalicular bile acid transporter; cMOAT: canalicular multispecific organic anion transporter; MDR: multidrug resistance; MDR1, MDR3: *MDR1* and *MDR3* gene products, respectively (also known as P-glycoprotein); MRP: multidrug resistance-associated protein.

9.3.1. P-Glycoprotein

P-glycoprotein (P-gp or gp-170) is a multidrug resistance (*MDR*) gene product with a molecular weight of about 170 kDa, and functions as an ATP-dependent drug efflux pump in the membrane that lowers the intracellular concentrations of amphiphilic cytotoxic drugs (Silverman and Schrenk, 1997). P-gp is one of the causes of multidrug resistance of cancer cells to a wide range of chemotherapeutic agents. It is interesting that P-gp is also present in normal cells. Their locations are confined in the lumenal domains of cells in some (excretory) organs such as liver, intestine, kidney, and brain, and their physiological functions appear to be related to active transport of organic cations. There are two *MDR* gene products in normal human cells, i.e., MDR1 and MDR3, and three *mdr* gene products in rat and mouse, i.e., mdr1a, mdr1b, and mdr2. MDR1, mdr1a, and mdr1b confer drug resistance on otherwise drug-sensitive cells, but MDR3 and mdr2 do not. Recent studies suggested that in normal hepatocytes MDR1 and MDR3 mediate active biliary excretion of hydrophobic organic (cationic) compounds and phosphatidylcholine across the canalicular membranes of the hepatocytes, respectively (Smit *et al.*, 1995).

9.3.1.1. Common Physicochemical and Structural Characteristics of P-Glycoprotein Substrates

Substrates for P-gp, especially the MDR1 gene product, appear to be bulky organic cations at physiological pH. Important general physicochemical and structural features are:

1. Molecular weight >400.
2. Log P >1.
3. The presence of at least one planar aromatic ring moiety, which can interact with hydrophobic P-gp drug binding domains.
4. Preferably, cationic at physiological pH, e.g., through amine moiety of the molecule that can be protonated at physiological pH.

9.3.1.2. Overlapping Substrate Specificity of Cytochrome P450 and P-Glycoprotein

It has been found that there is a significant overlap of the substrates for cytochrome P450 3A (CYP3A) and P-gp, including a wide range of hydrophobic (chemo)therapeutic agents, and their tissue distribution patterns. These findings suggest that CYP3A and P-gp may play complementary roles in drug disposition via metabolism and active secretion of drugs, especially in the villi of the small intestine, where CYP3A and P-gp can act synergistically as a barrier against the oral absorption of drugs (Wacher *et al.*, 1995; Zhang *et al.*, 1998).

9.3.2. Multidrug Resistance-Associated Protein

Multidrug resistance-associated protein (MRP) was first detected in some multidrug-resistant cell lines, in which no overexpression of either P-gp nor mRNA

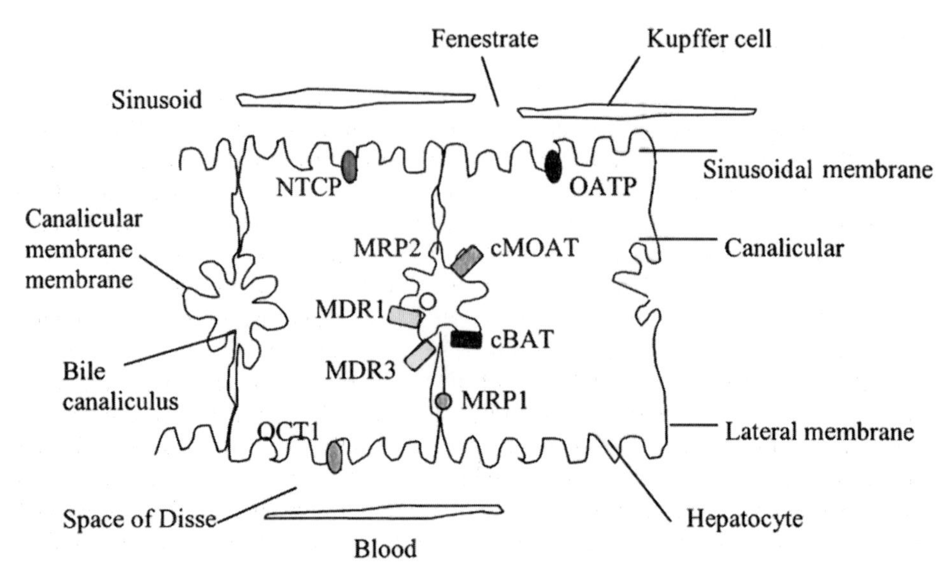

Figure 9.1. Current concept of carrier-mediated transport systems for organic cations and anions at the sinusoidal, the lateral, and the canalicular membrane domains of hepatocytes. cBAT: canalicular bile acid transporter; cMOAT: canalicular multispecific organic anion transporter; MDR1, MDR2: multidrug resistance gene products (in human hepatocytes), MRP1, MRP2: multidrug resistance-associated proteins; NTCP: Na^+-taurocholate co-transporting polypeptide; OATP: organic anion transportingtt polypeptide; OCT1: organic cation transporter.

from the encoding MDR1 gene could be detected. Later, it was found that MRP is a member of the ATP binding cassette (ABC) family of transporters along with P-gp, and the MRP gene can confer drug resistance in tumor cells. The molecular weight of MRP is about 190 kDa and it has a broad spectrum of substrates (Kusuhara *et al.*, 1998; Meijer *et al.*, 1997). MRP is also found in normal cells and its tissue distribution patterns are similar to those of P-gp, including the liver (canaliculus), erythrocyte membranes, heart, kidneys, intestinal brush border membranes, and lungs.

At least two MRP isoforms, MRP1 and MRP2, have been found in human and rodent hepatocytes. MRP1 is present in lateral membrane domains of normal hepatocytes at a very low level, whereas MRP2 is localized exclusively in canalicular membranes of hepatocytes (Fig. 9.1). MRP1 is known as an ATP-dependent glutathione S-conjugate transporter and functions as an ATP-dependent transporter for amphiphilic organic anions such as leukotriene C_4 (LTC_4), dinitrophenol S-glutathione, and calcein. It has been suggested that MRP1 functions as a transporter not only for glutathione S-conjugates but also for glucuronide and sulfate conjugates of various compounds. MRP2 is also known as the canalicular multispecific organic anion transporter (cMOAT), and it has been suggested that it functions as an ATP-dependent transporter for various amphipathic organic anions as well as organic cations in a glutathione-dependent manner.

REFERENCES

Hirom P. C. *et al.*, The physicochemical factor required for the biliary excretion of organic cations and anions, *Biochem. Soc. Trans.* **2**: 327–330, 1974.

Kusuhara H. *et al.*, The role of P-glycoprotein and canalicular multispecific organic anion transporter in the hepatobiliary excretion of drugs, *J. Pharm. Sci.* **87**: 1025–1040, 1998.

Meijer D. K. F. *et al.*, Hepatobiliary elimination of cationic drugs: the role of P-glycoproteins and other ATP-dependent transporters, *Adv. Drug Del. Rev.* **25**: 159–200, 1997.

Müller M. and Jansen P. L. M., Molecular aspects of hepatobiliary transport, *Am. J. Physiol.* **272**: G1285–G1303, 1997.

Silverman J. A. and Schrenk D., Expression of the multidrug resistance genes in the liver, *FASEB J.* **11**: 308–313, 1997.

Smit H. *et al.*, Multiple carriers involved in the biliary excretion of cationic drugs, *Hepatology* **22**: 309A, 1995.

Wacher V. J. *et al.*, Overlapping substrate specificities and tissue distribution of cytochrome P450 3A and P-glycoprotein: implications for drug delivery and activity in cancer chemotherapy, *Mol. Carcinog.* **13**: 129–134, 1995.

Yamazaki M. *et al.*, Recent advances in carrier-mediated hepatic uptake and biliary excretion of xenobiotics, *Pharm. Res.* **13**(4): 497–513, 1996.

Zhang Y. *et al.*, Overlapping substrate specificities of cytochrome P450 3A and P-glycoprotein for a novel cysteine protease inhibitor, *Drug Metab. Dispos.* **26**: 360–366, 1998.

10

Nonlinear Pharmacokinetics

In general, pharmacokinetic profiles of most drugs over therapeutic concentration ranges are concentration-independent. This concentration-independent pharmacokinetics is often referred as linear or first-order kinetics, implying that the rate of change in drug concentration is *proportional to the drug concentration* without saturation of any kinetic processes. At high concentrations, however, drug pharmacokinetics tends to change in a concentration-dependent way owing to saturation of certain processes (Jusko, 1989; Ludden, 1991). In this chapter, nonlinear pharmacokinetics will be defined and explained and the ways in which it can be elucidated from experimental data will be discussed.

10.1. DEFINITIONS

Any pharmacokinetic process of a drug, i.e., absorption, distribution, metabolism, and excretion, that cannot be adequately described with first-order (linear) kinetics of drug concentrations can be considered nonlinear kinetics. In other words, nonlinear pharmacokinetics implies deviations in the rate of change in the amount (or concentration) of a drug in any physiological or experimental system (whole body, organs, or compartments) from first-order kinetics in a dose (or concentration)- and/or a time-dependent manner.

10.1.1. Dose Dependency

Dose dependency in exposure generally indicates that the dose-normalized plasma drug concentrations or the dose-normalized AUC is not constant and depends on the dose levels. This can be due to transient saturation of any enzyme or carrier-mediated process such as metabolism or active transport at high doses (or concentrations). For instance, if clearance processes of a drug can be saturated at high concentrations, the dose-normalized AUC values after intravenous administration at high dose levels are higher than those at lower dose levels. Dose dependency can be viewed as transient and reversible at high drug doses (or concentrations), which reverse at low doses (or concentrations). Dose (or concentration)-dependent nonlinear kinetics is also called "capacity-limited" kinetics, and can be often described with the Michaelis–Menten equation.

10.1.2. Time Dependency

Time dependency in exposure implies that dose-normalized individual concentrations (or AUC) of a drug after multiple dosing or at a particular (reference) time after single dosing differ from those after single dosing or at a different time after single dosing at the same dose, respectively. A major distinctive feature of time dependency as opposed to dose (or concentration) dependency is that it results from actual physiological (or biochemical) changes or differences with time in the organ(s) associated with the corresponding disposition parameters of the drug. For instance, when a drug induces its own metabolism (autoinduction), its exposure levels decrease after multiple dosing as compared to those after single dosing owing to an actual increase in the amount of metabolizing enzymes with time. Another type of time-dependent pharmacokinetics is the so-called "chronopharmacokinetics," which describes the effects of rhythms in physiological or biochemical functions of the body on the pharmacokinetic behavior of a drug in a time-dependent manner (Levy, 1982).

10.2. MICHAELIS–MENTEN KINETICS

Generally, the initial rate of an enzyme reaction such as biosynthesis of an endogenous compound or metabolism of a xenobiotic is directly proportional to the concentration of the substrate at low concentrations. However, as the substrate concentration increases, the reaction reaches a certain maximum rate. In the simplest case, the overall processes of an initial enzymatic reaction can be described by following scheme:

$$E + S \underset{k_{-1}}{\overset{k_1}{\rightleftharpoons}} ES \xrightarrow{k_2} E + P$$

In brief, the substrate (S) binds to free enzyme molecule (E) and forms an enzyme–substrate complex (ES), which then can either dissociate back to E and S, or further break down to E and product (P); k_1 and k_{-1} are the association and dissociation rate constants, respectively; and k_2 is the rate constant for the production of P. During the early stages of the reaction, the reverse process, i.e., $E + P \rightarrow ES$, is assumed to be negligible, because the concentration of P is essentially zero. The equation describing the *initial* rate of reaction as a function of the substrate concentration can be derived from this scheme, and is known as the Michaelis–Menten equation [Eq. (10.1)], first proposed by Henri in 1903 and originally formulated from the equation describing a simple *in vitro* enzymatic reaction. This rather simple equation has been found to be extremely useful for describing apparent nonlinear plasma concentration–time profiles of many drugs in both *in vivo* and *in vitro* situations.

MICHAELIS–MENTEN EQUATION:

$$v = \frac{V_{max} \cdot C}{K_m + C} \tag{10.1}$$

C is the original concentration of the substrate ([S]) at the beginning of the reaction; v is the *initial* rate of reaction [substrate depletion ($-dC/dt$), or product formation (dP/dt)]; K_m is the Michaelis–Menten constant ($[E] \cdot C/[EC]$); and V_{max} is the maximum rate of the reaction ($k_2 \cdot [E]_t$), with $[E]_t$ being the total concentration of the enzyme.

In practice, consumption of no more than 5% of the original amount of substrate is considered acceptable for estimating the initial reaction rate with the original substrate concentration at the beginning of a reaction to satisfy the Michaelis–Menten equation. K_m is inversely related to the affinity between the substrate and the enzyme, i.e., the smaller the K_m value, the stronger the affinity between the substrate and the enzyme, and vice versa. K_m is also equal to the substrate concentration at which the rate of the process is half of V_{max}. V_{max} would be attained at infinite substrate concentration where all the enzymes are saturated with the substrate and present as ES, and is directly proportional to the total enzyme concentration. The important assumptions required for the Michaelis–Menten equation include:

1. Only a single substrate and a single enzyme–substrate complex with 1-to-1 stoichiometry between substrate and enzyme are involved, and the enzyme–substrate complex breaks down directly to form free the enzyme and the product.
2. The reaction rate is measured during the very early stages of the process so that the reverse reaction from the enzyme and the product to the enzyme–substrate complex ($E + P \rightarrow ES$) is negligible.
3. The substrate concentration is significantly higher than the enzyme concentration so that the formation of an enzyme–substrate complex does not alter the substrate concentration to any significant extent.

10.3. PHARMACOKINETIC IMPLICATIONS OF MICHAELIS–MENTEN KINETICS

For most drugs, the rate of change in the amount of drug in the body after dosing is governed primarily by enzymatic reactions such as metabolism or carrier-mediated transport including, e.g., absorption, biliary excretion, and renal secretion. Therefore, most pharmacokinetic processes can in principle be subject to nonlinear kinetics. Despite the fact that the Michaelis–Menten equation is most applicable to *in vitro* reactions, it can also be used to describe apparent nonlinear plasma concentration *vs.* time profiles of a drug *in vivo*, if the body behaves like a single compartment [Eq. (10.2)]. It is important to note that K_m and V_{max} estimates based on *in vivo* data are *apparent* values. This is because unlike *in vitro* experiments, numerous factors other than enzymatic reactions cause nonlinear kinetics in *in vivo* situations (Cheng and Jusko, 1988):

$$-\frac{dC_p(t)}{dt} = \frac{V_{max,app} \cdot C_p(t)}{K_{m,app} + C_p(t)} \tag{10.2}$$

$C_p(t)$ is the concentration of the drug in the plasma at time t; $-dC_p(t)/dt$ is the rate of drug disappearance in the plasma at time t over a short period of time, dt; $K_{m,app}$ is the apparent Michaelis–Menten constant; and $V_{max,app}$ is the apparent maximum rate of elimination of the drug. Under two different extreme conditions, Eq. (10.2) can be simplified into either first- or zero-order kinetics as shown below.

10.3.1. First-Order Kinetics

When drug concentrations are substantially lower than $K_{m,app}$ (usually $C_p(t) < 0.1 \times K_{m,app}$), the rate of change in the plasma drug concentration becomes a first-order function of the concentration, i.e.,

$$-\frac{dC_p(t)}{dt} = \left(\frac{V_{max,app}}{K_{m,app}}\right) \cdot C_p(t) = k \cdot C_p(t) \tag{10.3}$$

In this case, the rate of drug disappearance is directly proportional to its concentration, and, thus, the plasma concentration–time profile of the drug follows first-order kinetics with a rate constant of V_{max}/K_m.

10.3.2. Zero-Order Kinetics

When plasma drug concentrations are substantially higher than $K_{m,app}$ (usually $C_p(t) > 10 \times K_{m,app}$), the value at which the enzymatic process can be saturated, the rate of change of the drug concentration becomes independent of the concentration itself:

$$-\frac{dC_p(t)}{dt} = V_{max,app} \tag{10.4}$$

Under these conditions, the rate of drug disappearance is constant, i.e., zero-order kinetics.

10.3.3. Characteristics of the Plasma Concentration–Time Profile of a Drug Subject to Michaelis–Menten Kinetics

Figure 10.1 illustrates plasma drug concentration–time profiles on a semi-logarithmic scale after intravenous injection in a one-compartment body system, for which the slopes are shallower at initial high concentrations than at low concentrations. These apparent concentration-dependent changes in the slopes indicate that at high concentrations during the initial phase, drug elimination is saturated and governed by zero-order kinetics, but as the concentration decreases during the later phases, the drug elimination processes follow first-order kinetics.

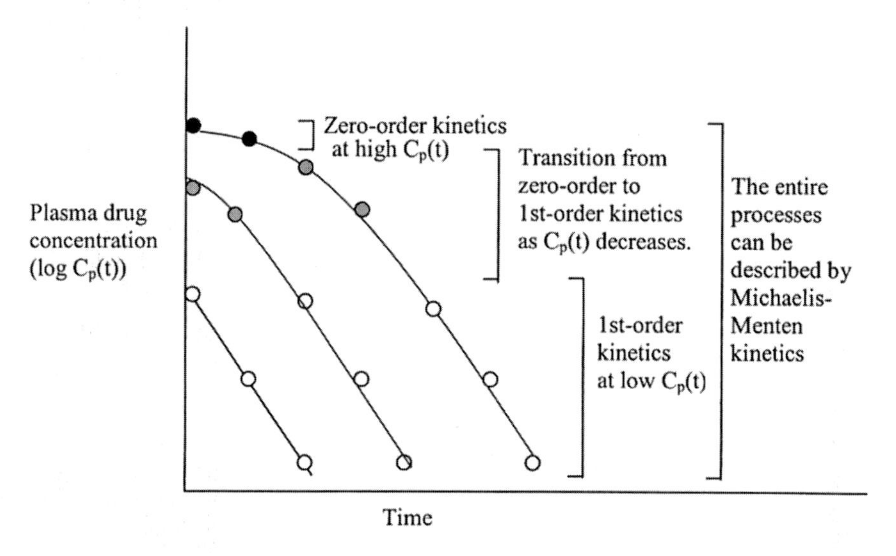

Figure 10.1. Semilogarithmic plasma concentration–time profiles of a hypothetical drug after intravenous injection at three different doses. Elimination of the drug is assumed to be mediated by an apparent single Michaelis–Menten process. A one-compartment system is assumed for drug disposition. At drug concentrations much higher than the apparent Michaelis–Menten constant ($K_{m,app}$) (●), the slopes of the plasma concentration–time plot are shallower and more variable than those at concentrations much lower than $K_{m,app}$ (○), which become steeper and constant. At the intermediate concentrations (●), the transition of shallow slopes at high concentrations toward the steeper slope is noticeable as the concentration decreases.

10.3.4. Estimating $V_{max,app}$ and $K_{m,app}$ from the Plasma Concentration–Time Profile *In Vivo*

The values of $V_{max,app}$ and $K_{m,app}$ can be estimated directly from log $C_p(t)$ *vs.* time profiles following the intravenous administration of a drug, when its disposition profiles can be adequately described with a one-compartment model and a single Michaelis–Menten equation, according to Eqs. (10.5) and (10.6):

$$\frac{V_{max,app}}{(2.303) \cdot K_{m,app}} = \frac{\log[C_0^*/C_p(t)]}{t} \tag{10.5}$$

and

$$K_{m,app} = \frac{C_0}{(2.303) \cdot \log(C_0^*/C_0)} \tag{10.6}$$

where C_0 is an estimated concentration at time zero obtained by backextrapolating the log $C_p(t)$ *vs.* time plot to time zero. C_0^* is a zero-time intercept of a plot extrapolated from the terminal log–linear portion of the log $C_p(t)$ *vs.* time plot (Fig. 10.2).

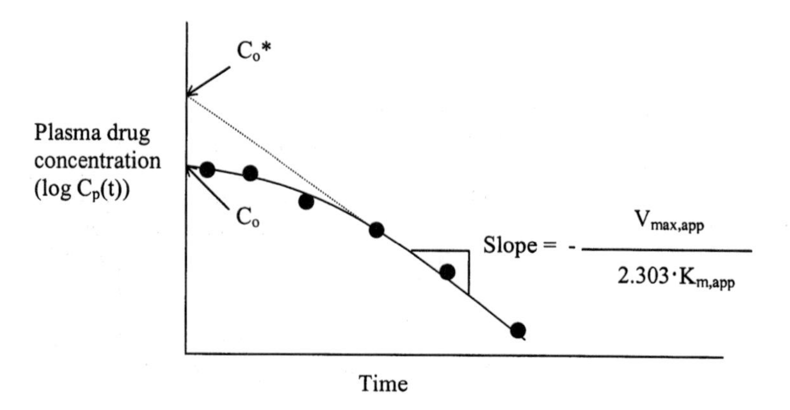

Figure 10.2. Estimate of $V_{max,app}$ and $K_{m,app}$ from a log $C_p(t)$ *vs.* time plot of a drug exhibiting a Michaelis–Menten type elimination after intravenous administration. The terminal slope of the plot equals $-V_{max,app}/(2.303 \cdot K_{m,app})$, and the estimates of C_0^* and C_0 by curve-fitting can be used to calculate $K_{m,app}$.

10.3.5. Systemic Clearance and Nonlinearity

When the elimination process of a drug can be described by simple Michaelis–Menten kinetics with a one-compartment model, based on Eq. (10.2), its systemic clearance (Cl_s) can be expressed as

$$-\frac{dA(t)}{dt} = \underbrace{\frac{V_{max,app} \cdot V}{K_{m,app} + C_p(t)}}_{\uparrow} \cdot C_p(t) = Cl_s \cdot C_p(t) \quad (10.7)$$

$$\boxed{Cl_s = \frac{V_{max,app} \cdot V}{K_{m,app} + C_p(t)}} \quad (10.8)$$

where A(t) is the total amount of the drug present in the body at time t $[A(t) = C_p(t) \cdot V]$ and V is the volume of drug distribution. Depending on the relative magnitudes of $C_p(t)$ and $K_{m,app}$, Cl_s can be concentration-dependent (nonlinear) or concentration-independent (linear),

10.3.5.1. First-Order Kinetics

Under the first-order kinetic conditions, i.e., at $C_p(t) \ll K_{m,app}$, Cl_s of the drug becomes concentration-independent and constant:

$$Cl_s = \frac{V_{max,app} \cdot V}{K_{m,app}} \; (= \text{constant}) \quad (10.9)$$

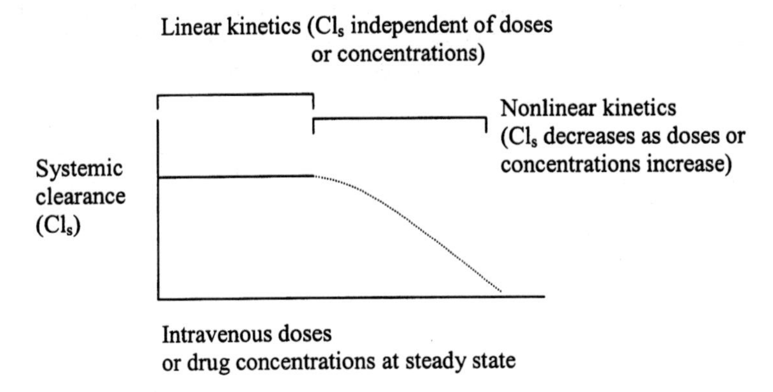

Figure 10.3. Potential changes in the systemic clearance of a drug as a function of an intravenous dose or plasma (or blood) drug concentration at steady state after continuous infusion, when elimination of the drug follows simple Michaelis–Menten kinetics, assuming a one-compartment body system.

10.3.5.2. Zero-Order Kinetics

Under the zero-order kinetic conditions, i.e., at $C_p(t) \gg K_{m,app}$, Cl_s of the drug is inversely related to $C_p(t)$ and becomes a function of drug dose or concentration:

$$Cl_s = \frac{V_{max,app} \cdot V}{C_p(t)} \tag{10.10}$$

Potential changes in Cl_s of a drug as a function of the intravenous dose levels (or concentrations at steady state after continuous infusion) are illustrated in Fig. 10.3.

10.3.6. Effects of Nonlinearity on Pharmacokinetic Parameters

Table 10.1 summarizes the potential effects of changes in various saturable pharmacokinetic processes at high doses or concentrations on several pharmacokinetic parameters, including clearance, volume of distribution, dose-normalized AUC, and terminal half-life of a drug.

10.3.7. Terminal Half-Life and Nonlinear Kinetics

Nonlinearity may or may not alter the terminal half-life ($t_{1/2}$) of a drug. If a drug's disposition profile can be readily described with simple Michaelis–Menten kinetics and a one-compartment system, its $t_{1/2}$ over the same concentration range should be the same regardless of the dose level. Thus, the differences in $t_{1/2}$ over the same concentration range at different dose levels can be indicative of the presence of nonlinearity in the system, which cannot be readily described with simple Michaelis–Menten kinetics, or a multicompartment system for the body. Assay sensitivity and product inhibition, which are two of the important factors in the apparent changes in $t_{1/2}$ at different dose levels, are discussed below.

Table 10.1. Potential Effects of Changes in Saturable Pharmacokinetic Processes on Systemic Clearance (Cl_s), Volume of Distribution (V_{ss}), Dose-Normalized AUC (AUC/dose), and Terminal Half-Life ($t_{1/2}$) of a Drug, as Doses or Concentrations Increase[a]

Pharmacokinetic parameter	Change at high doses	Potential effect on			
		Cl_s	V_{ss}	AUC/dose	$t_{1/2}$
Ratio of unbound and total drug concentrations in plasma	↑	↑	↑	↓	?
Intrinsic hepatic clearance	↓	↓	↔	↑ (likely)	↑
Dose-normalized rate of carrier-mediated intestinal absorption	↓	↔	↔	↓	↔
Concentration-normalized rate of carrier-mediated distribution into organs or tissues	↓	↔, ↓[a]	↔	↔, ↑	↔, ↑
Biliary or renal clearance	↓	↓	↔	↑ (likely)	↑

[a]Organ clearance can be affected by the rate of drug transport from the blood into the organ, when membrane transport of the drug is mediated by active carrier system.

10.3.7.1. Product Inhibition

If a metabolite(s) inhibits the biotransformation of its parent drug (product inhibition) *in vivo*, $t_{1/2}$ of the drug can increase with the dose. In order to verify product inhibition, metabolite in question can be identified and the initial drug elimination rates in the absence and presence of the metabolite can be compared in *in vitro* metabolism studies (Lin *et al.*, 1984; Perrier *et al.*, 1973).

10.3.7.2. Assay Sensitivity

In some cases, an apparent increase in a drug's $t_{1/2}$ with increasing doses may be due simply to limited assay sensitivity, which precludes detection of drug levels during the terminal phase at low doses in a multicompartmental body system. This is a rather common issue with drugs exhibiting multiexponential concentration–time profiles (Fig. 10.4). Important differences in definitions and terminology of linear and nonlinear pharmacokinetic processes are summarized in Table 10.2.

10.4. FACTORS CAUSING NONLINEAR PHARMACOKINETICS

Various factors for nonlinear kinetic behaviors in ADME processes and corresponding examples are summarized below.

Absorption

- Poor aqueous solubility and/or slow dissolution (griseofulvin).
- Site-specific absorption along the GI tract (phenytoin).

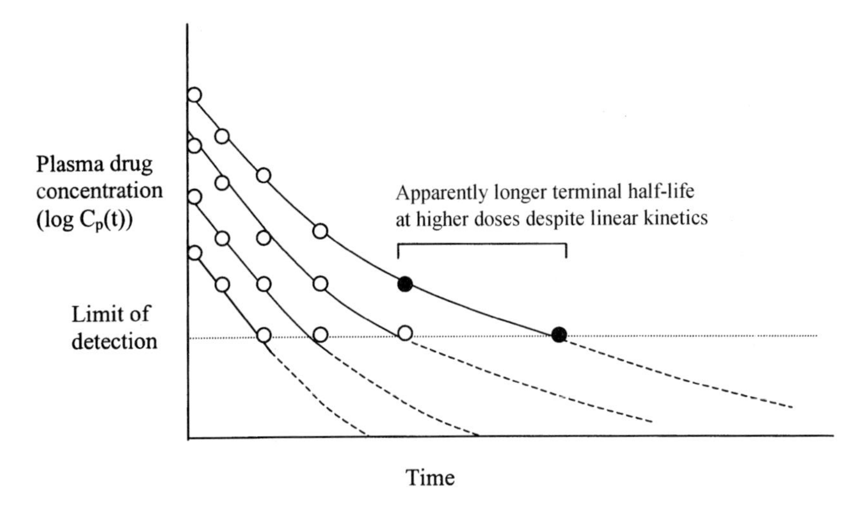

Figure 10.4. Limited assay sensitivity resulting in an apparent increase in the terminal half-life of a drug with increasing doses after intravenous administration in a multicompartment system. The dotted lines represent exposure levels of the drug below the limit of detection.

- Carrier-mediated absorption (riboflavin).
- P-glycoprotein efflux in intestinal epithelial cells (cyclosporin A).
- Saturable first-pass effect by the intestine and/or the liver (propranolol).
- Dose/time-dependent changes in gastrointestinal physiology including: (a) gastric emptying, (b) gastrointestinal motility, (c) gastrointestinal blood flow rate.

Distribution

- Nonlinear plasma protein binding (valproic acid).
- Carrier-mediated membrane transport (thiamine).
- Nonlinear tissue binding (prednisolone).

Metabolism

- Saturable metabolism (ethanol).
- Product inhibition (dicoumarol).
- Cosubstrate depletion (acetaminophen).
- Nonlinear plasma protein binding (prednisolone).
- Autoinduction.

NOTE: Capacity-limited metabolism is the most common and fundamentally important nonlinear mechanism in pharmacokinetics.

Excretion

- Nonlinear protein binding and/or glomerular filtration (naproxen).
- Carrier-mediated tubular secretion (cimetidine)/reabsorption(riboflavin).
- Carrier-mediated biliary excretion (iodipamide).

Table 10.2. A Summary of Definitions and Characteristics of Linear and Nonlinear Kinetics in a One-Compartment System

Kinetics	Definition	Elimination rate[a] $[-dC_p(t)/dt]$	Systemic clearance	Slope of plasma concentration–time profile after intravenous injection on a semilog scale	Synonyms
Linear	First-order kinetics	$k \cdot C_p(t)$ [proportional to $C_p(t)$]	Constant	$-k/2.303$	Dose (concentration- or time-) independent kinetics
Nonlinear[b]	Zero-order kinetics	k_0 (=constant)	Inversely related with $C_p(t)$	No definitive description available[c]	Dose (concentration- or time-) dependent, or capacity-limited kinetics
	Michaelis–Menten kinetics	$\frac{V_{max,app} \cdot C_p(t)}{K_{m,app} + C_p(t)}$	Constant at low $C_p(t)$ and inversely related to $C_p(t)$ at high $C_p(t)$	Changing from shallower slopes at high concentrations during the initial phase to steeper slopes reaching $-V_{max,app}/(2.303 \cdot K_{m,app})$ at low concentrations during the terminal phase	

[a] $C_p(t)$: concentration of drug in plasma at time t after intravenous injection; k, k_0: first-order or zero-order rate constants, respectively; $K_{m,app}$: apparent Michaelis–Menten constant; and $V_{max,app}$: apparent maximum rate of the process.
[b] Any process other than linear kinetics, e.g., zero-order or Michaelis–Menten kinetics.
[c] $C_p(t) = C_p(0) - k_0 \cdot C_p(0)$ is drug concentration at time 0 after intravenous injection.

Other

- Pharmacological effects of a drug that affect animal physiology, such as a decrease in blood flow rate after administration of antihypertensive agents.
- Chronopharmacokinetics: circadian rhythm or seasonal variability in kinetics
- Pathophysiological changes influencing pharmacokinetics of a drug, such as an increase in hepatic CYP2E1 content in diabetes or an increase in α_1-acid glycoprotein after surgery.

10.5. RECOGNIZING NONLINEAR PHARMACOKINETICS

When the system is linear, all the pharmacokinetic processes become concentration- and time-independent, and the pharmacokinetic parameters such as dose-normalized exposure levels (AUC/dose, AUMC/dose, and C_{max}/dose), t_{max}, $t_{1/2}$, Cl_s, V_{ss}, and MRT should remain constant regardless of the dose or concentration level. The most noticeable characteristic of nonlinearity of drug disposition is the lack of superimposibility of dose-normalized concentration *vs.* time profiles at different dose or concentration levels.

Principle of superposition (or dose proportionality): Superposition (also called "dose-proportionality") of drug disposition under linear conditions implies that dose-normalized drug concentrations or $AUC_{0-\infty}$ at different dose levels are the same. In other words, when the system follows linear kinetics, $C_p(t)$ or $AUC_{0-\infty}$ values at different doses become superimposible when they are dose-normalized:

$$\boxed{\frac{C_p(t) \text{ or } AUC_{0-\infty}}{\text{Dose}} = \text{Constant}} \tag{10.11}$$

On the other hand, nonlinearity can be viewed as "lack of superposition" of exposure levels at different dose levels. Virtually, all pharmacokinetic systems can be nonlinear at high dose (or concentration) levels owing to the saturation of various enzymatic or carrier-mediated processes. Figure 10.5 illustrates the relationship between dose normalized $AUC_{0-\infty}$ and dose levels under linear and nonlinear conditions.

In general, there is a more than dose-proportional increase in $AUC_{0-\infty}$ as a dose increases. Occasionally, however, an increase in $AUC_{0-\infty}$ with dose that is less than dose-proportional can be observed, which may be due to an increased systemic clearance associated with an increase in the unbound drug fraction at higher drug concentrations. In this case, a further increase in dose will eventually result in an overproportional increase in $AUC_{0-\infty}$, owing to saturation of the clearance mechanisms. Another example of an increase in $AUC_{0-\infty}$ after oral administration that is less than dose-proportional is impairment of absorption due to limited solubility of a drug or saturation of active transporter(s) in the intestinal membranes at high doses.

Pharmacokinetic profiles at a minimum of three different dose levels are required to adequately determine nonlinearity of a system. The dose-dependent changes of the following parameters are indicative of the presence of a nonlinear

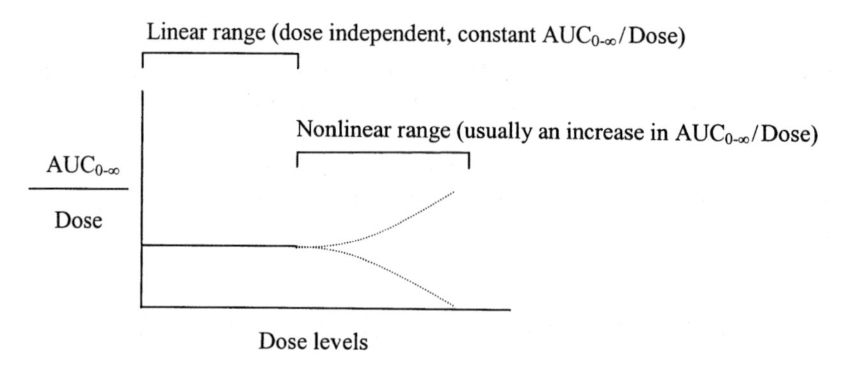

Figure 10.5. Schematic description of the relationship between dose-normalized $AUC_{0-\infty}$ and dose levels of a drug. At high dose levels, the dose-normalized $AUC_{0-\infty}$ can increase or decrease depending on the disposition properties of the drug.

pharmacokinetic system or process. Potential causes for the lack of superposition of those parameters are also discussed.

1. $C_p(t)$/dose: Some type of dose-dependence exists, but further study is needed to determine the causes for nonlinearity.
2. $AUC_{0-\infty}$/dose: Bioavailability or systemic clearance is nonlinear.
3. $AUMC_{0-\infty}$/dose: Absorption rate, Cl_s, or V_{ss} is nonlinear.

10.6. CHRONOPHARMACOKINETICS

Chronopharmacokinetics is defined as the circadian variations in pharmacokinetics of a drug, which are due to parallel changes in the physiological functions involved in drug absorption, distribution, metabolism, and excretion. Chronopharmacokinetic behaviors can result in apparent nonlinear kinetic profiles of a drug in a time-dependent manner (Bruguerolle, 1998; Lemmer and Bruguerolle, 1994). Circadian variations in different pharmacokinetic processes are summarized below (Labrecque and Belanger, 1991; Reinberg and Smolensky, 1982).

10.6.1. Absorption

It is known that there are circadian variations in gastric emptying time, gastric motility, and gastrointestinal blood flow, which in general are faster during the activity period than the sleeping period in both animals and humans. These circadian changes in physiological functions can alter the exposure profiles of orally dosed drugs.

10.6.2. Distribution

It has been reported that there are circadian variations in the blood albumin concentrations in both laboratory animals and humans, which in general are higher during the period of activity than during the sleep. The free drug concentration can be affected by these rhythmic changes in albumin levels resulting in time-dependent variations in drug distribution in addition to the circadian changes in blood flow.

10.6.3. Metabolism

Circadian variations in the activity of different metabolizing enzymes have been noted in some laboratory animals. In rats, the metabolizing enzyme activity usually increases at night and decreases while the animals are asleep during the daytime. In many cases, the circadian changes in the activities of both phase I and phase II enzymes are found to be positively correlated with the enzyme concentrations.

10.6.4. Excretion

Studies in both animals and humans have indicated that the glomerular filtration rate is highest in the middle of the activity period and lowest during sleep. This seems to be due to the circadian rhythms in the systemic blood pressure and the circulating vasoactive hormones producing time-dependent variations in renal hemodynamics. Circadian changes have also been found in urinary pH, which is more acidic at night than during the day. In rats, it is reported that the bile flow is higher in the middle of the activity period.

10.7. TOXICOKINETICS

Toxicokinetics (TK) is pharmacokinetic principles and techniques applied to concentration *vs.* time data generated at the (high) dose levels that are usually used in toxicity studies, in order to determine the rate, extent, and duration of exposure of the test compound in the animal species examined (Chasseaud, 1992; Clark and Smith, 1982; Smith *et al.*, 1990; Smith, 1993; Welling, 1995). The main objectives of TK for regulatory purposes are:

1. To establish a kinetic relationship between doses and exposure levels.
2. To evaluate the results of toxicity observed in the test animal species based on the rate, extent, and duration of exposure levels of the test compound at different dose levels.
3. To provide information on the relationship between exposure levels of the test compound and the extent of toxicity found in the test animal species for direct comparison with potential human exposure to the test compound.
4. To support clinical study designs and data interpretation by assessing the safety margin, i.e., no observed adverse effect level (NOAEL) in the test animal species divided by a target therapeutic exposure level in a clinical setting.

Owing to the high dose levels used in TK studies, it is not uncommon to observe nonlinearity in drug disposition. Usually three dose levels are examined in short- or long-term TK studies, i.e., low (close to target clinical dose levels), medium, and high (producing toxicity in animals). Although a thorough understanding of the causes of nonlinearity in exposure can be important for the evaluation of toxicity of the compound, the main purposes of TK studies as stated above are to determine NOAEL and the absolute exposure levels of the test compound associated with toxicity in the animals. The purposes of toxicity studies (drug safety evaluation) in animals [usually one rodent (rat) and one nonrodent species (dog); sometimes mice and monkeys can be used] in addition to those of TK studies are:

1. To ascertain toxicity in the test animal species.
2. To identify the organ(s) or tissue(s) associated with the toxicity.
3. To characterize types of toxicity in conjunction with exposure levels of the compound.

REFERENCES

Bruguerolle B., Chronopharmacokinetics, current status, *Clin. Pharmacokinet.* **35**: 83–94, 1998.

Chasseaud L. F., The importance of pharmacokinetic/toxicokinetic and metabolic information in carcinogenicity study design, *Drug Inform. J.* **16**: 445–455, 1992.

Cheng H. and Jusko W. J., Mean residence time concepts for pharmacokinetic systems with nonlinear drug elimination described by the Michaelis-Menten equation, *Pharm. Res.* **5**: 156–164, 1988.

Clark B. and Smith D. A., Pharmacokinetics and toxicity testing, *CRC Crit. Rev. Toxicol.* **12**: 343–385, 1982.

Jusko W. J., Pharmacokinetics of capacity-limited systems, *J. Clin. Pharmacol.* **29**: 488–493, 1989.

Labrecque G. and Belanger P. M., Biological rhythms in the absorption, distribution, metabolism and excretion of drugs, *Pharmacol. Ther.* **52**: 95–107, 1991.

Lemmer B. and Bruguerolle B., Chronopharmacokinetics, are they clinically relevant? *Clin. Pharmacokinet.* **26**: 419–427, 1994.

Levy R. H., Time-dependent pharmacokinetics, *Pharmacol. Ther.* **17**: 383–397, 1982.

Lin J. H. *et al.*, Effect of product inhibition on elimination kinetics of ethoxybenzamide in rabbits: analysis by physiological pharmacokinetic model, *Drug Metab. Dispos.* **12**: 253–256, 1984.

Ludden T. M., Nonlinear pharmacokinetics; clinical implications, *Clin. Pharmacokinet.* **20**: 429–446, 1991.

Perrier D. *et al.*, Effect of product inhibition on kinetics of drug elimination, *J. Pharmacokinet. Biopharm.* **1**: 231–242, 1973.

Reinberg A. and Smolensky M. H., Circadian changes of drug disposition in man, *Clin. Pharmacokinet.* **7**: 401–420, 1982.

Smith D. A., Integration of animal pharmacokinetic and pharmacodynamic data in drug safety assessment, *Eur. J. Drug Metab. Pharmacokinet.* **18**: 31–39, 1993.

Smith D. A. *et al.*, Design of toxicokinetic studies, *Xenobiotica* **20**: 1187–1199, 1990.

Welling P. G., Differences between pharmacokinetics and toxicokinetics, *Toxicol. Pathol.* **23**: 143–147, 1995.

11

Pharmacodynamics and Pharmacokinetic/ Pharmacodynamic Relationships

Pharmacodynamics is the study that establishes and elucidates relationships between concentrations of a drug at the receptor or target organ (effect site) and the intensity of its pharmacological effect. Pharmacodynamic studies can provide a means for identifying important pharmacological and toxicological properties of a drug in animals and humans, including, e.g., efficacious target concentrations, drug safety margin, potential risk factors, and the presence of active metabolites (Holford and Sheiner, 1982). In many cases, it is difficult to measure drug concentrations right at the effect site, and thus therapeutic drug monitoring is often performed by measuring drug concentrations in plasma or other readily accessible body fluids, which may not be directly related to the intensity of the drug effect. Pharmacokinetic and pharmacodynamic modeling contributes to a better understanding of the relationship between drug concentrations in biological fluids, where concentrations are measured, and those in the effect site. In this chapter, pharmacodynamics and various pharmacokinetic/pharmacodynamic models and physiological and experimental factors that may complicate pharmacodynamic studies are discussed.

11.1. PHARMACODYNAMICS

11.1.1. Definition

Pharmacodynamics deals with the relationship between drug concentrations at the effect site or concentrations (usually unbound concentrations) in plasma in equilibrium with effect site concentrations, and the magnitude of the observed pharmacological effect of the drug.

11.1.2. Effect Site

The effect site (site of action) of a drug can be a target receptor/enzyme(s) or an organ(s) where the initial pharmacological responses to the drug are produced.

For instance, the effect site of antidepressant agents would be the brain, and thus drug concentrations in the cerebrospinal fluid might be more relevant in establishing pharmacodynamic relationships than the plasma or blood drug concentrations under nonsteady state conditions. This is because drug concentrations between the brain and the blood can differ significantly at nonsteady state owing to delay in drug transport across the blood–brain barrier. On the other hand, the blood can be considered an effect site for certain antibiotics that target blood pathogens. In many cases, the experimentally measured pharmacological effects of a drug *in vivo* are only remotely or indirectly related to the initial stimulus it produces at the effect site, because there is often a series of biological events after the initial stimulus that lead to the observed effects.

Drug concentration at the effect site: The most direct link between concentration and effect can be made by simultaneous measurements of drug concentrations at the effect site and the corresponding pharmacological effects. However, owing to experimental difficulties in simultaneous measurements, the pharmacodynamic relationship is sometimes investigated when drug concentrations at the effect site are considered to be in equilibrium with those in readily available biological fluids such as plasma or blood after steady state infusion or multiple dosing.

11.1.3. Pharmacological Effects

The pharmacological activity of a drug generally includes a series of sequential events, i.e., interaction of drug molecules with their action sites or receptors, induction of a stimulus to the effector systems, and subsequent production of the effect (observed pharmacological endpoints). For instance, although warfarin rapidly blocks the synthesis of the prothrombin complex activity P, it takes several days before the levels of circulating P are reduced sufficiently to decrease normal prothrombin activity, a pharmacological endpoint. Therefore, the intensity of the pharmacological response observed does not necessarily reflect the result of direct interaction of drug molecules with target receptors or enzymes, especially when the receptors and the effector systems are not located in the same organs or tissues.

11.1.3.1. Characteristics of Pharmacological Responses

The reliable and reproducible measurement of drug's pharmacological effects is probably the most important factor required to establish meaningful pharmacodynamic relationships. There are several criteria by which to define the various characteristics of pharmacological responses.

(a) All-or-None vs. Graded Responses. The fixed (all-or-none) response of a drug implies that there is or is not a response, e.g., mortality rate, whereas the graded response may vary with dose or drug concentration. The graded response can be further divided into absolute and relative responses, which can be measured in a definitive or relative manner, respectively. For instance, blood pressure or body temperature can be viewed as an absolute response, while percent muscle contractility is considered a relative response.

(b) Direct vs. Indirect Responses. For some drugs, the pharmacological end-points may directly reflect the capacity (intrinsic activity) and affinity (potency) of the interaction between a drug and specific receptors/enzymes. In some cases, however, the pharmacological response to a drug may not be directly related to its binding rate to receptors, but result from serial biological events after its initial interaction with receptors. For instance, the gradual decrease in blood pressure brought about by a diuretic can be viewed as an indirect response.

(c) Reversible vs. Irreversible Responses. A reversible response means that the baseline effect prior to treatment can be restored when the drug is no longer being administered. On the other hand, an irreversible response implies that the drug causes changes in certain *in vivo* physiological or biological conditions such that the baseline effect cannot be restored when the drug is withdrawn. For most drugs, the effect is reversible. The pharmacological effect of drugs such as antibiotics or antineoplastic agents can be viewed as irreversible.

11.1.3.2. Receptor Theory

The pharmacological effect of a drug is mediated by the initial interaction of the drug molecules with particular receptor(s) according to the following scheme:

$$\mathrm{R} \overset{+\mathrm{D}}{\rightleftarrows} \mathrm{R{-}D} \longrightarrow \mathrm{E}$$

Inactive receptor (*R*)	*Activated receptor by drug* (D?)	*Pharmacological effect* (E)

11.1.4. Differences among Pharmacokinetics, the Pharmacokinetic/Pharmacodynamic Relationship, and Pharmacodynamics

When dealing with the overall time course of a dose–effect relationship, there are three different stages to be considered, as indicated in Fig. 11.1 (Holford and Sheiner, 1981).

STAGE 1: The relationship between the dose and the time course of drug concentrations in biological fluids (*pharmacokinetics*).

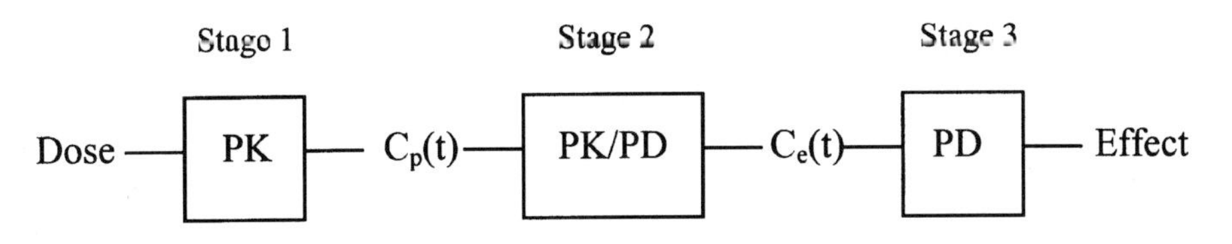

Figure 11.1. Overall relationships between the administered dose and the intensity of the measured pharmacological effects at the three different stages $C_e(t)$: concentration of drug at the effect site at time t, $C_p(t)$: concentration of drug in biological fluids such as plasma or blood at time t, PD: pharmacodynamics, PK: pharmacokinetics, PK/PD: pharmacokinetic/pharmacodynamic relationship.

STAGE 2: The time-dependent relationship between the drug concentrations in biological fluids such as plasma and at the effect site, which can be established by linking pharmacokinetics and pharmacodynamics of the drug with PK/PD modeling approaches (*pharmacokinetic and pharmacodynamic relationship*).

STAGE 3: The relationship between drug concentration at effect site and the observed pharmacological effects (*pharmacodynamics*).

The direct pharmacodynamic (PD) relationship can be established between plasma (or blood) concentrations and the observed pharmacological effects of a drug without pharmacokinetic/pharmacodynamic (PK/PD) modeling when: (1) drug concentrations at the effect site have reached equilibrium with concentrations in the plasma, especially free drug concentrations, after steady state infusion or multiple doses; or (2) the plasma (or blood) pool is the drug action site such as is the case with antibiotics. Except for these two cases, PK/PD relationships of a drug may have to be examined to establish its true PD profiles. PK/PD modeling can offer:

1. A better understanding of pharmacological behaviors of a drug (duration of action, delayed effect, circadian rhythm, biofeedback, etc).
2. Opportunity for recognizing the presence of active (or inhibitory) metabolite(s).
3. A better strategy for a therapeutic dose regimen.
4. A better understanding of potential drug–drug interactions.

11.1.5. Important Factors in Pharmacodynamic Study Designs

A thorough understanding of PK profiles is a prerequisite for elucidating reliable PD profiles. Factors in three different areas, i.e., animals, materials, and researchers, that have to be considered for reliable PK and PD experiments are noted below (Levy, 1985).

11.1.5.1. Animals

- Maintain the same strain and age from the same supply source.
- Use at least three animals/sex for study.
- If needed, carry out the same treatment on control groups, such as administration of dosing vehicle with no drug or sham operation.
- Preferentially, perform the studies at the same time of day to reduce any variability due to the circadian rhythm in animal physiology.
- Preferentially, avoid coprophagy when studies are performed in rodents.
- Do not draw too much body fluid from animals for sampling, especially in small animals. In general, less than 10% of the total blood volume can be drawn from small animals in a week.

11.1.5.2. Materials

- Identify polymorphism of crystallinity of the drug dosed, especially for suspension formulation for oral dosing. In general, an amorphous drug shows an apparently

higher aqueous solubility, which can lead to faster and more extensive oral absorption compared to the more stable crystalline forms.
- Maintain homogeneity of dosing formulations, especially when precipitation of the drug in the GI tract or injection site is expected.
- Use an optimal dosing volume (e.g., <5 or <10 ml/kg of oral dosing volume in rats when fed or fasted, respectively, to avoid potential overflow of dosing solution from the stomach to the lungs).

11.1.5.3. Researcher

- Determine the pharmacological response in *in vivo* experiments by the same researcher, especially when subjective quantification methods of drug effects are used.

11.1.6. Effects of Protein Binding on Pharmacodynamics

It has been suggested that the *in vivo* pharmacological efficacy of a drug is mediated solely by the unbound drug in plasma, because only unbound drug molecules from plasma proteins are available for interaction with target receptors. The extent of protein binding is especially important in assessing the pharmacodynamics of drugs in disease states, in which the amount of plasma protein content may be altered. For instance, epilepsy in individuals with renal failure receiving phenytoin is adequately controlled at a lower total concentration of the drug as compared to those with normal renal function, because the unbound drug concentrations are the same between the two groups owing to reduced protein binding in patients with renal failure (du Souich *et al.*, 1993).

11.2. PHARMACODYNAMIC MODELS

11.2.1. Definition

Pharmacodynamic models are mathematical schemes based on classical receptor theory for an empirical description of the intensity of a pharmacological response to a drug as a function of its concentrations at *the effect site.*

11.2.2. Implications of Pharmacodynamic Models

Pharmacodynamic models are useful for describing the apparent pharmacodynamic profiles of a drug and also for gaining some insight into its underlying physiological or biological processes (Holford and Sheiner, 1982; Ritschel and Hussain, 1984; Schwinghammer and Kroboth, 1988). Several pharmacodynamic models have been developed based on two assumptions: (a) drug response is reversible and, (b) there is only one type of receptor with one binding site for a drug.

11.2.3. Types of Pharmacodynamic Models

11.2.3.1. Linear Model

The linear pharmacodynamic model is useful when the efficacy of a drug is proportional to its concentrations at the effect site. The linear model can be derived from the E_{max} model, when its concentrations at the effect site is significantly lower than EC_{50} (see the E_{max} model).

$$E = S \cdot C_e + E_0 \tag{11.1}$$

C_e is the drug concentration at the effect site, E is the intensity of the effect, E_0 is the baseline effect in the absence of the drug, and S is the slope.

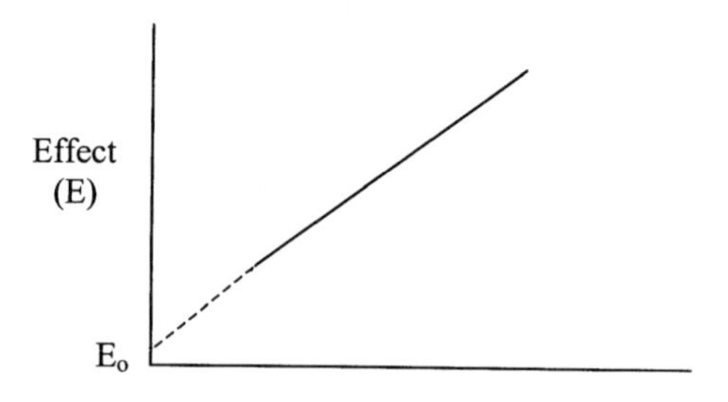

Figure 11.2. Linear model.

11.2.3.2. Log–Linear Model

The log–linear model is based on the empirical observation that a plot of effect *vs.* log concentration of many drugs exhibits a linear approximation between 20 and 80% of the maximum effect. Like the linear model, the concentration–effect relationship within this range can be analyzed using linear regression:

$$E = S \cdot \log C_e + I \tag{11.2}$$

S is the slope and I is an empirical constant with no physiological meaning.

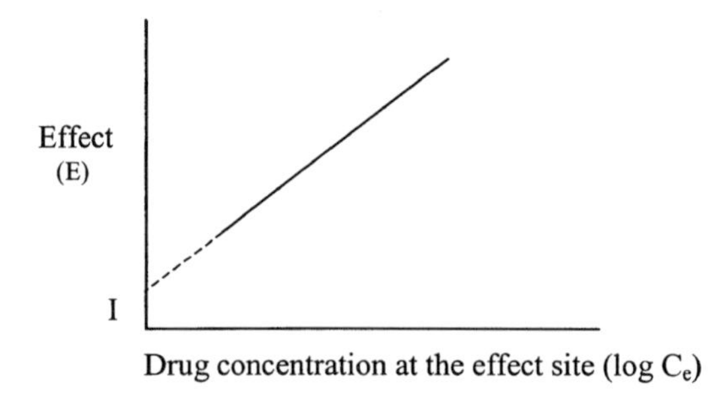

Figure 11.3. Log–linear model.

11.2.3.3. E_{max} *Model*

The E_{max} model can describe the concentration–effect curve over the full range from the baseline effect to the maximum effect of a drug. Accurate measurement of both E_{max} and EC_{50} are critical in the E_{max} model. In fact, only a few drugs have been shown to have this relationship *in vivo* mainly owing to the difficulties involved in conducting studies over a wide range of concentrations, especially at high concentrations because of the concomitant potential toxicity:

$$E = \frac{E_{max} \cdot C_e}{EC_{50} + C_e} (+E_0) \quad (11.3)$$

where E_{max} is the maximum effect, EC_{50} is the drug concentration that produces 50% of the maximum effect, and E_0 is the baseline effect in the absence of any drug.

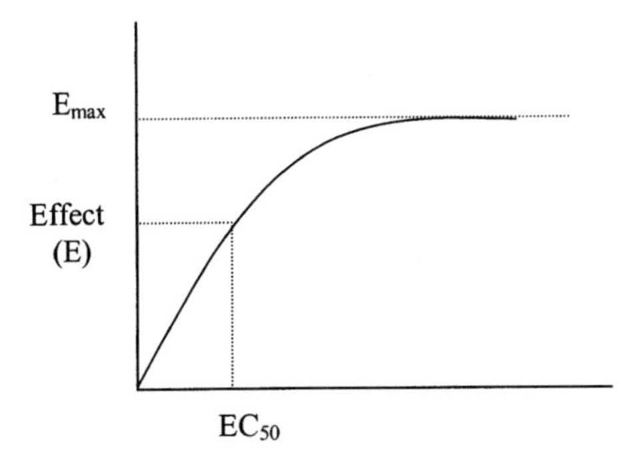

Figure 11.4. E_{max} model.

11.2.3.4. Sigmoid E_{max} *Model*

The sigmoid E_{max} model can be used when the concentration–effect curve exhibits more an S-shape pattern than a simple hyperbola (the E_{max} model), and is steeper or shallower than predicted by the E_{max} model. The sigmoid function originally proposed by Hill (1910) is often called the Hill equation:

$$E = \frac{E_{max} \cdot C_e^n}{EC_{50}^n + C_e^n} (+E_0) \quad (11.4)$$

where n is the Hill coefficient that affects the slope of the curve: at $n > 1$ the curve becomes sigmoid with a steeper slope than the E_{max} model at concentrations around EC_{50}; at $n = 1$: the curve is identical to the usual hyperbola of the E_{max} model; and at $n < 1$ the curve is steeper at low concentrations, but shallower at high concentrations than the E_{max} model.

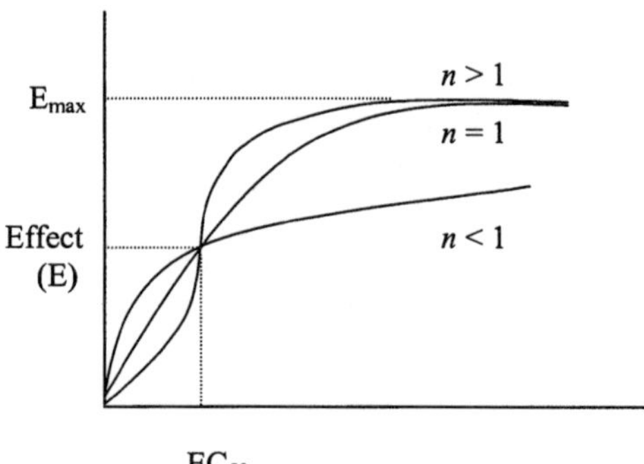

Figure 11.5. Sigmoid E_{max} model.

11.2.3.5. Inhibitory E_{max} *Model*

The same pharmacodynamic models discussed earlier can be used to describe the inhibitory effect of a drug by arranging the equations so that the observed effect becomes the difference between the drug's baseline effect (E_0) and inhibitory effect. For instance, the E_{max} model can be adapted to express the inhibitory response of a drug with maximum inhibitory effect (I_{max}) and IC_{50}, at which 50% of I_{max} is produced. This model is useful for investigating the inhibitory effects of a drug without transforming the data:

$$E = E_0 - \frac{I_{max} \cdot C_e}{IC_{50} + C_e} \tag{11.5}$$

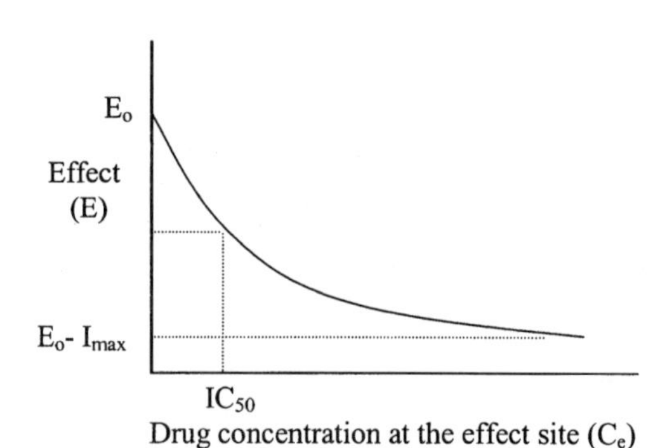

Figure 11.6. Inhibitory E_{max} model.

11.2.3.6. Models for Multiple Receptors

Only pharmacodynamic models applicable for drugs binding to a single type of receptor with one binding site have been described so far. For an increasing number of drugs, however, it is becoming evident that there may be either multiple receptors

Table 11.1. A Summary of Various Pharmacodynamic Models

Model	E	Characteristics
Linear	$S \cdot C_e + E_0$	Predicts the baseline effect when the concentration is zero Unable to define the maximum effect at high concentrations Error-prone at high or low drug concentrations
Log–linear	$S \cdot \log C_e + I$	Suitable for predicting drug effects over 20–80% of the maximum effect Unable to define the baseline and the maximum effects
E_{max}	$\frac{E_{max} \cdot C_e}{EC_{50} + C_e} + E_0$	Able to describe the pharmacodynamic relationship over a wide range of drug concentrations Predicts the baseline and the maximum effects
Sigmoid E_{max}	$\frac{E_{max} \cdot C_e^n}{EC_{50}^n + C_e^n} + E_0$	Able to describe an S-shape pattern of effect curve by adjusting n values Predicts the baseline and the maximum effects

and/or multiple binding sites with different activities (Campbell, 1990; Paalzow *et al.*, 1985). These can cause cascade- or U-shape concentration–effect curves. For instance, the analgesic effect of clonidine is mediated by two agonistic effects, both of which can be described by different E_{max} models, producing a stepwise increase in effect with concentration [Eq. (11.6), Fig. 11.7A]. On the other hand, for blood pressure, clonidine exhibits a hypotensive activity at low concentrations, but a hypertensive activity at higher concentrations owing to concentration-dependent binding to presynaptic α_2 and postsynaptic α_1 receptors, respectively, producing a U-shape concentration–effect curve [Eq. (11.7), Fig. 11.7B].

DUAL AGONISTIC RECEPTORS (cascade shape concentration-effect curve, Fig. 11.7A):

$$E = \underbrace{\frac{E_{max,1} \cdot C_e^n}{EC_{50,1}^n + C_e^n}}_{E_1} + \underbrace{\frac{E_{max,2} \cdot C_e^m}{EC_{50,2}^m + C_e^m}}_{E_2} \tag{11.6}$$

BIPHASIC RECEPTORS (agonist–antagonist, U- or bell-shape concentration–effect curve, Fig. 11.7B):

$$E = \underbrace{\frac{E_{max,1} \cdot C_e^n}{EC_{50,1}^n + C_e^n}}_{E_1} - \underbrace{\frac{E_{max,2} \cdot C_e^m}{EC_{50,2}^m + C_e^m}}_{E_2} \tag{11.7}$$

where $E_{max,1}$ and $E_{max,2}$ are, respectively, the maximum effect for each of two different receptors, $EC_{50,1}$ and $EC_{50,2}$ are the drug concentrations that produce 50% of $E_{max,1}$ or $E_{max,2}$, and n and m are the Hill coefficients.

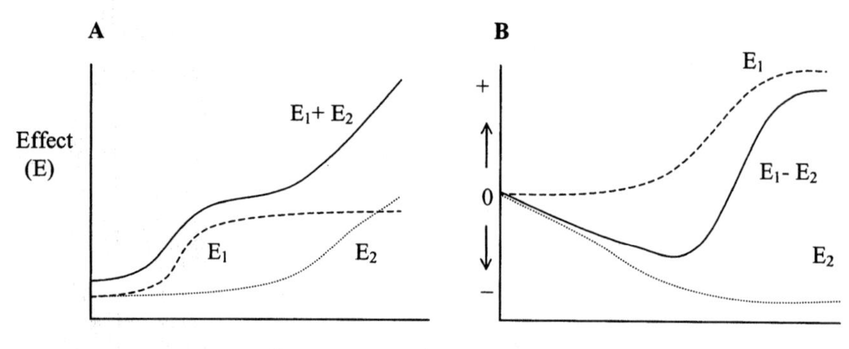

Figure 11.7. Drug concentration at the effect site(s) *vs.* effect for dual agonistic receptors (A) and agonistic–antagonistic receptors (B).

11.2.4. Model Selection

The initial decision to select one particular pharmacodynamic model(s) over another for a certain data set should be based on: (a) the need for estimating particular parameters (EC_{50}, E_{max}, E_0, etc), and (b) the ability of the model and availability of data suitable to provide estimates of those parameters. When more than one model is chosen and there are no obvious deficiencies in the models for parameter estimation, the choice of a model can be made according to certain statistical indications (Schwinghammer, 1988) including: (a) the smaller sum of squares of residuals (the differences between the observed and predicted effect values), (b) the smaller asymptotic standard deviations, (c) the smaller 95% confidence interval, (d) the smaller Akaike's Information Criterion (AIC) value, and the greater r^2 values (see Chapter 2).

11.2.5. Difficulties in Pharmacodynamic Modeling

In many cases, it is difficult to measure drug concentrations at the effect site because of limited sample availability or inaccessibility of the effect site, or to quantitatively assess the pharmacological effect owing to the complicated mechanisms and physiological variations involved (Levy, 1985; Oosterhuis and van Boxtel, 1988; Schwinghammer and Kroboth, 1988). In addition, there are numerous endogenous and exogenous factors that may have confounding effects on the study of pharmacodynamics of a drug in *in vivo* systems as summarized in the following:

- Lack of correlation between the plasma drug concentrations and concentrations at the effect site.
- Difficulties in reliable and reproducible measurements of pharmacological effects.
- Difficulties in quantifying direct pharmacological effects of a drug.

- Complicated pharmacological effects: (a) multiple receptors or target tissues; (b) concentration-dependence in effect (e.g., agonist at low concentrations, but antagonistic at high concentrations).
- The presence of endogenous agonists or antagonists at variable concentrations.
- Difficulties in distinguishing pharmacodynamic variability from pharmacokinetic variability: Pharmacokinetic issues such as active (or inhibitory) drug metabolites can complicate the isolation of pharmacological effects related solely to the parent drug.
- Development of tolerance or sensitization after the prolonged exposure of a drug: (a) tolerance (down-regulation), i.e., decrease in the density of receptors related to a drug after long-term administration (tachyphylaxis); (b) sensitization (up-regulation), i.e., increase in the density of receptors causing an overshoot effect after withdrawal of a drug.
- Physiological homeostatic responses (biofeedback mechanisms): administration of a drug can influence the concentration of endogenous agonists acting on the same receptor.
- Disease states: disease-induced pharmacodynamic alterations may not necessarily be the result of changes in receptor characteristics or effector mechanisms.
- Interindividual or intraindividual variability in pharmacodynamics owing to genetic or environmental factors.
- Stereoisomerism: drug concentration–effect relationships can be further complicated when the drug is a mixture of racemates or diastereomers and its isomers have different activities, e.g., labetolol and propranolol. If a stereospecific assay for a drug is not available or there is interconversion among isomers with different activities, establishing a meaningful relationship between drug concentration and effect becomes especially difficult.
- Drug–drug interaction in pharmacodynamics by the coadministered drug.

11.3. PHARMACOKINETIC/PHARMACODYNAMIC MODELING

11.3.1. Definition

Pharmacokinetic/pharmacodynamic (PK/PD) mathematical modeling to elucidate the relationship between drug concentrations in plasma and at the effect site by examining the plasma drug concentration *vs.* effect relationship, thus revealing the true pharmacodynamic profiles of a drug, i.e., the relationship between drug concentrations at the effect site and its pharmacological effects.

11.3.2. Implications of Pharmacokinetic/Pharmacodynamic Modeling

Simultaneous measurement of drug concentrations at the effect site [$C_e(t)$] and its pharmacological effect [$E(t)$] is the most desirable approach to reveal the true pharmacodynamic (PD) profiles of a drug. It is, however, seldom feasible to measure $C_e(t)$. In many *in vivo* studies, PD analyses are frequently performed based on drug concentrations in plasma [$C_p(t)$], assuming that $C_p(t)$ is in rapid equilibrium with $C_e(t)$, and that unbound drug concentrations are the same in both the plasma and

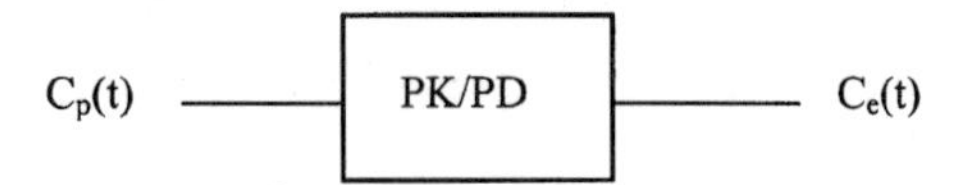

Figure 11.8. Pharmacokinetic/pharmacodynamic modeling to elucidate the relationship between drug concentrations in plasma [$C_p(t)$] and at the effect site [$C_e(t)$] under nonsteady state conditions.

at the effect site. If $C_p(t)$ is not in rapid equilibrium with $C_e(t)$, i.e., there is a delay in drug distribution from the plasma to the effect site or vice versa, or the system is subject to time-dependent kinetic and/or dynamic changes, the pharmacological response will take time to develop, and there will be no apparent relationship between $C_p(t)$ and E(t). PK/PD modeling can elucidate the relationship between $C_p(t)$ and $C_e(t)$ by integrating available information on $C_p(t)$ *vs.* E(t), and reveal the true PD of a drug between $C_e(t)$ and E(t) without establishing equilibrium between $C_p(t)$ and $C_e(t)$ (Fig. 11.8). A schematic description of PK/PD model developing processes when there is a time-delay discrepancy in the relationship between plasma drug concentrations and observed pharmacological effects under nonsteady state conditions is shown in Fig. 11.9.

11.3.3. Types of Pharmacokinetic/Pharmacodynamic Models

11.3.3.1. Effect Compartment (or Link) Model

The effect compartment model assumes that the differences between $C_p(t)$ and $C_e(t)$ are due to a delay in drug distribution from the plasma (the central compartment) to the effect site (the effect site compartment). The model is designed to eliminate the time-related discrepancy between $C_p(t)$ and E at different time points

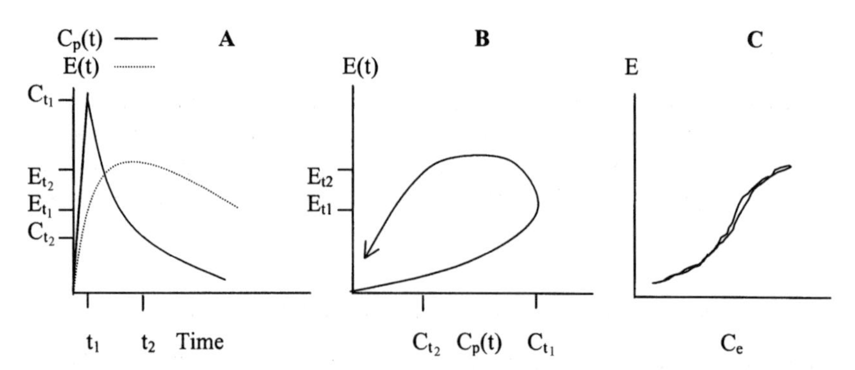

Figure 11.9. Schematic representations of pharmacokinetic/pharmacodynamic (PK/PD) modeling processes: (A) Plasma drug concentration–time and effect–time profiles after a short intravenous infusion up to time t_1. (B) Plasma drug concentration–effect plot showing an apparent discrepancy between the plasma drug concentration levels (C_{t_1} and C_{t_2}) and the intensity of the effect at different time points (t_1 and t_2) (see Counterclockwise hysteresis). (C) The PK/PD model predicts $C_e(t)$ based on $C_p(t)$, and reveals the true PD relationships between C_e and E by eliminating the time delay between $C_p(t)$ and E(t). $C_p(t)$ and C_e: drug concentration in plasma at time t and at the effect site, respectively, and E: the intensity of effect.

caused by *distributional delay*, by introducing the so-called equilibration rate constant, k_{eo} (Holford and Sheiner, 1981). Figure 11.10 is a schematic representation of the effect compartment model connected to the central compartment of a multicompartment model system after intravenous bolus injection of a drug into the central compartment.

The following steps are taken for the effect compartment model approaches:

STEP 1: Derive an equation describing $C_e(t)$ from $C_p(t)$ based on a PK model for $C_p(t)$ and k_{eo}. For instance, when the plasma drug exposure is adequately described with a single-compartment model and the drug is given intravenously, $C_e(t)$ can be expressed as a function of $C_p(t)$ and microconstants, k_{el} and k_{eo}:

$$C_e(t) = \frac{D \cdot k_{eo}}{V \cdot (k_{eo} - k_{el})} \cdot (e^{-k_{el} \cdot t} - e^{-k_{eo} \cdot t}) \tag{11.8}$$

and

$$C_p(t) = \frac{D}{V} \cdot e^{-k_{el} \cdot t} \tag{11.9}$$

where V is volume of distribution of the drug.

STEP 2: Select a proper PD model and incorporate $C_e(t)$ obtained from step 1 into it. For instance, if an E_{max} model is chosen, E(t) can be written as a function of $C_e(t)$:

$$E(t) = \frac{E_{max} \cdot C_e(t)}{EC_{50} + C_e(t)} \tag{11.10}$$

Equations (11.8)–(11.10) can be simultaneously fitted to experimental data on $C_p(t)$ and E(t) to estimate k_{eo}, E_{max}, and EC_{50}, using an appropriate nonlinear regression method.

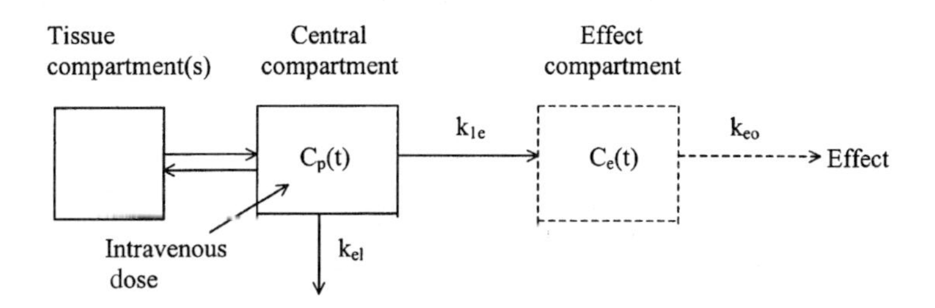

Figure 11.10. Schematic description of the effect compartment model. The hypothetical effect compartment is connected to the central compartment, usually, a plasma pool. When more than one compartment is required for describing the pharmacokinetics of a drug, the tissue compartments can be connected to the central one. Drug concentration at the effect site is governed by drug concentrations in the central compartment, and the extent of k_{eo}. $C_p(t)$ and $C_e(t)$: drug concentrations in plasma and at the effect site, respectively; E(t): effect; k_{le}, k_{el}, and k_{eo}: distribution rate constant from the central to the effect compartment, elimination rate constant from the central compartment, and equilibration rate constant, respectively.

11.3.3.2. Indirect Response Model

The premise of the indirect response model is that a slow onset of drug response or a slow return to baseline, i.e., time-related discrepancies in the $C_p(t)$–E relationship, can be due to indirect responses to the drug stemming from drug-related stimulation or inhibition, respectively, of the factors controlling either the production or the loss of the effect (Fig. 11.11) (Bellissant *et al.*, 1998; Dayneka *et al.*, 1993).

The rate of change of the effect in the absence of the drug can be described by

$$\frac{dE(t)}{dt} = k_{in} - k_{out} \cdot E(t) \tag{11.11}$$

where k_{in} and k_{out}, are, respectively, the zero-order and the first-order rate constant for production and loss of an effect. The following steps can be taken with the indirect response model.

STEP 1: Identify an equation for describing the rate of change of the effect (or response) over time in the presence of the drug. There are four different equations available, depending on whether the production or loss of response variables is stimulated or inhibited:

$$\frac{dE(t)}{dt} = k_{in} \cdot S(t) - k_{out} \cdot E(t), \text{ when the drug stimulates the effect}$$

$$= k_{in} \cdot I(t) - k_{out} \cdot E(t), \text{ when the drug inhibits the effect}$$

$$= k_{in} - k_{out} \cdot S(t) \cdot E(t), \text{ when the drug stimulates loss of the effect}$$

$$= k_{in} - k_{out} \cdot I(t) \cdot E(t), \text{ when the drug inhibits loss of the effect}$$

The stimulation function S(t), and the inhibition function I(t), can be expressed as follows:

$$S(t) = 1 + \frac{E_{max} \cdot C_p(t)}{EC_{50} + C_p(t)} \tag{11.12}$$

$$I(t) = 1 - \frac{C_p(t)}{IC_{50} + C_p(t)} \tag{11.13}$$

where EC_{50} and IC_{50} are, respectively, the drug concentrations producing 50% of the maximum stimulation and 50% of maximum inhibition achieved at the effect site.

STEP 2: The equation chosen is fitted to the experimental data on $C_p(t)$ and E(t) to estimate those parameters using an appropriate nonlinear regression method.

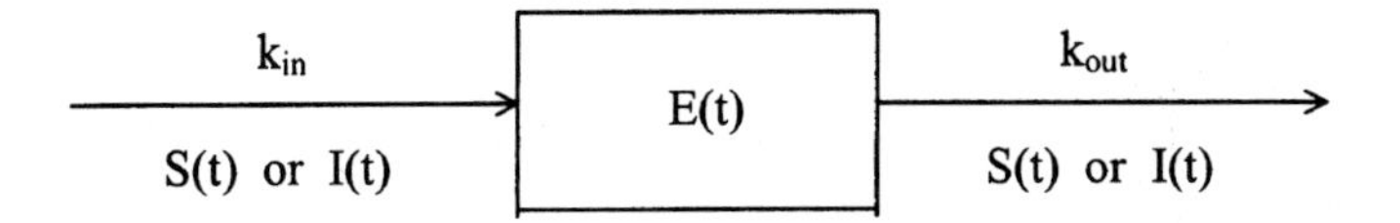

Figure 11.11. Schematic representation of the indirect response model. E(t): Effect; S(t) and I(t): stimulation and inhibition functions, respectively; k_{in} and k_{out}: the zero-order rate constant for production and the first-order rate constant for loss of the effect, respectively.

11.4. PROTERESIS OR HYSTERESIS

When $C_p(t)$ after a single dose or multiple doses is plotted against the corresponding E(t) and connected in a time sequence, the intensity of the effect is sometimes different at different time points even at the same concentrations (Campbell, 1990; Oosterhuis and van Boxtel, 1988), and it is the pattern of this relationship as a function of time that defines proteresis or hysteresis in pharmacodynamics.

Proteresis (Clockwise Hysteresis) and Hysteresis (Counterclockwise Hysteresis): If the intensity of the drug effect is higher at the earlier time points than at the later

Table 11.2. Factors Causing a Proteresis or Hysteresis Loop in a Plasma Drug Concentration [C_p(t)] *vs.* Effect [E(t)] Plot

Pattern of a C_p(t) *vs.* E(t) plot connected in time sequence	Factors causing the time-related discrepancies in the apparent concentration–effect relationship
Proteresis	Development of tolerance: a dampened response to the drug, after prolonged or repeated exposure (diazepam, morphine)
	Formation of antagonistic metabolites: antagonistic metabolites competing with the drug for the same binding sites on the receptor (pentobarbital)
	Down-regulation: decrease in the number of receptors after the prolonged exposure of the drug (isoproterenol)
	Biofeedback regulation (almitrine, nifedipine, tyramine)
Hysteresis	Distribution delay[a]: the nonequilibrium condition of drug concentrations between the plasma and the effect site owing to slow drug distribution from the plasma, sampling site, to the effect site (9-tetrahydrocannabinol, thiopental)
	Response delay: the delayed pharmacological response resulting from a series of biological events upon the initial stimulus by the drug interacting with a specific receptor at the effect site (corticosteroids, warfarin)
	Sensitization of receptors (angiotensin, propranolol)
	Formation of agonistic (active) metabolites (fenfluramine, camazepam, midazolam)
	Up-regulation: increase in the number of receptors after the prolonged exposure of the drug (propranolol)

[a]The most common cause for the hysteresis.

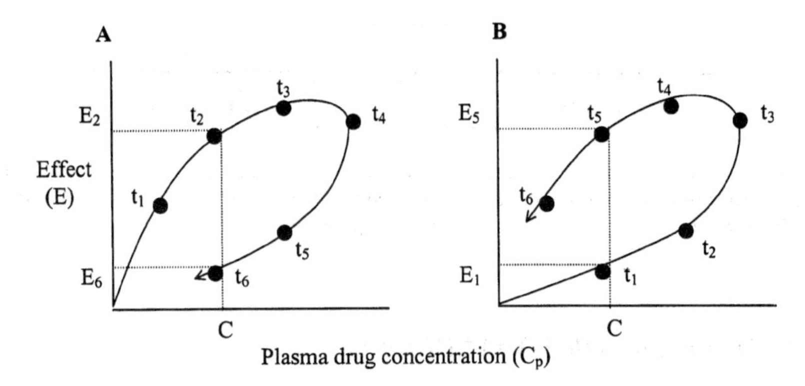

Figure 11.12. Proteresis (A) and hysteresis (B) relationships between plasma drug concentrations (●) from time t_1 to t_6 and the corresponding effect levels of a drug. A: The effect (E_2) at the earlier time point (t_2) is more pronounced than that (E_6) at the later time point (t_6) despite the same plasma concentration (C). B: The effect (E_1) at t_1 is lower than that (E_5) at t_5 at the same concentration (C).

ones at the same plasma drug concentrations, the clockwise pattern of the loop connecting concentration *vs.* effect data points in a time sequence reflects proteresis (or clockwise hysteresis) (Fig. 11.12A).

If the intensity of the drug effect is higher at the later time points than at the earlier ones at the same plasma drug concentrations, this counterclockwise pattern of the loop connecting data points in a time sequence or an apparent delay in effect is referred to as (counterclockwise) hysteresis (Fig. 11.12B).

In both proteresis and hysteresis, the pharmacological response is considered to be direct and reversible. These time-related discrepancies in the apparent concentration–effect relationship revealed by proteresis or hysteresis loops can be due to a lack of equilibrium in drug concentrations between the plasma and the effect site, but also can be due to many other factors (Table 11.2).

REFERENCES

Bellissant E. *et al.*, Methodological issues in pharmacokinetic-pharmacodynamic modelling, *Clin. Pharmacokinet.* **35**: 151–166, 1998.

Campbell D. B., The use of kinetic-dynamic interactions in the evaluation of drugs, *Psychopharmacology* **100**: 433–450, 1990.

Dayneka N. L. *et al.*, Comparison of four basic models of indirect pharmacodynamic responses, *J. Pharmacokinet. Biopharm.* **21**: 457–478, 1993.

du Souich T. *et al.*, Plasma protein binding and pharmacological response, *Clin. Pharmacokinet.* **24**: 435–440, 1993.

Hill. A. V., The possible effects of the aggregation of the molecules of hemoglobin on its dissociation curves, *J. Physiol.* (*London*) **40**: 4–7, 1910.

Holford N. H. G. and Sheiner L. B., Understanding the dose-effect relationship: clinical application of pharmacokinetic-pharmacodynamic models, *Clin. Pharmacokinet.* **6**: 429–453, 1981.

Holford N. H. G. and Sheiner L. B., Kinetics of pharmacological response, *Pharmacol. Ther.* **16**: 143–166, 1982.

Levy G., Variability in animal and human pharmacodynamic studies, in M. Rowland *et al.* (eds.), *Variability in Drug Therapy: Description, Estimation, and Control*, Raven Press, New York, 1985, pp. 125–138, 1985.

Oosterhuis B. and van Boxtel C. J., Kinetics of drug effects in man, *Ther. Drug Monitor.* **10**: 121–132, 1988.

Paalzow L. K. *et al.*, Variability in bioavailability: concentration versus effect, in M. Rowland *et al.* (eds.), *Variability in Drug Therapy: Description, Estimation, and Control*, Raven Press, New York, 1985, pp. 167–185, 1985.

Ritschel W. A. and Hussain A., Review on correlation between pharmacological response and drug disposition, *Exp. Clin. Pharmacol.* **6**: 627–640, 1984.

Schwinghammer T. L. and Kroboth P. D., Basic concepts in pharmacodynamic modeling, *J. Clin. Pharmacol.* **28**: 388–394, 1988.

12

Predicting Pharmacokinetics in Humans

Prediction of pharmacokinetic profiles of new chemical entities in humans based on *in vitro* or *in vivo* preclinical data is an extremely useful tool for drug discovery and development to identify compounds with desirable pharmacokinetic properties. There are basically two different approaches for the prediction of the pharmacokinetics of compounds in humans: allometry and the physiologically based method.

12.1. ALLOMETRY

In vivo pharmacokinetic data from experiments with various laboratory animals can be used to predict pharmacokinetic profiles in humans with allometry. Allometric extrapolation of pharmacokinetics among different species is based on the underlying anatomical, physiological, and biochemical similarities among animals (Boxenbaum, 1982; Boxenbaum and D'Souza, 1990; Dedrick, 1973).

12.1.1. Definition

Allometry, literally "by a different measure," is the study of empirical relationships between observations such as physiological functions or the pharmacokinetics of a drug on the one hand, and the size, shape, body surface area, and/or life span of animals on the other, without necessarily examining the underlying mechanisms. It has been found that physiological functions, e.g., energy and oxygen consumption, metabolism, cardiac output, and heart rate, are quantitatively related to an animal's body weight and/or size. It is, therefore, assumed that the pharmacokinetic parameters of a drug, such as clearance, volume of distribution, and half-life, that are governed by physiological functions, e.g., organ blood flow, glomerular filtration, volume of blood, and tissue mass, can be related to body weight and/or size as well.

12.1.2. Applications of Allometry for Predicting Pharmacokinetics in Humans

The major premise behind interspecies allometric scaling is that there is a positive correlation between the physiological functions affecting drug disposition and the body weight of animals. For instance, the hepatic blood flow rate and the weight of the liver in different species can be expressed as $55.4 \cdot BW^{0.89}$ ml/min ($r = 0.993$) and $37.0 \cdot BW^{0.85}$ kg ($r = 0.997$), respectively, where BW is the body weight (kg) of the animal of interest (Boxenbaum and D'Souza, 1990). In fact, the hepatic blood flow rate in all species is about 1.5 ml/min/g liver weight. For a drug highly extracted by the liver, clearance will be primarily governed by the hepatic blood flow rate and it can thus be readily described allometrically. Another example is creatinine clearance, which reflects the capacity of the glomerular filtration rate. Renal clearances of a drug via glomerular filtration among different species can be expressed by an allometric equation with body weight.

The allometric approach can be useful, preferably when the following conditions are met in all the included species: (1) linear kinetics of drug disposition; (2) low or similar extent of protein binding; (3) drug elimination predominantly via physical or mechanical processes (e.g., hepatic blood flow or glomerular filtration); and (4) a sufficient number of animal species and enough experimental data for linear regressions of allometric equations. There are, however, plenty of examples in which these criteria are not met, yet a reasonable prediction of pharmacokinetics in humans can be made, and vice versa.

12.1.2.1. Allometric Equations

The most commonly used allometric equation for interspecies extrapolation of pharmacokinetic parameters is (Mordenti, 1986):

$$Y = \alpha \cdot X^{\beta} \tag{12.1}$$

where Y is the pharmacokinetic parameter of interest, such as clearance and volume of distribution, and X is the physiological parameter, usually body weight. The estimates of the allometric coefficient (α) and the allometric exponent (β) can be obtained, respectively, from an intercept and a slope of a log–log plot of Eq. (12.1) (Fig. 12.1).

$$\log Y = \log \alpha + \beta \cdot \log X \tag{12.2}$$

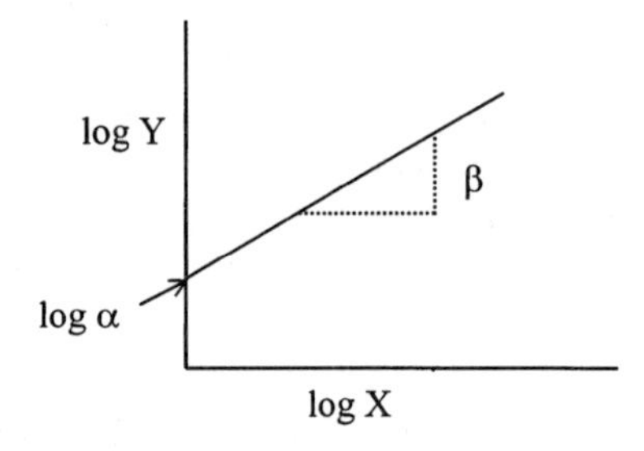

Figure 12.1. Allometric equation on a log–log scale.

Table 12.1. Experimentally Determined Systemic Plasma Clearance (Cl_s) and Volume of Distribution (V_{ss}) of a Hypothetical Drug in Rats, Monkeys, and Dogs and Predicted Values for Those Parameters in Humans Based on Allometry

	Experimentally determined values in			Predicted values in humans by allometry
	Rat	Monkey	Dog	
Body weight (kg)	0.25	5	10	70
Cl_s (ml/min)	10.4	95.3	175.7	*737*
V_{ss} (liters)	0.5	11.7	18	*140*

For instance, if a compound of interest shows a systemic clearance (Cl_s) and a volume of distribution at steady state (V_{ss}) in rats, monkeys, and dogs as summarized in Table 12.1, allometric equations for human predictions are $30 \cdot BW^{0.75}$ and $2 \cdot BW^1$, for Cl_s and V_{ss}, respectively based on Eq. (12.2) (Fig. 12.2).

Table 12.2 summarizes some of the known allometric relationships between physiological parameters and body weight among mammals (Boxenbaum and D'Souza, 1990; Mordenti, 1985, 1986).

12.1.2.2. Allometric Scaling of Clearance

(a) Hepatic Clearance. For drugs with high hepatic extraction, a reasonable allometric extrapolation of clearance between different species can be made. It is, however, difficult to apply allometry for drugs with intermediate to low hepatic clearance because clearance of those drugs is governed not only by hepatic blood flow rate, but also by other physiological and biological factors, including the

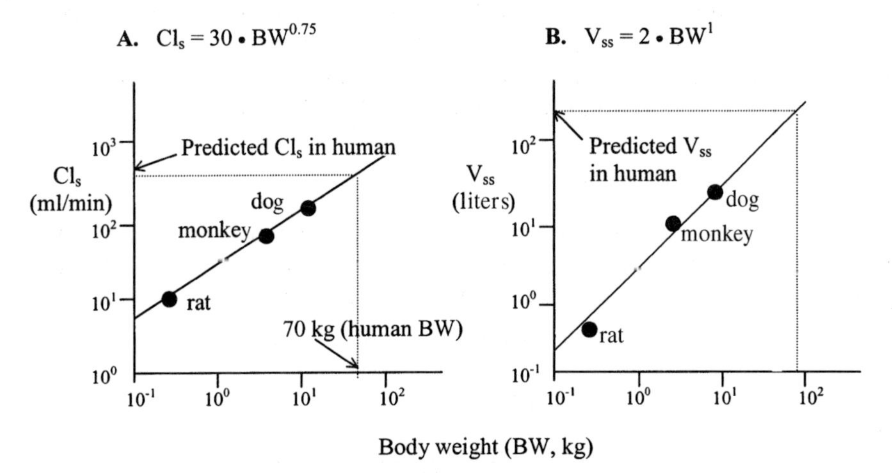

Figure 12.2. Allometric scaling for systemic clearance (Cl_s) and volume of distribution at steady state (V_{ss}) of a drug in humans. Allometric equations and plots for Cl_s (A) and V_{ss} (B) on a log–log scale are based on the animal data summarized in Table 12.2.

Table 12.2. Allometric Relationship, $Y = \alpha \cdot X^{\beta}$, between Physiological Parameters (Y) and Body Weight (X, kg) among Mammals

Physiological parameters	α	β	r^b
Cardiac output (ml/min)	166	0.79	—
Hepatic blood flow rate (ml/min)	55.4	0.89	0.993
Renal blood flow rate (ml/min)	43.06	0.77	—
Time for circulation of blood volume (min)	0.35	0.21	0.98
Cytochrome P450 weight (mg)	33.1	0.84	—
Nephra number	1.88×10^5	0.62	—

[a] Data taken from Boxenbaum and D'Souza (1990) and Mordenti (1985, 1986).
[b] r is the correlation coefficient.

activities of metabolizing enzymes, to which it is difficult to apply allometry. For drugs eliminated mainly by the hepatic cytochrome P450 enzymes and classified as low-clearance drugs, the clearance predicted in humans with allometry tends to be higher than the measured value. In this case, physiologically based methods may be more appropriate.

(b) Renal Clearance. A relatively reasonable interspecies extrapolation can be done for renal clearance with allometry if the extent of protein binding is similar among the different species. This is because renal elimination of a drug is essentially via either physical filtration (glomerular filtration) or passive processes, with the exception of active tubular secretion. The passive processes are determined in part by the physicochemical properties of the drug and its protein binding (Ritschel *et al.*, 1992).

12.1.2.3. Allometric Scaling of the Volume of Distribution

A reasonable prediction of the volume of distribution at steady state in humans can be made with allometric extrapolation if the extent of protein binding in the plasma is similar. This is because the volume of blood and the tissue mass are allometrically related to the size of the animals, and the extent of protein binding in tissues appears to be similar among different species (Mahamood and Balian, 1996*a*).

12.1.2.4. Allometric Prediction of Terminal Half-Life

In general, the terminal half-life ($t_{1/2}$) of a drug with plasma exposure showing a monophasic decline on a semilog scale tends to be proportional to body weight [Eq. (12.3)], based on allometric equations for Cl_s, and V_{ss} (Ings, 1990):

$$\underset{(\text{ml/min})}{Cl_s} = \alpha_1 \cdot \underset{(\text{kg})}{BW^{0.75}}, \qquad \underset{(\text{liters})}{V_{ss}} = \alpha_2 \cdot \underset{(\text{kg})}{BW^1}$$

and thus

$$\boxed{t_{1/2} \propto BW^{0.25}} \tag{12.3}$$

12.1.2.5. Neoteny

There are a number of examples where simple allometry based on body weight is not adequate for prediction of systemic clearance. This apparent deviation of allometric extrapolation can be improved by introducing the concepts of neoteny into the allometric equations. Neoteny means the retention of formerly juvenile characters by adult descendants produced by retardation of somatic development, i.e., a sort of sustained juvenilization or slow development of animals to adulthood. Related it to humans, neoteny is the preservation in adults of shapes and growth rates that characterize juvenile stages of ancestral primates (Boxenbaum and D'Souza, 1990). For instance, humans tend to reach puberty at about 60% of their final body weight, whereas most other mammals reach puberty at about 30% of their ultimate body weight. The concept of maximum life span (MLP), which appears to be an allometric function of brain and body weights, has been introduced to correct for such an evolutionary development in allometric scaling:

$$\text{MLP (years)} = (185.4) \cdot \text{Br}^{0.636} \cdot \text{BW}^{-0.225} \tag{12.4}$$

The scale of both brain weight (Br) and body weight (BW) is kg (Mahamood and Balian, 1996*b*; Sacher, 1959). For instance, the MLP in humans based on 70 kg BW is approximately 88.3 years. So, e.g., the allometric relationship between Cl_s and BW can be adjusted by MLP as follows:

$$Cl_s = \frac{\alpha \cdot \text{BW}^{\beta}}{\text{MLP}} \tag{12.5}$$

Sometimes, the allometric relationship for low-clearance drugs can be better established with Br, rather than with MLP and BW (Ings, 1990):

$$Cl_s = \alpha \cdot \text{Br}^{\beta_1} \cdot \text{BW}^{\beta_2} \tag{12.6}$$

12.2. PHYSIOLOGICALLY BASED APPROACH

Another method for pharmacokinetic prediction in humans is based on actual physiological, anatomical, and biochemical factors important in drug disposition such as:

1. Organ blood flow rates.
2. Organ size.
3. Tissue and fluid volumes.
4. Blood-to-plasma and tissue-to-plasma drug concentration ratios.
5. Protein binding.
6. Metabolizing enzyme activities, etc.

This method offers a mechanistic approach to the extrapolation of *in vitro* experimental findings to *in vivo* pharmacokinetic parameters within the same species.

12.2.1. Predicting Systemic Clearance of a Drug in Humans from *In Vitro* Data

When metabolism is the major elimination pathway of a drug in animals and is expected to be so in humans as well, simple allometric scaling may not be adequate for prediction of clearance in the latter owing to a substantial interspecies variability in metabolism (Calabriese, 1986). In recent years, substantial progress has been made in *in vitro* metabolism studies of xenobiotics, using, e.g., purified enzymes, subcellular fractions (liver microsomes and S9), whole cells (primary hepatocytes), and liver slices (Miners *et al.*, 1994). With the sensitivity and quantitative capability of modern analytical chemistry, the utility of these *in vitro* methodologies to predict *in vivo* hepatic clearance in humans becomes of great interest (Iwatsubo *et al.*, 1997).

The well-stirred and parallel-tube models are the two most commonly used clearance models for describing *in vivo* hepatic clearance (Cl_h) of drugs.

WELL-STIRRED MODEL:

(12.7)
$$Cl_h = \frac{Q_h \cdot f_u \cdot CL_{i,h}}{Q_h + f_u \cdot CL_{i,h}}$$

PARALLEL-TUBE MODEL:

(12.8)
$$Cl_h = Q_h \cdot (1 - e^{-f_u \cdot Cl_{i,h}/Q_h})$$

where Q_h, f_u, and $Cl_{i,h}$ are hepatic blood flow rate, the ratio of the unbound to total drug concentrations in blood, and the intrinsic hepatic clearance, respectively. As indicated in these equations, Cl_h of a drug is affected by these three factors. It is thus possible to calculate Cl_h if those parameters are estimated or measured experimentally *in situ* or *in vitro*. In general, Q_h values reported in literature can be used. The value for f_u can be experimentally determined. The key information needed to estimate Cl_h is $Cl_{i,h}$ for the metabolic activities in the liver. It has been demonstrated that $Cl_{i,h}$ estimated from various *in vitro* metabolism studies can be used for *in vivo* Cl_h prediction with a high degree of success, if the major elimination pathway of a drug is via hepatic metabolism (Houston 1994; Houston and Carlie, 1997).

A three-step strategy to extrapolate *in vitro* metabolism data of a drug to *in vivo* hepatic clearance in animals is as follows.

Measure V_{max} and K_m or $t_{1/2}$	step 1 →	$Cl_{i,\,in\ vitro}$	step 2 → scaling factors	$Cl_{i,h}$	step 3 → Q_h, f_u	Cl_h
In vitro metabolism studies		*Estimate in vitro intrinsic hepatic clearance*		*Estimate in vivo intrinsic hepatic clearance with proper scaling factors*		*Estimate hepatic clearance with* $Cl_{i,h}$, Q_h, *and* f_u *base on hepatic clearance models*

Step1: Estimate the *in vitro* intrinsic hepatic clearance ($Cl_{i,\,in\ vitro}$) based on V_{max} (the maximum rate of metabolism) and K_m (the Michaelis–Menten constant for the drug–enzyme interaction), or half-life of drug disappearance ($t_{1/2}$) in *in vitro* metabolism studies.

Step 2: Extrapolate $Cl_{i,\,in\ vitro}$ to *in vivo* intrinsic hepatic clearance ($Cl_{i,h}$) with proper scaling factors.

Step 3: Calculate *in vivo* Cl_h, according to hepatic clearance models with estimates of $Cl_{i.h}$, Q_h, and f_u.

Extrapolation processes from *in vitro* metabolism data to *in vivo* hepatic clearance will be discussed based mainly on metabolic stability studies in liver microsomes and hepatocytes; however, the same approaches can be applied to other *in vitro* systems, such as liver S9 or slices. In order for extrapolation to be successful, an *in vitro* study should be performed under linear conditions in terms of incubation time, and enzyme and substrate concentrations. A detailed description of each step is given below.

12.2.1.1. *Estimation of In Vitro Intrinsic Hepatic Clearance* ($Cl_{i,\,in\,vitro}$)

The true *in vivo* intrinsic hepatic clearance combines the activities of the metabolizing enzymes and hepatobiliary excretion. There are, however, no reliable methods for extrapolating *in vitro* or *in situ* data to *in vivo* biliary excretion in humans. Thus, intrinsic drug clearance measured in liver microsomes or hepatocytes that only represent metabolic activities in the liver [referred to, hereafter, as *in vitro* intrinsic hepatic clearance ($Cl_{i,\,in\,vitro}$)] will be considered for *in vivo* extrapolation, so estimates of Cl_h based on $Cl_{i,\,in\,vitro}$ that do not incorporate biliary excretion may be too low.

The utility of these *in vitro* experiments for predicting clearance is based on the assumption that there is a correlation between the rate of drug disappearance (or the rate of metabolite formation) and the *in vivo* metabolic clearance of the drug. Rate data generated from certain *in vitro* systems may underestimate the true metabolic clearance *in vivo* owing to limitations in discerning metabolic capability. For instance, isolated microsomes, which are a subcellular fraction, contain only a portion of the total range of drug-metabolizing enzymes present in the original tissue *in vivo*, and thus microsomal incubations may underestimate the true metabolic clearance if the compound undergoes metabolism via cytosolic or mitochondrial enzymes *in vivo*. Nevertheless, recent studies have demonstrated that a successful quantitative extrapolation of *in vitro* metabolism data to *in vivo* clearance could be achieved under carefully designed conditions with a few assumptions (Houston 1994; Houston and Carlie, 1997).

(a) Definition of $Cl_{i,\,in\,vitro}$. $Cl_{i,\,in\,vitro}$ can be viewed as the intrinsic ability of the hepatic metabolizing enzymes present in *in vitro* systems, such as liver microsomes, S9, hepatocytes, or liver slices, to eliminate a drug from the incubation medium (12.9). This is because the rate of drug disappearance (or metabolite formation) in *in vitro* systems depends solely on the activities of metabolizing enzymes without influences from other factors that are present *in vivo*, e.g., blood flow, drug binding to blood components, and cofactor supply. (Rane *et al.*, 1977).

$$Cl_{i,\,in\,vitro} = \frac{\text{Initial rate of drug disappearance (or metabolic formation) from the incubation medium}}{\text{Unbound drug concentration available to the metabolizing enzymes}}$$

(12.9)

Although the drug disappearance rate can be replaced by the rate of formation of all the metabolites of the drug, for simplicity hereafter only the drug disappearance rate will be used to describe $Cl_{i,\,in\,vitro}$. Disappearance of a drug via the metabolizing enzymes can be viewed as an enzymatic reaction describable by Michaelis–Menten kinetics:

$$\text{(12.10)} \qquad \text{Rate of drug disappearance} = \frac{V_{max} \cdot C_{1,u}}{K_m + C_{1,u}}$$

where V_{max} and K_m are the maximum *initial* disappearance rate of a drug and the apparent Michaelis–Menten constant of total enzymes involved in its elimination in the medium, respectively. There is usually more than one enzyme metabolizing a drug, and each has its own V_{max} and K_m for the substrate, which contributes to the *apparent* overall V_{max} and K_m values measured. $C_{1,u}$ is the concentration of drug not bound to microsomal proteins or macromolecules in hepatocytes, which is available to the metabolizing enzymes. From Eqs. (12.9) and (12.10) $Cl_{i,\,in\,vitro}$ can be expressed as

$$\text{(12.11)} \qquad Cl_{i,\,in\,vitro} = \frac{(V_{max} \cdot \cancel{C_{1,u}})/(K_m + C_{1,u})}{\cancel{C_{1,u}}} = \frac{V_{max}}{K_m + C_{1,u}}$$

When $C_{1,u}$ is less than 10% of K_m (linear condition), $Cl_{i,\,in\,vitro}$ can be viewed simply as the ratio between V_{max} and K_m [Eq. (12.12)], and becomes independent of the drug concentration (Chenery *et al.*, 1987):

$$\text{(12.12)} \qquad \boxed{Cl_{i,\,in\,vitro} = V_{max}/K_m \text{ (under linear conditions)}}$$

NOTE: *IN VIVO* INTRINSIC HEPATIC CLEARANCE ($Cl_{i,h}$). $Cl_{i,h}$ is a measure of the intrinsic ability of all the metabolizing enzymes and biliary excretion mechanisms to eliminate a drug, which is not influenced by other physiological factors that affect Cl_h.

(b) Unit of $Cl_{i,\,in\,vitro}$. In general,

- μl/min/mg protein: flow rate (ml/min) normalized by protein concentration when the study is performed in liver microsomes or
- μl/min/10^6 hepatocytes: normalized by number of cells when hepatocytes are used.

(c) Estimating $Cl_{i,\,in\,vitro}$. There are two different methods for estimating $Cl_{i,\,in\,vitro}$ from *in vitro* metabolism studies with liver microsomes or hepatocytes, based on: (1) estimation of V_{max} and K_m of the metabolizing enzymes in the systems, or (2) half-life of substrate disappearance during the initial phase at a single substrate concentration.

(i) Estimation of $Cl_{i,\,in\,vitro}$ *based on* V_{max} *and* K_m *of the metabolizing enzymes.* $Cl_{i,\,in\,vitro}$ can be estimated by measuring V_{max} and K_m of the metabolizing

enzymes responsible for the disappearance of a drug in *in vitro* systems under linear conditions, according to Eq. (12.12).

• *Definition and unit of* V_{max}. V_{max} is a theoretical maximum initial disappearance rate of a drug via all the metabolizing enzymes when the initial substrate concentration at the beginning of the reaction approaches infinity. The magnitude of V_{max}, therefore, is dependent on the amount of active enzymes available for metabolism of the drug. When the drug concentration in the incubating medium is μM, the unit of V_{max} is usually: (1) pmol/min/mg protein, i.e., the amount of drug metabolized per unit time (pmol of drug/min) normalized by microsomal protein concentration when the study is performed in liver microsomes (mg protein); or (2) pmol/min/10^6 hepatocytes, normalized by the number of cells in the incubation medium when hepatocytes are used (10^6 hepatocytes).

• *Definition and unit of* K_m. K_m is the *apparent* Michaelis–Menten constant for the interaction between a drug and all the enzymes responsible for its metabolism. K_m can be also viewed as the initial concentration of the substrate required to reach half of V_{max}. The true K_m of an enzyme (or apparent K_m for multiple enzymes) is constant for the given enzyme, and is independent of the amount of enzyme present in the reaction medium. The unit of K_m is usually μM (or mM).

• *Estimating of* V_{max} *and* K_m. V_{max} and K_m can be estimated by fitting the Michaelis–Menten equation to a plot of the initial disappearance rates of a drug *vs.* its concentrations, or by fitting the equations transformed from the Michaelis–Menten equation, such as the Lineweaver–Burk plot, to the data.

Michaelis–Menten kinetics: V_{max} and K_m can be estimated by fitting the Michaelis–Menten equation to the initial disappearance rates of the substrate at different drug concentrations [Eq. (12.13) and Fig. 12.3].

$$V = \frac{V_{max} \cdot C}{K_m + C} \tag{12.13}$$

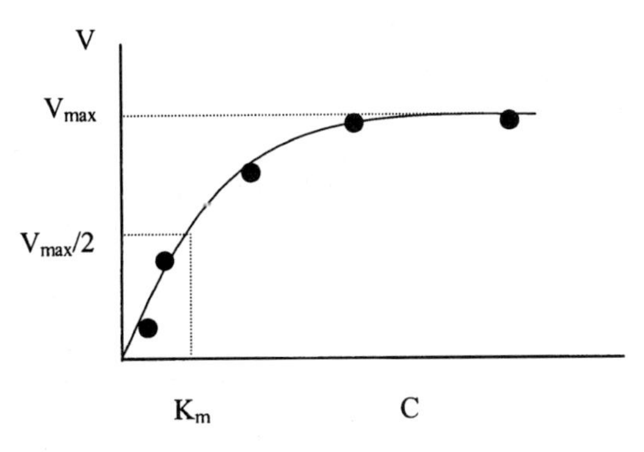

Figure 12.3. Michaelis–Menten plot (sometime called a "substrate saturation curve") illustrating the changes in initial disappearance rates (V) as a function of drug concentration (C). K_m is the same as the concentration at which V is a half of the theoretical maximum initial velocity (V_{max}).

where V represents the *initial* disappearance rates of the drug at given drug concentrations (C) with a fixed amount of enzyme, i.e., a fixed amount of microsomal protein or hepatocytes.

Construction of Michaelis–Menten Plot: Several important experimental conditions are required for a proper estimate of V_{max} and K_m from a Michaelis–Menten plot:

1. Concentration of the drug in the reaction medium should be substantially higher than the enzyme concentration.
2. Reaction conditions are established such that reliable measurement of the *initial rate* of disappearance of the drug at a given concentration is warranted. In general, initial reaction rates should be determined when less than 10% of the drug has been converted to primary metabolites in order to maintain linearity of the reaction rate. In this way, further biotransformation of metabolites to secondary or tertiary metabolites can be minimized.
3. Reaction rate should be determined at a minimum of at least five different drug concentrations (at least two and three concentrations below and above K_m, respectively).
4. For reliable estimates of V_{max} and K_m, in general, determinations should be conducted over drug concentrations ranging from 0.1- to 10-fold of K_m.

When *initial* disappearance rates of a drug are measured, it is important to ensure that drug loss from the incubating medium equates solely with the metabolism rather than with the adsorption of the drug to apparatus or cellular components. Three important factors in determining the incubation conditions for reliable measurement of initial rates of the metabolism of a drug in *in vitro* experiments are: (a) the range of drug concentrations, (b) microsomal protein concentration (or number of hepatocytes), and (c) assay sensitivity for the drug (or its metabolites). In order to construct a Michaelis–Menten plot, several preliminary experiments have to be performed to determine optimum ranges for the drug concentration and the amount of protein.

Determination of substrate and enzymatic protein concentrations: The lowest and highest drug concentrations have to be determined during the preliminary studies for the Michaelis–Menten plot. The former can be selected based on assay sensitivity for the drug and should be substantially lower than K_m of the enzyme(s). The latter can be chosen based on the aqueous solubility of the drug. A small amount of organic cosolvent may be used to enhance solubility of lipophilic drugs, but in this case, the effects of the cosolvent on enzyme activities should be monitored and minimized (see Chapter 8).

It is also important to select a protein concentration of microsomes (or a number of hepatocytes) that yields a reliable estimate of initial drug disappearance rates. Let us assume that the disappearance profiles of a hypothetical drug at the highest concentration, 0.5 μM, are examined after incubation at three different protein concentrations of liver microsomes (Fig. 12.4). At the lowest protein concentration, the rate of drug disappearance (differences in substrate concentration

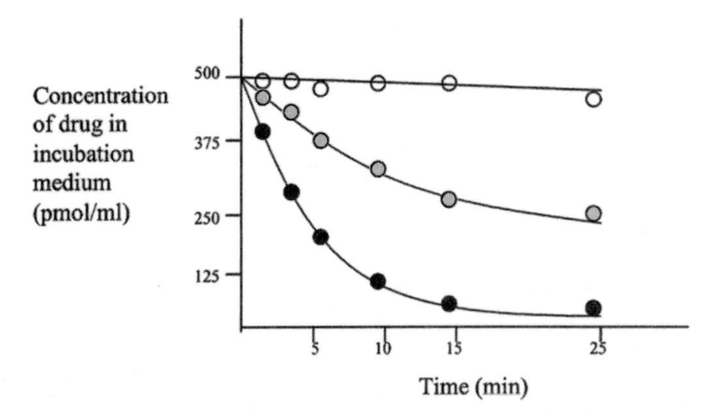

Figure 12.4. Hypothetical plot showing the changes in concentration over time after incubation of a drug at 0.5 μM in 1 ml of microsomes at three different protein concentrations (○ low, ● medium, ● high). The slope of the curve represents the drug disappearance rate.

over an incubation period) is consistent up to 25 min of incubation. However, consumption of the drug is less than 15% of its initial concentration, which may be difficult to distinguish reliably from the normal assay variability (usually 10–15%). At a medium protein concentration, the initial drug disappearance rate is maintained up to 10 min with approximately a 30% decrease in drug concentration. At the highest protein concentration, the drug disappearance rate is no longer the same as the initial rate shortly after incubation, which requires highly accurate time measurements and instantaneous complete mixing of drug, cofactor, and microsomes at the beginning of the experiment. From these preliminary experiments, it appears that the medium microsomal protein concentration would be the most suitable for measuring initial drug disappearance rates at 0.5 μM.

Reasons for measuring initial rates: The initial rate of drug disappearance in *in vitro* metabolism should be measured for the Michaelis–Menten plot for the following reasons:

1. The concentration of a drug after incubation is lower than at the beginning of the incubation. As it is drug disappearance rates measured after incubation that are used to construct the Michaelis–Menten plot against drug concentrations at the beginning of the incubation, it is important to minimize any discrepancy caused by the difference in drug concentrations at the start and at the time of measurement.
2. Any prolonged incubation in a closed *in vitro* system such as isolated liver microsomes can cause formation of metabolites from the primary metabolites of a drug, which does not occur *in vivo*, owing to the lack of phase II metabolism and other elimination pathways.
3. Inactivation or denaturation of enzymes can become significant over time in *in vitro* systems. Thus, it is desirable to determine the initial rates during the early time points, when a significant decrease in enzyme activity is expected over time.

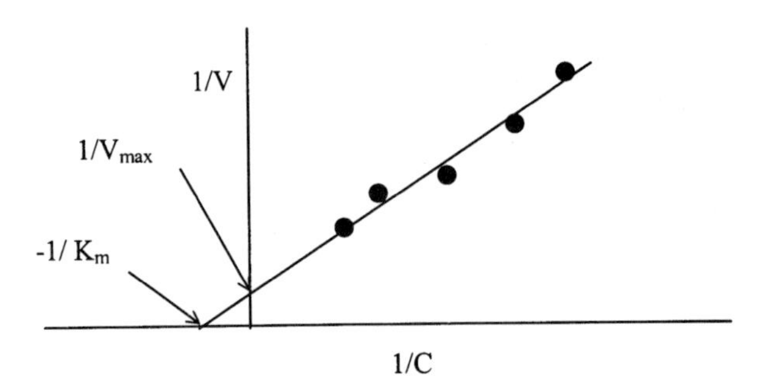

Figure 12.5. Lineweaver–Burk plot. V and C are the initial drug disappearance rate and drug concentration, respectively. The x- and y-axis intercepts of the Lineweaver–Burk plot are $-1/K_m$ and $1/V_{max}$, respectively.

Formation rate of metabolites: When the initial formation rates of metabolites are used, the sum of the V_{max}/K_m terms for each metabolic pathway can be used to estimate overall $Cl_{i,\, in\, vitro}$ of a drug:

(12.14) $$Cl_{i,\, in\, vitro} = \frac{V_{max,1}}{K_{m,1}} + \frac{V_{max,2}}{K_{m,2}} + \frac{V_{max,3}}{K_{m,3}} + \cdots$$

where $V_{max,1,2,3\ldots}$ and $K_{m,1,2,3\ldots}$ are maximum formation rates of each metabolite and Michaelis–Menten constants of the enzymes for the corresponding metabolites, respectively.

Lineweaver–Burk plot: It is often difficult to properly define Michaelis–Menten kinetics at high or low drug concentrations owing to limited aqueous solubility and assay limitation of a drug. The Lineweaver–Burk plot consists of reciprocals of V and C, and transforms the hyperbola of the Michaelis–Menten plot into a straight line (Fig. 12.5). It can be used to estimate V_{max} and K_m based on data at low drug concentrations.

NOTE: RELATIONSHIPS BETWEEN MICHAELIS–MENTEN AND LINEWEAVER–BURK EQUATIONS. The Lineweaver–Burke equation is inversely related to the Michaelis–Menten equation:

MICHAELIS–MENTEN EQUATION

$$V = \frac{V_{max} \cdot C}{K_m + C}$$

LINEWEAVER–BURK EQUATION

(12.16) $$\frac{1}{V} = \frac{K_m}{V_{max}} \cdot \frac{1}{C} + \frac{1}{V_{max}}$$

Kinetics for multiple enzymes: When more than one enzyme is involved in the metabolism of a drug, the Lineweaver–Burk plot may become curvilinear, whereas it is less noticeable in the Michaelis–Menten plot. Thus, a Lineweaver–Burk plot (or Eadie–Hofstee plot [V *vs.* V/C)] is considered to be more reliable for assessing the involvement of multiple enzymes. Once those plots indicate a multienzyme system, suitable equations, shown below, can be fitted to the Michaelis–Menten plot.

MICHAELIS-MENTEN KINETICS FOR ONE ENZYME AND LINEAR KINETICS FOR THE OTHER:

$$V = \frac{V_{max,1} \cdot C}{K_{m,1} + C} + Cl_{i,2} \cdot C \tag{12.17}$$

TWO SETS OF MICHAELIS–MENTEN EQUATIONS FOR TWO DIFFERENT ENZYMES:

$$V = \frac{V_{max,1} \cdot C}{K_{m,1} + C} + \frac{V_{max,2} \cdot C}{K_{m,2} + C} \tag{12.18}$$

How to Determine V_{max} *and* K_m *Experimentally and Estimate* $Cl_{i,\,in\,vitro}$:

1. Measure the disappearance rates (V) of the drug over a linear range of the disappearance curve at more than five drug concentrations in liver microsomes (or hepatocytes). For instance, if a drug concentration decreases from 1 to 0.9 μM after 5-min in a 0.5 ml incubation buffer at a microsomal protein concentration of 1 mg/ml, an initial drug disappearance rate at 1 μM of the drug is 20 pmol/min/mg protein:

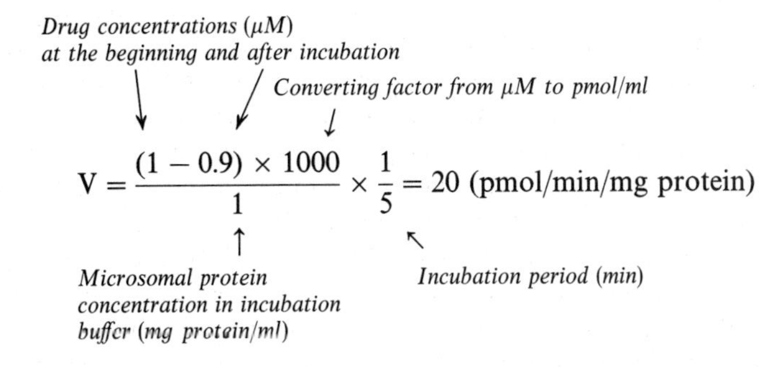

The initial rate of drug disappearance (V) is usually expressed as pmol/min/mg microsomal protein or number of hepatocytes ($\times 10^6$).

2. Fit the Michaelis–Menten equation to the drug disappearance rate *vs.* concentration profile (the Michaelis–Menten plot) determined at five different substrate concentrations with a nonlinear regression program (e.g., PCNONLIN) for estimating of V_{max} and K_m.

3. Calculate $Cl_{i, \textit{in vitro}}$ by dividing V_{max} by K_m [Eq. (12.12)]. For instance, if V_{max} and K_m are 100 pmol/min/mg protein and 5 μM, respectively, $Cl_{i, \textit{in vitro}}$ is 0.02 ml/min/mg protein:

$$Cl_{i,\ \textit{in vitro}} = \frac{\overset{V_{max}\ (\text{pmol/min/mg protein})}{\downarrow} \atop 100}{\underset{\nearrow K_m\ (\mu\text{M}) \quad \nwarrow \textit{Converting factor from } \mu M \textit{ to pmol/ml}}{5 \times 1000}} = 0.02\ (\text{ml/min/mg protein})$$

(ii) Estimating $Cl_{i, \textit{in vitro}}$ *based on the half-life of drug disappearance.* $Cl_{i, \textit{in vitro}}$ can be also estimated from the half-life of initial drug disappearance at a single drug concentration substantially lower than K_m of the enzyme [Eq. (12.19)]. This approach is useful, especially when it is difficult to carry out the Michaelis–Menten kinetics experiments over a wide range of concentrations (Chenery *et al.*, 1987):

$$Cl_{i,\ \textit{in vitro}} = \frac{0.693}{t_{1/2} \cdot \text{microsomal protein concentration}} \tag{12.19}$$

$t_{1/2}$ is the half-life of drug disappearance:

$$t_{1/2} = \frac{(0.693) \cdot (t_2 - t_1)}{\ln (C_1/C_2)} \tag{12.20}$$

C_1 and C_2 are drug concentrations at time t_1 and t_2 after incubation, respectively. Certainly, this is an easier and faster method than Michaelis–Menten kinetics for estimating $Cl_{i, \textit{in vitro}}$, as long as several assumptions can be made.

First, the drug concentration at which the half-life of drug disappearance is measured is substantially lower than K_m ($<10\%$ of K_m) to ensure a linear condition. Since this method requires only one drug concentration, it is important to choose an appropriate concentration and the right incubation conditions for linear kinetics. A concentration that is too high is not suitable because of potential substrate inhibition and a limitation in solubility, and a concentration that is too low is also inappropriate owing to assay limitation. Second, $t_{1/2}$ of drug disappearance should be measured over a period where the disappearance rate is linear (usually $<10\%$ of the disappearance rate of the initial concentration of the drug). In the case of low-clearance drugs, measuring $t_{1/2}$ during the period when up to 30% of the drug disappears may be acceptable. Owing to these restrictions, an estimate of $Cl_{i, \textit{in vitro}}$ based on this method tends to be lower than that from V_{max} and K_m estimated from the Michaelis–Menten plot. Nevertheless, its experimental simplicity offers an advantage over other more complicated methods, especially during drug discovery when a crude rank order of a large number of compounds in potential metabolic clearance in humans is sufficient for compound selection.

Table 12.3. Physiological and Biochemical Parameters Important for Scaling up *In Vitro* Intrinsic Hepatic Clearance to *In Vivo* Intrinsic Hepatic Clearance and for Estimating Hepatic Clearance in Humans

Physiological parameters	Values from literature	Scaling factors[a]
Liver weight	25.7 g liver/kg body weight[b]	—
Hepatic blood flow rate	20.6 ml/min/kg body weight[b]	—
Hepatocyte number	120×10^6 cells/g liver	3100
P450 contents in hepatocytes	0.14 nmol/10^6 cells	—
Microsomal protein	52.5 mg protein/g liver	1350
	77 mg protein/g liver	1980
P450 contents in microsomes	0.32 nmol/mg microsomal protein	—
	0.296 nmol/mg microsomal protein	—

[a]Data taken from Bäärnhielm *et al.* (1986), Davies and Morris (1993), and Iwatsubo *et al.* (1997).
[b]Scaling factors for estimating *in vitro* intrinsic hepatic clearance normalized by kg body weight in humans (ml/min/kg body weight), when *in vitro* intrinsic hepatic clearance is expressed as ml/min/mg microsomal protein (or ml/min 10^6 hepatocytes).
[c]Assuming that an average human body weight is 70 kg.

For instance, if $t_{1/2}$ of drug disappearance is 30 min over a linear range of the drug concentration–time profile at an initial concentration of 2 μM in 0.5 ml of incubation buffer at a microsomal protein concentration of 1 mg/ml, $Cl_{i,\,in\,vitro}$ becomes 0.02 ml/min/mg protein:

$$Cl_{i,\,in\,vitro} = \frac{\overset{\text{Constant}}{0.693}}{\underset{\text{Half-life (min)}}{30} \times \underset{\text{Microsomal protein concentration in incubation buffer (mg protein/ml)}}{1}} = 0.02 \text{ ml/min/mg protein}$$

12.2.1.2. *Extrapolation from* $Cl_{i,\,in\,vitro}$ *to In Vivo Intrinsic Hepatic Clearance* ($Cl_{i,h}$)

Extrapolation of $Cl_{i,\,in\,vitro}$ to $Cl_{i,h}$ can be achieved by scaling up the microsomal protein (or P450) concentration (or, where relevant, the number of hepatocytes) used in *in vitro* studies to that of the whole liver. Proper scaling factors in humans and rats are summarized in Tables 12.3 and 12.4, respectively. For instance, if an estimated $Cl_{i,\,in\,vitro}$ is 0.02 ml/min/mg microsomal protein measured after incubating a drug in human liver microsomes, $Cl_{i,h}$ of the drug in human liver *in vivo* can be estimated by multiplying $Cl_{i,\,in\,vitro}$ by a scaling factor of 1350 (or 1980) (Table 12.3), i.e.,

$$Cl_{i,h} = \underset{Cl_{i,\,in\,vitro}\ \text{(ml/min/mg microsomal protein)}}{0.02} \times \underset{\text{Scaling factor}}{1350} = 27 \text{ (ml/min/kg body weight)}$$

Table 12.4. Physiological and Biochemical Parameters Important for Scaling up *In Vitro* to *In Vivo* Intrinsic Clearance and for Estimating Hepatic Clearance in Rats[a]

Physiological parameters	Published values	Scaling factors[b]
Liver weight	45 g liver/kg body weight[c]	—
Hepatic blood flow rate	81 ml/min/kg body weight[c]	—
Hepatocyte number	135×10^6 cells/g liver	6100
Microsomal protein	45 mg proteins/g liver	2000
	54 mg protein/g liver	2400
P450 contents in microsomes	0.98 nmol/mg microsomal protein[d]	—

[a]Data taken from Bäärnhielm *et al.* (1986) and Houston (1994).
[b]Scaling factors for estimation of *in vivo* intrinsic hepatic clearance in rat normalized by kg body weight (ml/min/kg body weight), when *in vitro* intrinsic hepatic clearance is expressed as ml/min/mg microsomal protein (or ml/min 10^6 hepatocytes).
[c]Assuming that an average body weight of rat is 0.25 kg.
[d]Bäärnhielm *et al.*, 1986.

where the scaling factor is 52.5 mg microsomal protein/g liver multiplied by 25.7 g liver/kg body weight in humans which is equal to 1350 mg microsomal protein/kg body weight. Thus, the unit of $Cl_{i,h}$ becomes ml/min/kg body weight.

12.2.1.3. *Utilization of Hepatic Clearance Models to Estimate In Vivo Hepatic Clearance of a Drug*

Estimates of *in vivo* hepatic clearance (Cl_h) can be achieved by incorporating estimates of $Cl_{i,h}$, Q_h, and f_u into the hepatic clearance model(s). For instance, if 90% of a drug is bound to blood components ($f_u = 0.1$) and $Cl_{i,h}$ is estimated at 27 ml/min/kg body weight, Cl_h of the drug in humans *in vivo* is then 2.66 or 2.53 ml/min/kg body weight, according to the well-stirred or parallel-tube models for hepatic clearance, respectively.

Table 12.5. Concentration of Cytochrome P450 and the Amount of Microsomal Protein per Gram of Liver in Rats, Dogs, and Humans (Mean ±SD)

Species	Cytochrome P450 (nmol/mg protein)	Microsomal protein concentration (mg/g liver)	Liver weight (g/kg body weight)
Rat	0.980 ± 0.10[a]	54[a], 45[b]	42.4[a], 45[b]
Dog	0.474 ± 0.080[a]	43[a]	25.6[a]
	0.78 ± 0.08[c]	—	—
Human	0.296 ± 0.093[a], 0.32[d]	77[a], 52.5[d]	20.2[a]

[a]Data taken from Bäärnhielm *et al.* (1986).
[b]Data taken from Houston (1994).
[c]Data taken from Duignan *et al.* (1987).
[d]Data taken from Iwatsubo *et al.* (1997).

WELL-STIRRED MODEL:

$$\overset{ml/min/kg\ body\ weight}{} \qquad \overset{no\ unit}{} \qquad \overset{ml/min/kg\ body\ weight}{}$$

$$Cl_h = \frac{Q_h \cdot f_u \cdot Cl_{i,h}}{Q_h + f_u \cdot Cl_{i,h}} = \frac{20.6 \times 0.1 \times 27}{20.6 + 0.1 \times 27}$$

$$= 2.66 \text{ ml/min/kg body weight}$$

PARALLEL-TUBE MODEL:

$$Cl_h = Q_h \cdot (1 - e^{-f_u \cdot Cl_{i,h}/Q_h}) = 2.53 \text{ ml/min/kg body weight}$$

Differences in the estimates of Cl_h from these two models are less important for low-clearance drugs, i.e., drugs with $f_u \cdot Cl_{i,h} \ll Q_h$. When the extraction ratio of a drug is greater than 0.7, the difference in the clearance estimate between the models becomes more apparent. Despite its assumptions and simplifications, the well-stirred model, the least sophisticated of the hepatic clearance models, appears to suffice for extrapolation of *in vitro* metabolism data to *in vivo* clearance. Similarly, the simple Michaelis–Menten kinetics is considered to be adequate for describing kinetic behaviors of metabolizing enzymes, despite the known complexities of enzyme families and reaction patterns.

12.2.1.4. Comparison of Various In Vitro Methods for Predicting In Vivo Hepatic Clearance

In general, primary hepatocytes (freshly isolated hepatocytes) seem to provide more reliable *in vitro* predictions of Cl_h than liver microsomes or slices. Studies using liver microsomes or slices tend to underestimate Cl_h as compared to hepatocytes. It has been found that the rate of drug metabolism in hepatocytes is approximately three times as fast as that in microsomes when normalized by microsomal protein concentration (Houston and Carlie, 1997; Miners *et al.*, 1994). The slower metabolism rate in microsomes can be the result of various factors. For instance,

1. Limited metabolic capability of microsomes: Hepatocytes can perform both phase I and phase II metabolism, whereas microsomes have phase I metabolism or only limited phase II metabolism such as UDP-glucuronosyl transferase activity depending on reaction conditions.
2. Damage to microsomes during preparation: Structural integrity of microsomes may be damaged to a certain extent during isolation owing to the disruptive nature of the preparation processes, and subsequent reestablishment of study conditions for metabolism may not be physiologically optimal.
3. Inactivation of enzymatic activities in microsomes: When the half-life of drug disappearance is used to estimate $Cl_{i,\,in\ vitro}$ rather than the Michaelis–Menten kinetics for separate V_{max} and K_m estimates, liver microsomes may not be able to maintain full activity over a sufficiently long incubation period for an accurate measurement of the half-life.

4. Nonspecific binding of drug to microsomal proteins: In microsomes, any nonspecific binding of a drug to microsomal proteins (usually 10 to 60% in rat liver microsomes at 1 mg microsomal protein) may cause the concentration of the drug available to the enzyme to be lower than that initially added in the reaction medium. On the other hand, the extent of nonspecific binding of a drug to intracellular protein in hepatocytes may be lower than in microsomal preparations, when normalized by enzyme activities.

Comparison between microsomes and hepatocytes for prediction in vivo clearance: Extrapolations from studies with hepatocytes tend to be more reliable than those from microsomes. For low-clearance drugs ($Cl_{i,\, in\, vitro} < 20\ \mu l/min/mg$ microsomal protein), prediction is reasonable; however, for high-clearance drugs ($Cl_{i,\, in\, vitro} > 100\ \mu l/min/mg$ microsomal protein), predicted values of Cl_h based on microsomal data are substantially lower than those from the corresponding hepatocyte data or measured Cl_h.

Using liver slices for predicting in vivo clearance: Liver slices have not been used as extensively as microsomes or primary hepatocytes for predicting Cl_h for the following reasons:

1. Difficulties in preservation and limited availability of fresh liver slices: Only limited success in cryopreservation of slices.
2. Lack of population approach: Difficulties in phenotyping liver slices from different individuals prior to the studies.
3. Diffusion (transport)-limited metabolism: Because of the limited diffusibility of a drug across the multilayers of hepatocytes in slices (approximately five layers of hepatocytes in about 200-μm-thick slices), drug metabolism takes place mainly in the outer layers of a liver slice. In other words, drug availability to hepatocytes in liver slices is rather limited as compared to that in isolated hepatocytes.

12.1.2.5. Effects of Nonspecific Binding of a Drug on In Vitro Metabolism

Michaelis–Menten kinetics for enzyme reactions such as metabolism is based on the interaction between *unbound* drug and *unbound* enzyme molecules in the medium, forming enzyme–drug intermediates, which then break into metabolites and *unbound* enzymes:

Unbound drug molecule (D) *interacting with unbound enzyme* (E)	*Forming enzyme-drug intermediate*	*Producing metabolite* (M) *and releasing* E

$$D + E \underset{k_{-1}}{\overset{k_1}{\rightleftharpoons}} E - D \xrightarrow{k_2} E + M$$

The initial rate of drug disappearance according to Michaelis–Menten kinetics [Eq. (12.10)] is a function of the *unbound*, not the total, drug concentration actually

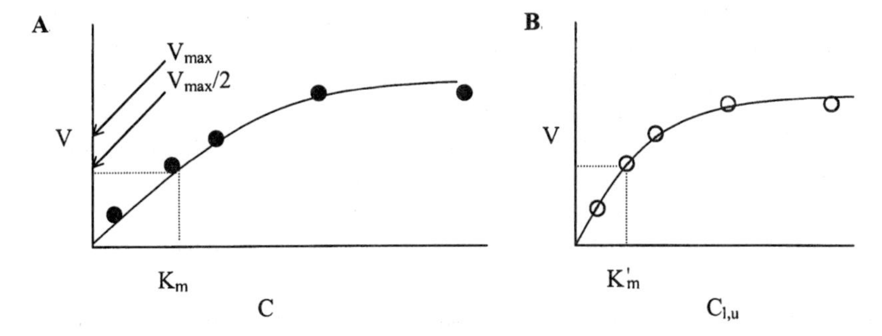

Figure 12.6. Differences in Michaelis–Menten constants obtained based on the relationship between initial disappearance rates (V) of a drug and its total (C, ●) or unbound ($C_{l,u}$, ○) concentrations in liver microsomes. K_m and K'_m are indicated as total (A) and unbound (B) drug concentrations, at which V is half of V_{max}. V_{max} is independent of nonspecific binding of the drug to microsomal proteins.

available to the metabolizing enzymes. K_m is equal to the *unbound* drug concentration in the medium at which the initial rate of drug disappearance is half of V_{max}, and thus should be determined from the relationship between the initial rate of drug disappearance and the unbound drug concentrations in the medium. V_{max} should not be affected by the extent of drug binding in the medium (Bäärnhielm *et al.*, 1986; Ito *et al.*, 1998; Obach, 1997).

For instance, when metabolism is studied in microsomes, the total concentration of the drug added to the medium may not be equal to its unbound concentration owing to the nonspecific binding to microsomal proteins. In fact, the values of K_m estimated based on total drug concentrations in the reaction medium can be significantly greater than that based on the unbound drug concentrations (Fig. 12.6). The K_m value corrected for the extent of nonspecific protein binding (K'_m) can be obtained by multiplying the value estimated based on total concentrations (i.e., K_m) by the ratio of unbound and total drug concentrations in the microsomes ($f_{u,m}$):

$$K'_m = f_{u,m} \cdot K_m \tag{12.21}$$

$f_{u,m}$ can be estimated through protein binding studies such as equilibrium dialysis without cofactors at a single drug concentration in a reaction medium containing microsomes, if the extent of protein binding is independent of drug concentration. Nonspecific binding to microsomal proteins has been found to be 10 to 60%, i.e., $f_{u,m}$ of 0.4 to 0.9, at a microsomal protein concentration of 1 mg/ml for several drugs.

For drugs with extensive nonspecific binding, the estimate for $Cl_{i,\, in\ vitro}$ using K_m can be too low. If extensive nonspecific binding of a drug is expected, it is important to consider $f_{u,m}$ for estimating the true $Cl_{i,\, in\ vitro}$:

$$Cl_{i,\, in\ vitro} = \frac{V_{max}}{K'_m} = \frac{V_{max}}{f_{u,m} \cdot K_m} \tag{12.22}$$

Discrepancies between the predictions from in vitro data and in vivo findings: It is not uncommon to find that hepatic clearance predicted on the basis of *in vitro* data can under- or overestimate the true *in vivo* hepatic clearance (Houston and Carlie, 1997). Several potential causes for these discrepancies are as follows:

1. Underestimation of *In Vivo Hepatic Clearance.* In many cases, *in vivo* hepatic clearance (Cl_h) predicted from $Cl_{i,\ in\ vitro}$ tends to be lower than the measured Cl_h, especially for high-clearance drugs ($Cl_{i,\ in\ vitro} > 100\ \mu l/min/mg$ protein). This discrepancy might be due to the following limitations of *in vitro* experiments:

- Limited metabolic activities: This becomes more apparent in liver microsome studies.
- Product inhibition by metabolites on enzyme activities: This is of concern especially in microsome studies owing to the lack of active phase II conjugating enzymes.
- Variability of *in vitro* systems (sample collection, storage condition, etc): The disruptive nature of the microsomal preparation can result in the loss of activities or variability in activities of different isoforms. In most cases, approximately 20% of the primary hepatocytes isolated using current methods are not viable.
- Poor penetration of drug into multilayers of hepatocytes in liver slices: Owing to poor penetration of the drug, drug metabolism takes place mainly in hepatocytes in the outer layer of liver slices.
- Nonspecific binding of a drug to microsomal proteins and/or experimental apparatus: Since only unbound drug is subject to metabolism, nonspecific binding of a drug in *in vitro* experiments can reduce the amount of drug available for metabolism, resulting in apparently lower $Cl_{i,\ in\ vitro}$ compared to $Cl_{i,h}$ when normalized by protein concentration.
- Inability to predict biliary clearance in *in vitro* metabolic systems.

2. Overestimation of *In Vivo* Hepatic Clearance: Occasionally, a predicted Cl_h based on $Cl_{i,\ in\ vitro}$ is higher than the measured Cl_h, which can be due to:

- A slow equilibrium of drug between sinusoidal blood and hepatocytes: When intrinsic hepatic clearance (metabolism and biliary excretion) of a drug is much faster than the efflux clearance from the hepatocytes to the sinusoidal blood, the apparent $Cl_{i,h}$ becomes governed by drug uptake from the sinusoidal blood into the hepatocytes. In this case, $Cl_{i,\ in\ vitro}$ estimated in liver microsomes, which does not account for a slow drug uptake from the blood into the hepatocytes *in vivo*, can lead to an overestimate of Cl_h (Kwon and Morris, 1997).

3. Other Factors:

- The presence of active transporters in the sinusoidal membranes of hepatocytes: When a drug is subject to active transporters located in the sinusoidal membranes, its unbound drug concentration within the hepatocytes can be different than it is in blood. The effects of active drug transport on Cl_h are often difficult to assess

Table 12.6. Known Standard Values for Volumes of Different Body Fluids in Animals and Humans

	Volumes (liters/kg)[a]		
Species	V_p	V_e	V_r
Rat, guinea pig	0.031	0.266	0.371
Rabbit	0.044	0.206	0.466
Monkey	0.045	0.163	0.485
Dog	0.052	0.225	0.328
Human	0.043	0.217	0.340

[a] V_p: plasma volume; V_e: extracellular fluid volume minus plasma volume; and V_r: physical volume into which the drug distributes minus the extracellular space (intracellular fluid volume).

solely from *in vitro* data and further experiments, such as liver perfusion, are needed.

- Interindividual variability in metabolism: Genetic polymorphism and environmental factors can cause significant variability in hepatic metabolic ability and capacity among different individuals. It is, therefore, often difficult to predict Cl_h representing population from *in vitro* experiments with liver samples obtained from a few individuals.

12.2.2. Predicting the Volume of Distribution of a Drug in Humans

Equation (12.23) developed by Φie and Tozer can be used to predict the volume of drug distribution at steady state in humans using the known values for actual physiological volumes and the fractions of drug not bound to plasma components determined experimentally (Φie and Tozer, 1979). An estimate of the ratio of unbound and total drug concentrations in intracellular space (f_{ut}) of a drug in humans can be obtained by averaging f_{ut} values calculated from at least two different species [Eq. (12.24)], assuming that f_{ut} is similar across species. The known standard values for V_p, V_e, and V_r in humans are shown in Table 12.6.

$$V_{ss} = (2.4) \cdot V_p + f_u \cdot V_p \cdot [(V_e/V_p) - 1.4] + (f_u/f_{ut}) \cdot V_r \qquad (12.23)$$

$$f_{ut} = \frac{f_u \cdot V_r}{V_{ss} - (2.4) \cdot V_p - f_u \cdot V_p \cdot [(V_e/V_p) - 1.4]} \qquad (12.24)$$

f_u is the ratio of unbound and total drug concentrations in plasma, f_{ut} is the ratio of unbound and total drug concentrations in intracellular space (tissues), V_p is the plasma volume (liters/kg), V_e is the extracellular fluid volume minus the plasma volume (liters/kg), V_r is the physical volume into which the drug distributes minus the extracellular space (liters/kg), and V_{ss} is the volume of distribution at steady state (liters/kg).

REFERENCES

Bäärnhielm C. *et al.*, *In vivo* pharmacokinetics of felodipine predicted from *in vitro* studies in rat, dog and man, *Acta. Pharmacol. et Toxicol.* **59**: 113–122, 1986.

Boxenbaum H., Interspecies scaling, allometry, physiological time and ground plan of pharmacokinetics, *J. Pharmacokinet. Biopharm.* **10**: 201–227, 1982.

Boxenbaum H. and D'Souza R. W., Interspecies pharmacokinetic scaling, biological design and neoteny. *Adv. Drug Res.* **19**: 139–196, 1990.

Calabriese E. J., Animal extrapolation and the challenge of human heterogeneity, *J. Pharm. Sci.* **75**: 1041–1046, 1986.

Chenery R. J. *et al.*, Antipyrine metabolism in cultured rat hepatocytes, *Biochem. Pharmacol.* **36**: 3077–3081, 1987.

Davies B. and Morris T., Physiological parameters in laboratory animals and humans. *Pharm. Res.* **10**: 1093–1095, 1993.

Dedrick R. L., Animal scale-up, *J. Pharmacokinet. Biopharm.* **1**: 435–461, 1973.

Duignan D. B. *et al.*, Purification and characterization of the dog hepatic cytochrome P450 isozyme responsible for the metabolism of 2,2′,4,4′,5,5′-hexachlorobiphenyl, *Arch. Biochem. Biophys.* **255**: 290–303, 1987.

Houston J. B., Utility of *in vitro* drug metabolism data in predicting *in vivo* metabolic clearance, *Biochem. Pharmacol.* **47**: 1469–1479, 1994.

Houston J. B. and Carlie D. J., Prediction of hepatic clearance from microsomes, hepatocytes, and liver slices, *Drug Metab. Rev.* **29**: 891–922, 1997.

Ings R. M. J., Interspecies scaling and comparisons in drug development and toxicokinetics, Xenobiotica **20**: 1201–1231, 1990.

Ito K. *et al.*, Quantitative prediction of *in vivo* drug clearance and drug interactions from *in vitro* data on metabolism, together with binding and transport, *Ann. Rev. Pharmacol. Toxicol.* **38**: 461–499, 1998.

Iwatsubo T. *et al.*, Prediction of *in vivo* drug metabolism in the human liver from *in vitro* metabolism data, *Pharmacol. Ther.* **73**: 147–171, 1997.

Kwon Y. and Morris M. E., Membrane transport in hepatic clearance of drugs. I: Extended hepatic clearance models incorporating concentration-dependent transport and elimination processes, *Pharm. Res.* **14**: 774–779, 1997.

Mahamood I. and Balian J. D., Interspecies scaling: a comparative study for the prediction of clearance and volume using two or more than two species. *Life Sci.* **59**: 579–585, 1996*a*.

Mahamood I. and Balian J. D., Interspecies scaling: predicting clearance of drugs in humans: three different approaches, *Xenobiotica* **26**: 887–895, 1996*b*.

Miners J. O. *et al.*, *In vitro* approaches for the prediction of human drug metabolism, *Ann. Rep. Med. Chem.* **29**: 307–316, 1994.

Mordenti J., Forecasting cephalosporin and monobactam antibiotic half-lives in humans from data collected in laboratory animals, *Antimicrob. Agents Chemother.* **27**: 887–891, 1985.

Mordenti J., Man versus beast: pharmacokinetic scaling in mammals, *J. Pharm. Sci.* **75**: 1028–1040, 1986.

Obach R. S., Nonspecific binding to microsomes: impact on scale-up of *in vitro* intrinsic clearance to hepatic clearance as assessed through examination of warfarin, imipramine, and propranolol, *Drug Metab. Dispos.* **25**: 1359–1369, 1997.

Φie S. and Tozer T. N., Effect of altered plasma protein binding on apparent volume of distribution, *J. Pharm. Sci.* **68**: 1203–1205, 1979.

Rane A. *et al.*, Prediction of hepatic extraction ratio from *in vitro* measurement of intrinsic clearance, *J. Pharmacol. Exp. Ther.* **200**: 420–424, 1977.

Ritschel W. A. *et al.*, The allometric approach for interspecies scaling of pharmacokinetic parameters, *Comp. Biochem. Physiol.* **103C**: 249–253, 1992.

Sacher G. A., Relationship of life span to brain weight and body weight in mammals, in G. E. W. Wolstenholm and M. O'Connor (eds.), *CIBA Foundation Colloquium on Aging*, Churchill, London, 1959, pp. 115–133.

13

Animal Physiology

A thorough understanding of animal physiology is a prerequisite for conducting pharmacokinetic studies and interpreting data in an appropriate way. This chapter presents data relating to animal physiology and recommendations for dosing in common laboratory animals. Table 13.1 summarizes physiological and biochemical parameters in animals, which can be used when designing studies in pharmacokinetics and metabolism with various laboratory animals and humans, and interpreting the results. Tables 13.2 and 13.3 summarize the information on drug administration in laboratory animals in terms of needle size and maximum dosing volume depending on the route of administration. Table 13.4 details the physiology of the rat, which is the most commonly used species in pharmacokinetic studies, and Table 13.5 can be used as a guideline for anesthesia in rats. Figure 13.1 illustrates the chemical composition of extracellular and intracellular fluids and the physiological differences between them, which can be of help in understanding the pharmacokinetic behavior of endogenous and exogenous substances at the cellular level.

Table 13.1. Important Physiological and Biochemical Parameters in Laboratory Animals and Humans

Species	Mouse	Hamster	Rat	Guinea pig	Rabbit	Cat	Monkey	Dog	Man	Note[a]
Body weight (kg, male)	0.02–0.06	0.11–0.14	0.25–0.4	0.6–1	2.5–6	2–4	5	10	70	1
Body surface area (m^2)	0.008	—	0.023	—	0.17	—	0.32	0.51	1.85	2
	0.004	0.026	0.033	0.06–0.07	0.13–0.30	—	—	—	—	3
Body temperature (°C)	36.5–38.0	37–38	37.5–38.5	38–40	38.5–39.5	38–39.5	36–40	38–39	37.0	3
Daily consumption										
Food (g/kg)	100–200	100	100	30–50	25–50	40–80	50–60	25	—	3
Water (ml/kg)	150	80–100	100–120	100	60	50	—	25–40	—	
Gestation (days)	17–21	15–16	20–23	58–75	30–35	58–71	150–183	53–71	—	
	avg. 19	avg. 15.5	avg. 21	avg. 68	avg. 31	avg. 65	avg. 165	avg. 63	—	
Birth weight (g)	0.5–1.5	2–3	5	70–100	30–100	90–130	450–500	200–500	—	3
Life span (yr)	1–2	2–3	2–3	5–6	5–6	10–17	20–30	10–15	—	3
Organ weight (g)										
Adrenals	0.004	0.027	0.05	—	0.5	—	1.2	1	14	2, 4, 5
Brain	0.36	—	1.8	—	14	—	90	80	1400	italic for
Fat	—	—	—	—	—	—	—	—	10,000	organ
Heart	0.08	0.46	1.0	—	5	—	18.5	80	330	volume
Intestine	*1.5*	*12.23*	*11.25*	—	*120*	—	*230*	*480*	2100	(ml)
Kidneys	0.32	0.96	2.0	—	13	—	25	50	310	
Liver	1.75	6.05	10.0	—	77	—	150	320	1800	
Lung	0.12	0.74	1.5	—	18	—	33	100	1000	
Marrow	*0.6*	—	—	—	*47*	—	*135*	*120*	*1400*	
Spleen	0.1	*0.54*	0.75	—	1	—	8	25	180	
Body Fluids										
Blood (ml/kg)	74.5	72.0	58.0	74.0	69.4	84.6	75.0	92.6	77.8	2
Hematocrit (%)	45	37	46	48	36	44	41	42	44	
Blood pH	—	7.39	7.38	7.35	7.35	7.35	—	7.36	7.39	6
Plasma (ml/kg)	48.8	45.5	31.3	38.8	43.5	47.7	44.7	53.8	47.9	2

Plasma protein (mg/ml plasma)										
Plasma albumin	32.7	—	31.6	—	38.7	—	49.3	26.3	41.8	2, 6
Plasma α-1-ACG	12.5	—	18.1	—	1.3	—	2.4	3.7	1.8	
Total	62	—	67	47	57	—	88	90	74	
Cerebrospinal fluid										
Volume (μl/g tissue)	—	—	—	—	—	—	—	4.9	22.4	7
Flow (μl/min)	—	—	2.2	—	10.1	22	—	47	429	
Interstitial fluid (ml/kg)	—	—	—	—	—	—	—	—	157.1	8
Intracellular fluid (ml/kg)	—	—	—	—	—	—	—	—	400.0	
Total body water (ml/kg)	—	—	—	—	—	—	—	—	600.0	
Blood flow (ml/min)										
Adrenal	—	—	—	—	—	—	—	—	100	
Adipose	—	—	0.4	—	32	—	20	35	260	2, 5
Brain	—	—	1.3	—	—	—	72	45	700	italics for
Heart	0.28	*0.14*	3.9	—	16	—	60	54	240	plasma
Hepatic artery	0.35	—	2.0	—	37	—	51	79	300	flow rate
Intestine	1.5	*5.3*	7.5	—	111	—	125	216	1100	
Kidneys	1.3	*5.27*	9.2	—	80	—	138	216	1240	
Liver	1.8	*6.5*	13.8	62	177	—	218	309	1450	
Lung	—	*28.4*	—	—	—	—	—	—	1400	
Marrow	*0.17*	—	—	—	*11*	—	*23*	*20*	*120*	
Muscle	0.91	—	7.5	—	155	—	90	250	750	
Portal vein	1.45	—	9.8	—	140	—	167	230	1150	
Skin	0.41	—	5.8	—	—	—	54	100	300	
Spleen	0.09	*0.25*	0.63	—	9	—	21	25	77	
Cardiac output	8.0	—	74.0	—	530	—	1086	1200	5600	
Blood in tissue (μl/g tissue)										
Adrenal gland	30	—	—	—	—	—	—	—	—	9
Bone	110	—	45	—	—	—	—	—	—	

Table 13.1 (Cont')

Table 13.1. Continued

Species	Mouse	Hamster	Rat	Guinea pig	Rabbit	Cat	Monkey	Dog	Man	Note[a]
Blood in tissue										
Bone marrow	—	—	—	55	—	—	—	—	—	
Brain	30	—	11	—	—	30	—	11	—	
Heart	—	—	60	—	—	84	—	66	—	
Intestine	90	—	28	—	—	—	—	41	—	
Kidney	340	—	92	—	—	93	—	81	—	
Liver	360	—	99	—	—	52	—	147	—	
Lung	490	—	111	—	—	147	—	301	—	
Skeletal muscle	30	—	4	—	—	27	—	11	—	
Skin	30	—	20	—	70	—	—	—	—	
Spleen	170	—	86	133	—	195	—	510	—	
Testis	60	—	6	—	—	—	—	—	—	
Heart rate (beats/min)	300–800	250–500	300–500	230–380	130–325	100–120	100–150	80–150	—	3
Bile flow	100	90	90	230	120	—	25	12	5	10
(ml/kg/day)	—	—	48–92	—	130	—	19–32	19–36	2.2–22.2	11
Bile pH	—	—	—	—	—	—	—	—	7.4–8.5	(hepatic duct bile)
									5.4–6.9	(gall bladder bile)
Bile concentration (relative to human)	0.05	0.06	0.6	0.02	0.04	—	0.2	0.23	1	10
Urine flow (ml/kg/day)	50	—	200	—	60	—	75	30	20	10
Urine concentration (relative to human)	0.4	—	0.1	—	0.34	—	0.26	0.66	1	10
Urine pH	—	—	—	—	—	—	—	—	6.3, 4.5–8.0	12
GFR (ml/min)	0.28	—	1.31	—	7.8	—	10.4	61.3	125	
	0.2	—	2.3	—	12	—	10.0	40	126	13
Number of glomeruli ($\times 10^5$/kg)	5.9	—	2.9	—	1.6	—	—	0.9	0.29	13

Lymph flow										
(ml/kg/day)										
Cervical duct	—	—	—	1.8	—	—	—	2.6	—	14
Heart	—	—	—	—	—	—	—	7.7	—	
Intestine	—	—	96	—	—	37	—	29	—	
Kidney	—	—	—	—	—	—	—	0.2, 2	—	
Leg	—	—	—	—	—	—	—	2.6	—	
Liver	—	—	7.7	—	—	18, 14	—	26, 36	—	
Lung	—	—	—	—	—	—	—	7, 10	—	
Right lymph duct	—	—	—	—	—	—	—	15, 21	—	
Thoracic duct	960	—	40, 96	39	30, 88	8, 40	—	55, 132	17, 144	
Thyroid gland	—	—	—	—	—	—	—	1.8	—	
Gastrointestinal (GI) tract										
Half gastric emptying	—	—	—	—	—	—	—	4–5	8–15	11, 15, 16
time (min)			(<15)					(180)	(77–130)	
Intestinal transit										
time (min)										
Small intestine	—	—	88	—	—	—	—	111	238	2, 15, 17
									(275, >600)	
Whole intestine	—	—	—	—	—	—	—	770	2350	2
								360–480	1200–1800	11
Intestine length (m)										
Small intestine	0.04	—	0.1–0.15	—	3.56	—	—	4.14	7	11, 18
Large intestine	—	—	0.02–0.03	—	2.26	—	—	0.74	—	
Intestine volume (ml)										
Whole intestine	1.5	12.23	11.25	—	120	—	230	480	2100	5
Gut lumen	1.5	—	8.8	—	—	—	230	—	2100	
Capacity (liters)										
Stomach volume	—	—	—	—	—	0.34	0.1	1	1–1.6	11
							(0.008)		(0.024)	basal
Small intestine	—	—	—	—	—	0.11	—	—	—	
Large intestine	—	—	—	—	—	0.12	—	—	—	
GI pH										
Saliva	—	—	—	—	—	—	—	—	6.0–7.0	19
Gastric juice	—	—	—	—	—	—	—	—	1.0–3.5	

Table 13.1 (Cont')

Table 13.1. Continued

Species	Mouse	Hamster	Rat	Guinea pig	Rabbit	Cat	Monkey	Dog	Man	Note[a]
GI pH										
Stomach	3.1–4.5	2.9–6.9	3.0–3.8	—	—	—	—	0.9–2.5	1.3–2.1	15, 20, 21
(fasted)	—	—	—	—	—	—	—	—	1.5–3.5	
(postprandial)	—	—	2.3–4.5	—	—	—	—	0.5–5.0	2.5–7.5	11, 22
Intestine	—	6.1–7.1	—	—	6.0–8.0	—	—	6.5–7.5	5.5–6.5	15, 20, 23
(fasted)	—	—	6.9–7.8	—	—	—	—	—	5–7	11, 22
(postprandial)	—	—	—	—	—	—	—	2–7	5–6	15
Pancreatic juice	—	—	—	—	—	—	—	—	8.0–8.3	19
Small intestinal secretion	—	—	—	—	—	—	—	—	7.5–8.0	
Large intestinal secretion	—	—	—	—	—	—	—	—	7.5–8.0	
Bile	—	—	—	—	—	—	—	—	7.8	
Feces	—	—	6.9	—	7.2	—	—	—	7.0–7.5	20, 24
Daily secretion (ml/kg/day)										
Saliva	—	—	—	—	—	—	—	—	14	19
Gastric juice	—	—	—	—	—	—	—	—	21	
Pancreatic juice	—	—	—	—	—	—	—	—	14	
Bile	—	—	—	—	—	—	—	—	14	
Small intestinal secretion	—	—	—	—	—	—	—	—	26	
Large intestinal secretion	—	—	—	—	—	—	—	—	2.9	
Metabolic Activities										
β-glucuronidase activity (nmol/hr/g content)										
Liver	2000–4000	2200	15,000–30,000	4500	5000	—	—	—	3000	25

Kidney	1000–4500	1100	4000–6000	600–1700	300	—	—	—	2000	
Lung	—	—	5000	—	—	—	—	—	500	
Spleen	4000–11000	—	15,000–30,000	5500	—	—	—	—	6500	
Brain	100	—	150	300	150	—	—	—	Trace	
Stomach	—	—	—	—	40	—	—	—	200–1000	
Small intestine	—	—	3000–5000	—	800	—	—	—	—	
Small intentinal content	1200	—	304	2.7	2.4	—	—	—	0.02	11, 26
	(5015)	—	(1341)	(139)	(45.4)	—	—	—	(0.9)	
Large intestine	—	—	3500	—	300	—	—	—	—	25
Large intestinal content	—	—	3000	—	2000	—	—	—	—	
Cytochrome P450	—	—	0.98	—	—	—	—	0.474	0.296	27
(nmol/mg protein)	—	—	(54)	—	—	—	—	(43)	(77)	

[a]Notes: (1) Average body weight for adult male animals. (2) Davies B. and Morris T., 1993. (3) Havenaar R. *et al.*, 1993. (4) Frank D. W., 1944. (5) Gerlowke L. E. and Jain R. K., 1983. (6) Altman P. L. and Dittmer D. S., 1971*a*. (7) Altman P. L. and Dittmer D. S., 1971*b*. (8) For an average man with 70 kg body weight, the plasma volume is 3 liters, the blood volume is 5.5 liters, the extracellular fluid outside the plasma (interstitial fluid) is 11 liters, the intracellular fluid is about 28 liters, and the total body water is approximately 42 liters (Benet L. Z. and Zia-Amirhosseini P., 1995). (9) Altman P. L. and Dittmer D. S., 1971*c*. (10) Clark B. and Smith D. A., 1982. (11) Kararli T. T., 1995. (12) Guyton A. C. and Hall J. E., 1996*a*. (13) Lin J. H., 1995. (14) Altman P. L. and Dittmer D. S., 1971*d*. (15) Dressman J. B., 1986. (16) Gastric half-emptying time after ingestion of water or normal saline (solid meal). In some studies in human, half-emptying time of the order of 30 min after a small meal has been reported, whereas after a large meal, it lasted up to 180 min. (17) Mean intestinal transit time of a Heidelberg capsule under fasting condition (tablet with heavy meal). There is more interindividual variation in transit time in dogs (15 to 206 min) than in humans (180 to 300 min), suggesting that drug absorption may be more variable and incomplete in dogs than in humans. (18) The postmortem length of the organ without fixation. (19) Guyton A. C. and Hall J. E., 1996*b*. (20) Ilett K. F., 1990. (21) The fasted gastric pH. In humans, postprandial gastric pH is initially higher (up to 7) owing to the buffering effect of food than the fasted gastric pH; then as gastric acid is secreted in response to eating, the pH gradually decreases to premeal values over a period of 60 to 90 min. In dogs, the initial buffering effect of food is not observed with more variability in postprandial pH (0.5 to 3–5, mean of 2.1). (22) pH in Wistar rats under fed condition. (23) Intestinal pH is consistently 1 unit higher in dogs than in humans at the same time of observation. (24) Chang R., 1981. (25) Measured as phenolphthalein produced from phenolphthalein β-glucuronidase in 1 hr at 38°C, and expressed as μg/g of moist tissue (Calabrese E. J., 1986). (26) Nanomoles of phenolphthalein-β-glucuronide deconjugated by bacterial β-glucuronidase present in proximal (distal) small intestine. (27) Numbers in parenthesis indicate mg microsomal protein content per g liver (Bäärnhielm C. *et al.*, 1986.

Table 13.2. Recommended Needle Size and Maximum Volume for Different Routes of Administration in Laboratory Animals[a]

		Needle size/maximum volume (ml)[b]				
	Weight	PO[c]	IV	IM	IP	SC
Mouse	20 g	1.0/0.5	25G/0.3	26G/0.05	26G/1	25G/0.5
Hamster	100 g	1.0/1	25G/0.5	25G/0.1	25G/3	25G/1
Rat[d]	250 g	2.0/2.5	25G/0.5[e]	25G/0.2	24G/5	24G/2
Guinea pig	600 g	2.0/4	24G/0.5	24G/0.3	24G/10	22G/3
Rabbit	4 kg	5.0/10	22G/5	23G/1	21G/50	21G/20
Cat	3 kg	–/10	22G/5	23G/1	21G/30	21G/20
Monkey	5 kg	–/10	22G/5	23G/1	21G/30	21G/20
Dog	10 kg	–/20	22G/10	21G/2	21G/100	20G/50

[a] Data taken from Fleckwell (1995), Iwarsson *et al.* (1994), Zwart (1993), and Sharp and LaRegina (1998).
[b] If compounds injected are irritants, much smaller volume should be used.
[c] Blunt cannula (oral gavage); firm restraint; vertical posture, pass tube along palate into esophagus (diameter of gavage (mm)/maximum volume).
[d] Intratracheal 40 μl, intranasal 100 μl, or subplantar 100 μl/foot (usually inject in only one foot).
[e] Single injection volume should not exceed 10% of the circulating blood volume and continuous 24-hr intravenous infusion should be less than 4 ml/kg body weight/hr.

Cell membrane

	Extracellular fluid	Intracellular fluid
Na^+	140	14 mEq/l
K^+	4	140 mEq/l
Ca^{+2}	2.4	0.0001 mEq/l
Mg^{+2}	1.2	58 mEq/l
Cl^-	103	4 mEq/l
HCO_3^-	28	10 mEq/l mEq/l
HPO_4^{-2}	4	75 mEq/l
SO_4^{-2}	1	2 mEq/l
Glucose	90	0-20 gm/dl
Amino acids	30	200 mg/dl
Cholesterol, Neutral fat, Phospholipids	0.5	2-95 g/dl
pO_2	35	20 mmHg
pCO_2	46	50 mmHg
pH	7.4	7.0 (6.0-7.4)
Proteins	2	16 g/dl
Osmolarity	280	280 mOsm/l
Total osmotic pressure	5423	5423 mmHg

Figure 13.1. The chemical compositions of extracellular and intracellulaqr fluids and the physiological differences between them.

Table 13.3. Needle Sizes and the Corresponding Standard Wire Gauge[a]

Needle size (mm)[b]	0.35	0.45	0.55	0.7	0.9	1.25	1.65	2.10
Gauge (G)	28	26	24	22	20	18	16	14

[a]Data taken from Iwarsson *et al.* (1994).
[b]External diameter.

Table 13.4. Physiological Parameters of the Rat[a]

Basic biological parameters	
Gestation	21–23 days
Birth weight	5–6 g
Life span	2.5–3.5 years
Puberty	50 ± 10 days
Male body weight	450–520 g[b]
Female body weight	250–300 g[b]
Body temperature (rectal)	35.9–37.5°C
Body surface area	0.03–0.06 cm^2
Food intake	50–60 g/kg body weight/day
Water intake	100–120 ml/kg body weight/day
GI transit time	12–24 hr
Heart rate	330–480 beats/min
Cardiac output	10–80 ml/min (mean of 50 ml/min)
Body fluids	
Total body water	167 ml[b]
Intracellular fluid	92.8 ml[b]
Extracellular fluid	74.2 ml[b]
Blood volume	57.5–69.9 (mean of 64.1 ml/kg)
Plasma volume	36.3–45.3 ml/kg (mean of 40.4 ml/kg)
Red blood cell volume	36.3 ± 1.0 mg/kg
Red blood cell (RBC) count	$5–10 \times 10^9$/ml
White blood cell (WBC) count	$3–17 \times 10^6$/ml
Hemoglobin (HB)	110–190 mg/ml
Hematocrit	0.35–0.57
Plasma albumin	29–59 mg/ml
Cerebrospinal fluid (CSF) volume	250 ± 16 μl[c]
CSF formation rate	2.83 ± 0.18 μl/min (1.88 ± 0.17 μl/min[c])
CSF pressure	38 ± 4 mm Hg[c]
Urine volume	55 ml/kg body weight/day

[a]Data taken from Cocchetto and Bjornsson (1983) and Sharp and LaRegina (1998).
[b]Body weight may vary with stock or strain.
[c]30-day-old rats.

Table 13.5. Injectable Anesthetics Suitable for Surgical or Light Anesthesia in Rats[a]

Type	Anesthesia time (min)	Anesthetic agent	Dose (mg/kg)[b]	Sleep time (min)
Surgical	⩽5	Methohexital	10–15 IV	10
	⩽5	Propofol	10 IV	10
	⩽10	Thiopental	30 IP	15
	At least 20	Ketamine + xylazine	75–100 IP + 10 IP	120–240
Light	15	Pentobarbital	40–50 IV	120–240
	20–30	Ketamine + diazepam	75 IP + 5 IP	120
	20–30	Ketamine + midazolam	75 IP + 5 IP	120
	60	Chloral hydrate	300–400 IP	120–180
	—	Ketamine	100 IP	—
	—	Methohexitone	10 IV	—
	—	Thiopentone	30 IV	—
	—	Urethane	1000 IP	—

[a]Data taken from Sharp and LaRegina (1998).
[b]IP: intraperitoneal injection; IV: intravenous injection.

REFERENCES

Altman P. L. and Dittman D. S., *Respiration and Circulation*, Federation of American Society for Experimental Biology, Bethesda, 1971, (*a*) p. 225, (*b*) pp. 388–390, (*c*) pp. 383–385, (*d*) pp. 438–439.

Bäärnhielm C *et al.*, *In vivo* pharmacokinetics of felodipine predicted from *in vitro* studies in rat, dog and man, *Acta. Pharmacol. Toxicol.* **59**: 113–122, 1986.

Benet L. Z. and Zia-Amirhosseini P., Basic principles of pharmacokinetics, *Toxicol. Pathol.* **23**: 115–123, 1995.

Calabrese E. J., Animal extrapolation and the challenge of human heterogeneity, *J. Pharm. Sci.* **75**: 1041–1046, 1986.

Chang, R., *Physical Chemistry with Applications to Biological Systems, 2nd Ed.*, Macmillan, New York, 1981.

Clark B. and Smith D. A., Pharmacokinetics and toxicity testing, *CRC Drit. Rev. Toxicol.* **12**: 343–385, 1982.

Cocchetto D. M. and Bjornsson T. D., Methods for vascular access and collection of body fluids from the laboratory rat, *J. Pharm. Sci.* **72**: 465–492, 1983.

Davies B. and Morris, T., Physiological parameters in laboratory animals and humans, *Pharm. Res.* **10**: 1093–1095, 1993.

Dressman J. B., Comparison of canine and human gastrointestinal physiology, *Pharm. Res.* **3**: 123–131, 1986.

Flecknell P. A., Anaesthesia, in A. A. Tuffery (ed.), *Laboratory Animals: An Introduction for Experiments*, 2nd Ed., John Wiley & Sons, New York, 1995, (*a*) pp. 255–294, (*b*) pp. 324–325.

Frank D. W., Physiological data of laboratory animals, in E. C. Melby, Jr. and N. H. Altman (eds), *Handbook of Laboratory Animal Science, Vol. II.I*, CRC Press, Cleveland, 1974, pp. 23–64.

Gerlowke L. E. and Jain R. K., Physiologically based pharmacokinetic modeling: principles and applications, *J. Pharm. Sci.* **72**: 1103–1127, 1983.

Guyton A. C. and Hall J. E., *Textbook of Medical Physiology*, 9th Ed., W. B. Saunders Co., Philadelphia, (*a*) p. 386, (*b*) p. 817.

Havennar R. *et al.*, Biology and husbandry of laboratory animals, in L. F. M. van Zutphen, V. Baumans and A. C. Beynen (eds.), *Principles of Laboratory Animal Science: A Contribution to the Humane Use and Care of Animals and to the Quality of Experimental Results*, Elsevier, New York, 1993, pp. 17–74.

Ilett K. F., Metabolism of drugs and other xenobiotics in the gut lumen and wall, *Pharmacol. Ther.* **46**: 67–93, 1990.

Iwarsson K. *et al.*, Common non-surgical techniques and procedures, in P. Svendsen and J. Hau (eds.), *Handbook of Laboratory Animal Science, Vol. I; Selection and Handling of Animals in Biomedical Research*, CRC Press, London, (*a*) 1994, p. 233, (*b*) p. 231.

Karali T. T., Comparison of the gastrointestinal anatomy, physiology, and biochemistry of humans and commonly used laboratory animals, *Biopharm. Drug Dispos.* **16**: 351–380, 1995.

Lin J. H., Species similarities and differences in pharmacokinetics, *Drug Metab. Dispos.* **23**: 1008–1021, 1995.

Sharp P. E. and LaRegina M. C., *The Laboratory Rat*, CRC Press, New York, 1998; (*a*) p. 138, (*b*) pp. 1–19, (*c*) pp. 105–107.

Zwart P., Biology and husbandry of laboratory animals, in L. F. M. van Zutphen, V. Baumans and A. C. Beynen (eds.), *Principles of Laboratory Animal Science: A Contribution to the Humane Use and Care of Animals and to the Quality of Experimental Results*, Elsevier, New York, 1993, pp. 17–74.

Glossary

ABC (see ATP binding cassette).

Absolute bioavailability (see also Relative bioavailability): The fraction of a drug reaching systemic circulation upon extravascular administration as compared to the dose size of the drug administered intravenously. Absolute bioavailability after extravascular administration (F) can be estimated as follows:

$$F = \frac{AUC_{ex,0-\infty} \cdot D_{iv}}{AUC_{iv,0-\infty} \cdot D_{ex}}$$

where $AUC_{ex,0-\infty}$ and $AUC_{iv,0-\infty}$ are the AUC from time 0 to ∞ after extravascular or intravenous administration of drug, respectively; and D_{ex} and D_{iv}: Extravascular or intravenous doses of drug, respectively.

Accuracy of assay (see also Precision of assay): Relative error of assay. Accuracy of assay can be assessed by determining the experimental concentrations of the control samples by substituting detector responses into the regression equation. The relative differences between the experimental concentrations and theoretical (nominal) concentrations yield the relative error.

Active metabolite: A metabolite of a drug with significant pharmacological effects.

Active transport (see also Diffusion): A mechanism of drug transport across a membrane that is carrier-mediated and saturable, and requires consumption of energy. The net movement of a drug by active transport can be against its concentration gradient.

ADME: Pharmacokinetic profiles, i.e., absorption, distribution, metabolism, and excretion of a drug.

Akaiki information criterion (AIC): A statistical method of determining the appropriate pharmacokinetic model(s) for plasma exposure data of the test compound proposed by Akaike (Akaike H., A new look at the statistical model identification, *IEEE Trans. Automat. Control.* **19**: 716–723, 1974). Among different models, the one yielding the lowest AIC value (highest negative in the case of negative values) is the most appropriate model for describing the data:

$$\text{AIC value} = n \cdot \ln(\text{WSS}) + 2 \cdot m$$

where n and m are the number of data points and parameters used in the model, respectively; and WSS is the weighted sum of squares.

Albumin: The most abundant plasma protein (35–50 μg/ml plasma) with molecular weight of approximately 65,000. Acidic compounds commonly bind to albumin. The liver is the body's major producer of albumin.

Allele: One of two or more alternate forms of a gene occupying the same position (locus) in a particular chromosome and containing specific inheritable characteristics.

Allometry: The study designed to establish empirical relationships between the size, shape, surface area, and/or life span of animals and their consequences in regard to physiological functions of animals or pharmacokinetics of drugs without necessarily understanding the underlying mechanisms. The most frequently used allometric relationship in drug pharmacokinetics is

$$\boxed{Y = \alpha \cdot X^{\beta}}$$

where Y is the pharmacokinetic parameter of interest, e.g., clearance or volume of distribution, and X is the physiological parameter such as body weight. Estimates for the allometric coefficient (α) and the allometric exponent (β) can be obtained from an intercept and a slope of a log–log plot of the above equation, respectively:

$$\boxed{\log Y = \beta \cdot \log X + \log \alpha}$$

Alpha (α) phase (distribution phase): The initial portion of a plasma drug concentration *vs.* time profile after rapid intravenous injection. The decrease in drug concentration in plasma during this phase is usually steeper than in the later phase (β-, pseudoequilibrium, elimination, or terminal phase) mainly due to the distribution of drug molecules from the plasma pool into other tissues and organs in the body. Drug elimination may also occur during the alpha phase.

Amorphous drug form (see also Crystalline drug form): Solid drug particles without definite crystalline structure. In general, the amorphous form of a chemical is more soluble in water than its crystalline form, which can lead to more extensive absorption after oral administration than with the crystalline compound. However, the crystalline form is thermodynamically more stable than the amorphous form, so that in time the amorphous form will convert to the more stable crystalline form. Owing to this instability of the amorphous form, the crystalline form of a compound is preferred for manufacturing and quality control.

Antagonism: The apparent total effect of two different drugs after coadministration, which is less than the addition of the individual effects obtained from each drug (the additive effect) after separate administration of drugs.

APCI: Atmospheric pressure chemical ionization.

API: Atmospheric pressure ionization.

ATP binding cassette (ABC): The ATP binding cassette (ABC) family is one of the largest superfamilies of proteins among prokaryotes and eukaryotes. In eukaryotes, this family can

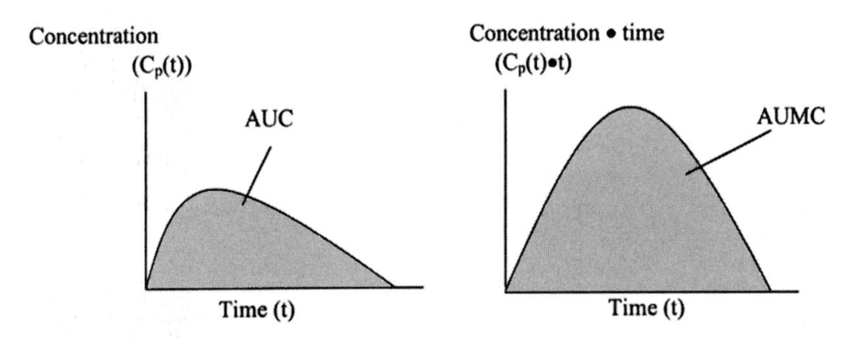

Figure G.1

be subdivided into four to six subclusters. Pharmacokinetically important ABC proteins include ATP-dependent active transporters including p-glycoprotein [multidrug resistance (*MDR*) gene products] and multidrug resistance-associated protein (MRP) responsible for active excretion of various organic cations and anions in the eliminating organs such as the liver.

AUC: The area under the plasma drug concentration *vs.* time curve. The unit of AUC is concentration multiplied by time (e.g., $\mu g \cdot hr/ml$).

AUMC: The area under the (first)-moment of a plasma drug concentration *vs.* time curve (area under the product of plasma drug concentration and time *vs.* time curve) on a linear scale. The unit of AUMC is concentration multiplied by time^2 (e.g., $\mu g \cdot hr^2/ml$) (Fig. G.1).

Autoinduction: A phenomenon in which a compound induces its own metabolism after single or multiple dosing. This can be due to induction and/or stabilization of the enzyme(s) metabolizing the compound. When a compound is subject to autoinduction, the extent of exposure such as AUC and/or C_{max} of the compound after multiple dosing is usually lower than that after single dosing.

BBB (see Blood Brain Barrier).

Beta (β) phase (pseudoequilibrium, postdistributive, elimination, or terminal phase): The later portion of plasma drug concentration *vs.* time profiles after rapid intravenous injection. The decrease in drug concentration in plasma during this phase is mainly due to the elimination of drug molecules from the plasma pool, and is usually shallower than in the earlier phase (alpha- or distribution phase). During the beta phase, the ratio between the total amount of drug present in the plasma and tissues remains constant.

Bioassay: Determination of the amount or pharmacological effectiveness of a biologically active substance by measuring the extent of its effects on a living organism.

Bioavailability (see also Absolute bioavailability and Relative bioavailability): The fraction of a drug reaching the systemic circulation unchanged following administration by any route. Since the availability of a drug in the systemic circulation after intravenous administration is usually unity, bioavailability (F) after extravascular administration, e.g., oral dosing, can be estimated as follows:

$$F = \frac{D_{iv} \cdot AUC_{po,0-\infty}}{D_{po} \cdot AUC_{iv,0-\infty}}$$

where D_{iv} and D_{po} are intravenous and oral doses, respectively; and $AUC_{iv,0-\infty}$ and $AUC_{po,0-\infty}$ are the areas under the plasma drug concentration *vs.* time curves after intravenous or oral administration, respectively, from time 0 to ∞.

Bioequivalence: Statistically comparable bioavailability (the extent as well as the rate of absorption) of a drug with different products or formulations at the same dose level. The same bioavailabilities after oral administration of two different formulations implies the same extent as well as the same rate of absorption. Therefore, even if two different formulations produce the same AUC after oral administration (the extent of absorption), the bioavailabilities of those formulations can be considered different (not bioequivalent), if the concentration–time profiles such as C_{max}, T_{max}, half-life, etc, are different from each other owing to the different rates of absorption.

Biophase: The site of action for the pharmacological effect of a drug in the body.

Biotransformation: The enzymatic or biochemical transformation of a drug in the body to other chemical forms, i.e., metabolites. In most cases, the biotransformation process of endogenous or exogenous compounds is irreversible. The term biotransformation is interchangeable with metabolism.

Blood brain barrier (BBB): The physiological and biochemical barrier between the blood and the brain that restricts endogenous and exogenous compounds in the blood from entering the brain by capillary endothelial cells with very tight junctions and no pinocytic vesicles. In general, lipid-soluble small molecules with a molecular weight under a 400–600 threshold are readily transported through the blood–brain barrier *in vivo* via lipid-mediated transport (Pardridge W. M., CNS drug design based on principles of blood–brain barrier transport, *J. Neurochem.* **70**: 1781–1792, 1998).

Bolus dose: An individual dose, usually given via rapid intravenous injection.

BSA: Bovine serum albumin.

Caco-2 cells: Cells originally derived from human adenocarcinoma colon cells. When grown on porous membranes, Caco-2 cells differentiate spontaneously to monolayers of polarized cells possessing the differentiated functions of intestinal enterocytes. The morphological and biochemical properties of the cells are closer to those of the small intestinal villus cells than to those of colonic cells, except for paracellular transport permeability, which is closer to that of the colonic epithelium. Owing to these characteristics, the Caco-2 cell line has been one of the most commonly used and extensively studied cell lines to assess membrane permeability of xenobiotics for intestinal absorption in humans.

Caplet: Capsule-shaped coated tablet.

Capsule: A solid dosage form in which the drug substance and other pharmaceutical adjuncts such as fillers are contained in either a soft or a hard shell, usually made of gelatin. In general, a drug substance from a capsule is released faster than from a tablet.

Cassette dosing (Cocktail dosing or N-in-1 dosing): Dosing a mixture of several compounds (N) in one dosing vehicle (1) to single animals as opposed to dosing individual compounds in one dosing vehicle to individual animals at a time.

Central compartment: The compartment representing the total plasma pool and well-perfused body organs and tissues for which drug concentration equilibrates instantaneously with that in plasma. It is assumed that any changes in the plasma drug concentration are quantitatively and qualitatively reflected by the drug concentration in the central compartment. The volume of the central compartment multiplied by the drug concentration in plasma at a given time indicates the total amount of drug present in the central compartment at that time. For many drugs, the volume of the central compartment is usually greater than the actual volume of the body's entire plasma pool.

Chelation: Intramolecular bonding or holding of a hydrogen or metal atom between atoms of a single molecule (e.g., chelation of iron with heme or chelation of calcium with EDTA).

Chirality (see also Enantiomer): Molecules that are not superimposable on their mirror images are called chiral. Chirality of a molecule is a necessary and sufficient condition for the existence of its enantiomer.

Chiral center (see also Enantiomer): A carbon atom to which four different groups are attached.

Chronopharmacology (chronergy): Rhythmic changes in pharmacological efficacy of drug as a function of time.

Chronopharmacokinetics: Rhythmic changes in pharmacokinetics of drug as a function of time.

Cirannual rhythm: A biological rhythm with a yearly cycle.

Circadian rhythm: A biological rhythm with a 24-hr cycle.

Clearance: The most general definition of clearance is the rate of elimination of a drug from the body (systemic clearance) or organ (organ clearance) normalized to its concentration in an appropriate reference body fluid such as plasma. Clearance can be also viewed as the apparent volume of the reference fluid completely cleared of the drug per unit time.

ClogP and **MlogP:** Log P values of a compound calculated based on the methods developed by the Medicinal Chemistry Department of Pomona College, CA, or by Moriguchi *et al.* (Moriguchi I. *et al.* Simple method of calculating octanol/water partition coefficient. *Chem. Pharm. Bull.* **40**: 127–130, 1992), respectively.

cMOAT [see also Multidrug resistance-associated protein (MRP)]: Canalicular multispecific organic anion transporter located in canalicular membranes of hepatocytes. cMOAT is believed to be MRP2.

Cocktail dosing (see Cassette dosing).

Compartment (see also Central compartment and Peripheral compartment): The imaginary space in the body representing a group of tissues, organs, and/or fluids that can be treated kinetically as a homogeneous unit. The kinetic homogeneity of a drug in the compartment does not necessarily imply that the drug concentration is equal in all the tissues within the compartment at any given time, rather that the times to reach distribution equilibrium between the plasma and each organ and tissue are similar.

Compartmental model: A pharmacokinetic approach to interpret the plasma drug concentration *vs.* time profile following drug administration, assuming that the body can be viewed as one or several different compartments. The number of compartments equals the number of exponential terms in a differential equation describing the plasma concentration–time profile. For instance, if the plasma concentration–time profile data are best fitted with a biexponential equation, a two-compartment model such as one shown in Fig. G.2 can be used.

Controlled-release dosage form (see also Sustained-release dosage form): A solid dosage form designed to release a drug substance over an extended period of time at a precisely controlled rate, usually zero-order rate, compared to the sustained release product. The advantage of controlled-release dosage forms is that they require less frequent drug administration than ordinary forms, yet maintain therapeutic drug concentrations, which leads to better patient compliance.

Creatinine (see also GFR): The endogenous muscle breakdown substance, which is produced in the body at a constant rate. Concentration of creatinine in plasma in humans is about 15 μg/ml. Owing to its negligible protein binding and almost exclusive excretion via glomerular filtration, the renal clearance of creatinine is considered to be the same as the glomerular filtration rate (GFR).

Crossover study: The study in which comparison treatments in animals (or humans) follow in a counterbalanced sequence, so that each animal (or human subject) receives both treatments, and thus serves as its own control.

Crystalline drug form (see also, Amorphous drug form and Polymorphism): Solid drug materials with definite identifiable crystalline shape. Some chemicals can exist in more than one crystalline form (polymorphism), depending on the conditions (temperature, solvent, time, pressure, etc) under which crystallization is induced.

CSF: Cerebrospinal fluid.

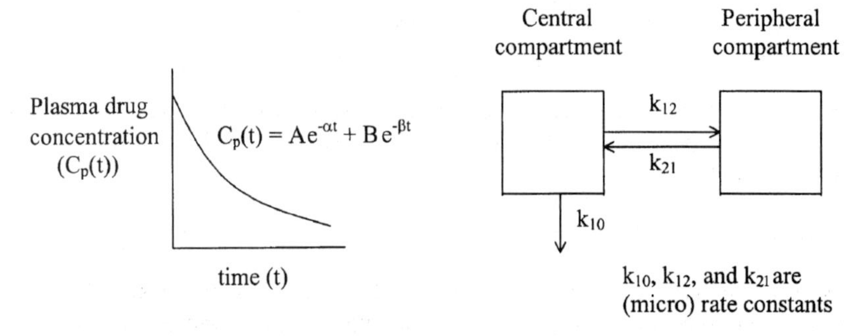

Figure G.2

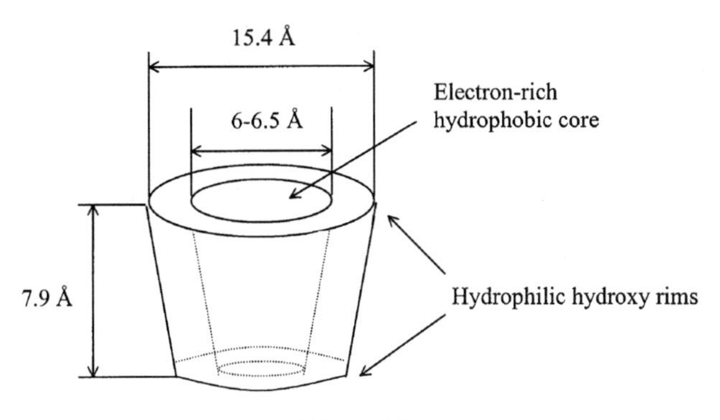

Figure G.3

Cyclodextrin (CD): Cyclic oligosaccharide. Natural cyclodextrins, i.e., α-, β-, and γ-cyclodextrins, are cyclic oligosaccharides of 6, 7, and 8 glucopyranose units, respectively, with aqueous solubility ranging from 150 to 230 mg/ml. Chemical substitution at the 2, 3, and 6 hydroxyl groups of the glucopyranose units of cyclodextrin, e.g., 2-hydroxypropyl β-cyclodextrin, can significantly increase the aqueous solubility of natural cyclodextrin up to more than 500 mg/ml. A three-dimensional structure of cyclodextrin resembles a truncated cone with a nonpolar electron-rich hydrophobic interior and a hydrophilic exterior (see Fig. G.3). Owing to these structural characteristics, cyclodextrins can enhance the aqueous solubility and stability of hydrophobic molecules and even biological macromolecules such as peptides and proteins by trapping the hydrophobic moiety of the compounds within their hydrophobic cores, reducing hydrophobic interactions between the compounds and the surrounding water molecules.

Deep compartment (see also Peripheral compartment): The compartment representing certain organs or tissues to which distribution of a drug from plasma is significantly slower than to other regions in the body. The prolonged drug elimination half-life often found with sensitive assay or following administration of a radiolabeled compound can be due to the slow release of a drug from the deep compartment. Identification of organs or tissues responsible for a deep compartment can be important because of the potential toxicity of a drug associated with its accumulation in those organs or tissues after multiple dosing.

Diastereomers (see also Chiral center and Enantiomers): Stereoisomers that are nonsuperimposable mirror images of each other, i.e., enantiomers. In general, diastereomers have more than one chiral center. A molecule can have only one enantiomer, but may have several diastereomers. The maximum number of possible stereoisomers of a molecule is equal to 2^n, where n is the number of chiral centers in the molecule. Diastereomers have similar chemical properties, but different physical properties such as solubility, melting points, and densities.

Diffusion (see also Active transport): Diffusive transport of a drug across a membrane. Facilitated diffusion is distinguished from simple diffusion by an enhanced rate and saturability of transport. Although facilitated diffusion is carrier-mediated transport, it is different from active transport in that the net movement of a substance is not against a concentration gradient, i.e., at equilibrium the concentration of the substance in the inside and on the outside of a cell is equal, and energy is not required for transport of the substrate.

Disposition: Disposition of a drug generally implies both distribution and elimination (metabolism and excretion) processes.

Distribution coefficient (also see Partition coefficient): Distribution coefficient (D or log D as generally described) is defined as the overall ratio of a compound, ionized and un-ionized, between organic and aqueous phases at equilibrium:

$$D = \frac{[\text{Un-ionized drug}]_{\text{organic phase at equilibrium}}}{[\text{Un-ionized drug}]_{\text{aqueous phase at equilibrium}} + [\text{Ionized drug}]_{\text{aqueous phase at equilibrium}}}$$

Distribution equilibrium: Condition at which the rate of change in the amount of a drug in the peripheral compartment is zero after intravenous bolus injection, when drug disposition can be adequately described with a two-compartment model. The volume of distribution of the drug at this time point is the so-called "apparent volume of distribution at steady state."

Distribution phase [see Alpha (α) phase].

DDS (see Drug delivery system).

Dose-dependent pharmacokinetics (see also Nonlinear kinetics): The phenomenon that the pharmacokinetic behaviors of a drug during absorption, distribution, metabolism, and excretion differ at different dose levels. The dose-dependent plasma concentration–time profiles of a drug at different dose levels are indicative of nonlinear kinetics of the drug in the body.

Dose proportionality (see Superposition).

Drug delivery system (DDS): The technology utilized to direct a drug substance to the desirable body site for release and absorption. For instance, a transdermal patch is a drug delivery system.

E (see Extraction ratio).

ED_{50}: The dose level of a drug that is efficacious in 50% of animals following administration.

EDTA: Ethylenediaminetetraacetic acid ($C_{10}H_{16}N_2O_8$, molecular weight 292.24), a chelating agent for heavy metals or divalent cations such as Ca^{+2}. EDTA can be also used as an anticoagulant or antioxidant in foods.

EHC (see Enterohepatic circulation).

Electrospray ionization: Ion formation from samples in solution by the use of electrospray. In an electrospray interface, the liquid chromatography (LC) effluent (a sample solution) is nebulized into small electronically charged aerosols into the atmospheric pressure region by applying an electric field (usually 3-kV potential difference) in a narrow-bore capillary or electrospray needle. Electrospray ionization is most suited to compounds that exist as preformed ions in the LC effluent or can be readily ionizable by altering the pH or to polar neutral molecules that can be associated with small ions such as NH_4^+.

Elimination phase (see Beta phase).

EM (see Extensive metabolizer).

Emulsifying agent: A chemical used to promote and maintain the dispersion of finely subdivided particles of a liquid in an immiscible liquid vehicle (e.g., polyoxyethylene 50 stearate, Cremophor El).

Emulsion: A dispersion dosage form in which small globules of a liquid-containing drug substance are distributed throughout an immiscible liquid vehicle. For instance, oil-in-water (o/w) emulsions refer to oil globules (internal phase) containing a drug dispersed in a water medium (external phase). Conversely, emulsions with an aqueous internal phase dispersed in an oily external phase are termed water-in-oil (w/o) emulsions.

Enantiomers: Two stereoisomers that exhibit nonsuperimposabl!e mirror images of each other. For instance, two enantiomers of lactic acid are shown in Fig. G.4. Enantiomers have the same physicochemical properties such as solubility, boiling point, melting point, and retention time on nonchiral analytical columns but not the ability to rotate the plane of polarized light, i.e., dextrorotatory (*d* or +, rotating right) or levorotatory (*l* or −, rotating left).

Encapsulating agent: A chemical used to form thin shells to enclose a drug substance or drug formulation for ease of administration.

Enteric coating: Coatings applied to a solid dosage form of drug substances such as tablets or capsules to permit safe passage of the dosage form through the acidic environment of the stomach to the intestine where dissolution of the dosage form takes place. The enteric coated tablets or capsules are suitable for drugs unstable at low pH or for targeting intestinal absorption sites or lesions.

Enterohepatic circulation (EHC): The phenomenon in which part of a drug excreted into the bile is reabsorbed from the intestine and becomes subject to the continuous recirculating process of excretion into bile and subsequent reabsorption from the intestine.

Extensive metabolizer (EM, see also Poor metabolizer): An individual with normal metabolizing activities for a certain drug.

Extracellular fluid: The fluid that lies outside of the cells, which consists of intravascular and interstitial fluids. There are about 14 liters of extracellular fluid in an average 70-kg man, which is approximately 20% of the total body weight or 33% of the total body fluid (42 liters in a 70-kg man).

Extraction ratio (E): The fraction of a drug extracted from each unit volume of blood per pass through the eliminating organ such as the liver or the kidney from inlet to outlet of blood flow. The extraction ratio (E) relates the organ clearance (Cl) to the blood flow rate (Q) into the organ, i.e., $Cl = Q \cdot E$.

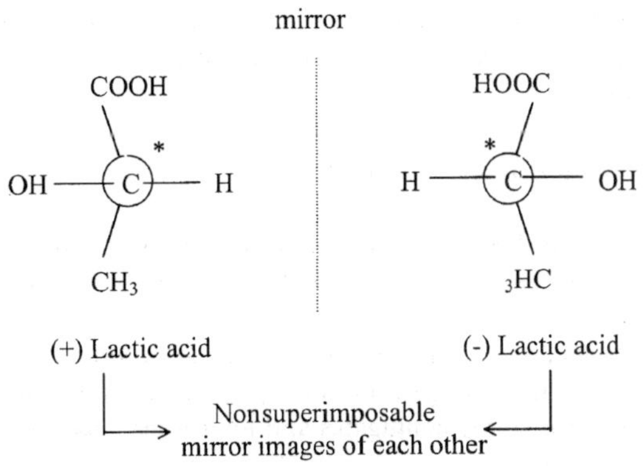

mirror

COOH, OH, C*, H, CH_3 — (+) Lactic acid

HOOC, H, C*, OH, $_3HC$ — (-) Lactic acid

Nonsuperimposable mirror images of each other

Figure G.4

Extravascular fluid: The fluid that lies outside the blood vessels, which is composed of interstitial and intracellular fluids.

Extravascular administration: All routes of administration of a drug other than those in which the drug is injected directly into the vascular blood stream, such as intraarterial or intravenous injection.

Ex vivo: Experiments conducted with biological samples such as blood or urine obtained from animals that have been pretreated with the compound(s) of interest.

FIM: First in man (see also Phase I study).

First-order kinetics: A kinetic process in which the rate of change in concentration [C(t)] or in the amount of drug with time is directly proportional to the drug concentration, i.e.,

$$\text{Rate of change of drug concentration} = k \cdot C(t)$$

where k is the first-order rate constant.

First-pass effect (presystemic elimination): Drug loss prior to reaching systemic circulation for the first time during absorption after oral administration, which is due to drug metabolism in the enterocytes (gastrointestinal membranes), drug metabolism and biliary excretion in the liver, and/or drug metabolism in the lung.

Flavin-containing monooxygenase (FMO): In addition to the cytochrome P450, hepatic microsomes contain a second class of monooxygenase, the flavin-containing monooxygenase (FMO). FMOs are present mainly in liver, kidney, and lung, and are considered to be important for metabolizing heteroatom-containing compounds, along with cytochrome P450. FMOs require NADPH and O_2 for oxidation of the nucleophilic nitrogen (N), sulfur (S), and phosphorus (P) heteroatoms of xenobiotics rather than direct oxidation at a carbon atom. FMOs are heat-labile and can be inactivated in the absence of NADPH by warming microsomes at 50°C for 1 min.

Flip-flop kinetics: The term describing the phenomenon observed when the rate of absorption of a drug from the site of administration is similar to or slower than the rate of its elimination from the body after it enters the systemic circulation. This can be often observed after oral administration of sustained-release drug formulations. When a drug is subject to flip-flop kinetics, the slope of the plasma concentration [$C_p(t)$] *vs.* time profile during the later phase after oral (P.O.) administration is shallower than that after intravenous (I.V.) bolus injection, as illustrated in Fig. G.5.

FMO (see Flavin-containing monooxygenase).

Free drug hypothesis: A hypothesis that the biological activity of a given hormone is governed by its unbound rather than its protein-bound concentration in plasma.

Futile cycling (see also Reversible metabolism): A phenomenon in which conjugated metabolites (glucuronide or sulfate conjugates) of endogenous or exogenous compounds undergo successive cycles of synthesis of the conjugates and hydrolysis back to the parent compounds by more than one enzyme.

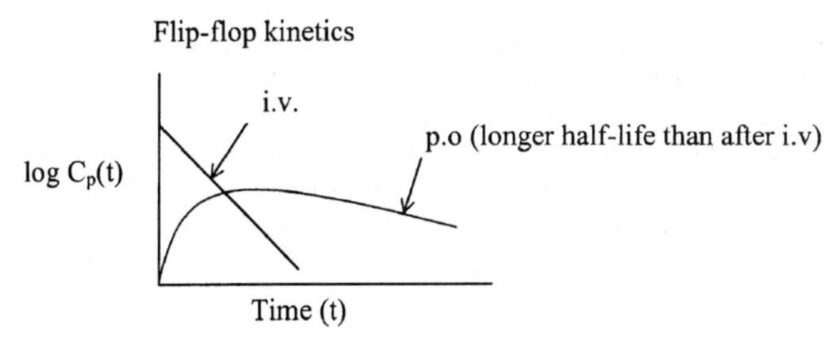

Figure G.5

GCP (see Good clinical practice).

Gene therapy: A therapeutic approach to the treatment of diseases (resulting mainly from genetic defects) by delivering a human gene to a target organ with the subsequent production of a "gene product" or protein such as an enzyme, which can act as a drug. The delivery of a gene can be achieved by, e.g., liposomes, DNA vectors, or recombinant retroviruses.

Genotype: The fundamental assortment of genetic information (genes) contained in chromosomes of an organism.

GFR (see Glomerular filtration rate).

GLP: Good laboratory practice.

Glomerular filtration rate (GFR): Glomerular filtration rate can be used as an index for a subject's renal function. In general, renal impairment causes a decrease in GFR.

Glutathione S-transferase (GST): Glutathione S-transferases represent an integral part of the phase II detoxification system. The glutathione conjugation reaction protects cells against oxidatively and chemically induced toxicity and stress by catalyzing the glutathione conjugation with an electrophilic moiety of hydrophobic and often toxic substrates. In the liver, the glutathione S-transferase accounts for up to 5% of the total cytosolic protein.

GMP: Good manufacturing practice.

Good clinical practice (GCP): Guidelines for planning and performing clinical trials.

gp170 (see P-glycoprotein).

GST (see Glutathione S-transferase).

Gunn rat: A mutant strain of Wistar rat that is genetically deficient in the conjugation of certain aglycones such as, e.g., bilirubin, planar, and bulky phenols.

Half-life: The time it takes a drug concentration in the blood or plasma (or the amount in the body) to decline to one-half of its reference value.

Heated nebulizer: A liquid chromatography (LC)–mass spectrometry (MS) interface, which consists of a pneumatic nebulizer combined with a heated desolvation tube (or chamber) for the introduction of a sample solution (LC column effluent) into an atmospheric pressure chemical ionization (APCI) source. In a heated nebulizer, the column effluent from LC is pneumatically nebulized at room temperature into a heated tube, where evaporation of the solvent in LC effluent (desolvation process) takes place. Owing to the heat required to volatilize a sample solution prior to APCI, a heated nebulizer is not suitable for thermolabile compounds.

Hematocrit: The fraction of the blood composed of red blood cells. Hematocrit can be determined by centrifuging blood in a "hematocrit tube," until the red blood cells become tightly packed in the bottom of the tube. The hematocrit values in healthy men and women are roughly 0.4 and 0.36, respectively.

Henderson–Hasselbalch equation: An equation elucidating the relationship between pH and pK_a of acids and bases:

$$\text{Acid (HA): } pH = pK_a + \log([A^-]/[HA])$$

$$\text{Base (B): } pH = pK_a + \log([B]/[BH^+])$$

Heterocedacity (see Homocedacity).

Homocedacity: When calculated residuals in linear regression are independent of each other and normally distributed with equal variances, the condition of the equal variance of the data is called homocedacity. When there is unequal variance, the condition of the unequal variance of the data is called heterocedacity (Fig. G.6).

Homeostasis: The physiological processes responsible for maintaining the constancy of the internal biochemical and biological functions and conditions in living organisms.

HPLC: High-pressure (or performance) liquid chromatography.

Hybrid rate constant (see also Compartment model): Apparent rate constant consisting of more than one microconstant. For instance, α and β, the exponents in a biexponential equation describing biphasic plasma concentration–time profiles of a drug after intravenous injection, i.e., $C_p(t) = A \cdot e^{-\alpha t} + B \cdot e^{-\beta t}$, are hybrid constants, because those are functions of k_{10}, k_{12}, and k_{21}, the microconstants in a two-compartment model.

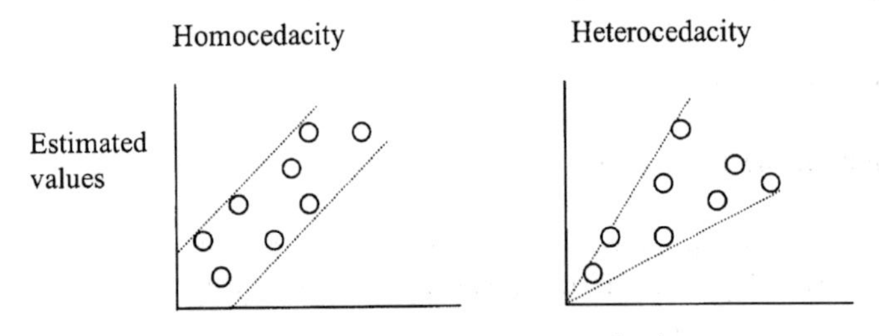

Figure G.6

Hypertonic solution: A solution with a higher osmolarity compared to the reference solution. For instance, solutions with more than 0.9% sodium chloride are considered hypertonic to the intracellular fluid with osmolarity of 280 mOsm/liter, which is equal to that of a 0.9% sodium chloride solution.

Hypotonic solution: A solution with a lower osmolarity compared to the reference solution. For instance, solutions with less than 0.9% sodium chloride are considered hypotonic to the intracellular fluid with osmolarity of 280 mOsm/liter, which is equal to that of a 0.9% sodium chloride solution.

Hysteresis: The time-related changes in the apparent relationship between the plasma concentration of a drug and its observed pharmacological effect. When the plasma drug concentration is plotted against the corresponding observed effect of the drug, and the points are connected in time sequence, the intensity of the observed effect at the same drug concentrations is sometimes different. If the effects are less pronounced during the later time points than those during the earlier time points at the same plasma concentrations, this concentration–effect relationship as a function of time is known as "clockwise hysteresis (or proteresis)." If the intensity of the effect is higher at the later time points than at the earlier ones for the same plasma concentrations, this phenomenon is known as "counterclockwise hyteresis (or hysteresis)."

I.A. injection: Intraarterial injection.

Investigator's brochure (IB): A document containing all pertinent information such as pharmacological and toxicological effects in animals and humans, if any data exist, known about the investigational new drug (IND) provided by the sponsor to all clinical investigators.

ICH: International committee of harmonization.

I.M. injection: Intramuscular administration (injection).

IND: Investigational new drug (see also Phase I).

Inducer: A compound that increases the concentration of metabolizing enzymes in tissues and thus enhances the rates of metabolism of endogenous or exogenous substrates.

Induction (or enzyme induction): The increase in enzyme content or activities. Induction of metabolizing enzymes usually results in faster metabolism of a drug, which can be due to the enhanced synthesis of enzymes and/or the stimulation of preexisting enzymes by inducers. Enhancement of enzyme synthesis by xenobiotics is known to be the most common mechanism for enzyme induction. If a drug stimulates its own metabolism, it is referred as autoinduction.

Inhibition: The decrease in enzyme content or activities. Inhibition of metabolizing enzymes usually results in slower metabolism of a drug, which can be due to the suppressed synthesis of enzymes and/or the inactivation of preexisting enzymes by inhibitors. Depending on the patterns of inhibitor interaction with the enzyme, there are three different types of inhibition, i.e., competitive, noncompetitive, and uncompetitive inhibition. (a) *Competitive inhibition*: Inhibitor competing for the same binding site of the enzyme with substrates: this type of inhibition causes an increase in K_m, the Michaelis–Menten constant, but no change in V_{max}, the maximum rate of enzymatic reaction, as compared to that in the absence of the inhibitor. (b) *Noncompetitive inhibition*: Inhibitor binding to both the free enzyme and the enzyme–substrate complex. This type of inhibition causes a decrease in V_{max}, but no change in K_m. (c) *Uncompetitive inhibition*: Inhibitor binding to the enzyme–substrate complex. This type of inhibition causes a decrease in both V_{max} and K_m.

In situ: Experiments conducted with intact organs or tissues.

Interstitial fluid: The fluid present in the space between cells, the so-called, interstitium (extracellular fluid minus plasma volume). There are about 11 liters of interstitial fluid in an average 70-kg man, which is approximately 16% of the total body weight or 26% of the total body fluid (42 liters in a 70-kg man). The interstitial fluid is derived from plasma by filtration and diffusion via capillary membranes and contains almost the same constituents as plasma, except for lower concentrations of plasma proteins.

Intracellular fluid: The fluid that resides within the cells. There are about 28 liters of intracellular fluid in an average 70-kg man, which is approximately 40% of the total body weight or 67% of the total body fluid (42 liters in a 70-kg man).

Intravascular fluid: The fluid that resides within blood vessels, i.e., plasma water.

Intrinsic clearance: In general, intrinsic clearance refers to the intrinsic ability of the eliminating organ, such as the liver, to eliminate drugs when there are no limitations in other physiological factors affecting *in vivo* drug clearance. For instance, intrinsic hepatic clearance implies the intrinsic ability of the liver to eliminate drugs via metabolism and/or biliary excretion, when, e.g., hepatic blood flow (supply of the drug to the liver), membrane transport (transport of the drug from the sinusoidal blood into hepatocytes), cofactor availability, and drug binding to blood components are not limiting factors for drug clearance.

In vitro: Experiments conducted in laboratory glassware.

In vivo: Experiments conducted in intact animals.

Iontophoresis: The process of transferring ionized molecules into the tissues, usually skin, by the use of a small electric current.

I.P. injection: Intraperitoneal injection.

Isoelectric point (p*I*): The pH at which the net charge of a molecule such as an amino acid, peptide, or protein is zero, i.e., at p*I*, the molecule is electronically neutral.

Isomers (structural isomer): Distinct molecular entities that have the same molecular formula and share a common characteristic in their chemical structure. There are two different kinds of structural isomers, i.e., constitutional isomers, which differ in the sequential arrangement of atoms, and stereoisomers, which have the same constitutions but differ in the spatial orientation of their atoms. For instance, 1-propranolol ($HO–CH_2–CH_2–CH_3$) and 2-propranolol ($CH_3–CHOH–CH_3$) are constitutional isomers, whereas (+)-lactic acid and (−)-lactic acid are stereoisomers (see Stereoisomers).

Isotonic solution: A solution with the same osmolarity as the reference solution. For instance, a 0.9% sodium chloride solution or a 5% glucose solution is isotonic to the intracellular fluid, since it produces the same osmolarity (280 mOsm/liter).

I.V. injection: Intravenous injection.

Lagrange method: A curve-fitting method with a cubic polynominal function to put a smooth curve through the data points such as plasma drug concentration $[C_p(t)]$ *vs.* time (t) plots. Usually, the four data points around the segment of interest in the plot are fitted with the cubic function shown below:

$$C_p(t) = A + B \cdot t + C \cdot t^2 + D \cdot t^3$$

Lag time: The time elapsed between the administration of a drug and its appearance in the systemic circulation or between its appearance in the systemic circulation and the manifestation of its pharmacological effects. For instance, the lag time between drug administration and drug exposure in blood or plasma can be often observed after oral administration of solid dosage forms such as tablets, capsules, and especially enteric-coated tablets. The delay is due to the slow disintegration and/or dissolution of dosage forms in the gastrointestinal tract before the drug is actually absorbed and can be anywhere from a few minutes to several hours.

LC: Liquid chromatography.

LD_{50}: The dose level of a drug that is lethal to 50% of animals after administration.

Linear conditions (see also Linear kinetics and Superimposibility): Conditions in which pharmacokinetic processes of a drug such as absorption, distribution, metabolism, and excretion can be properly described by the first-order kinetics. The linearity of the system can be recognized when the dose-normalized plasma concentration *vs.* time profiles of a drug at different dose levels are superimposible and can be properly described by the same first-order kinetics.

Linear kinetics (see also First-order kinetics): Pharmacokinetic processes, that can be described by first-order kinetics.

Liposome: A stable microscopic vesicle composed of one bilayer (unilamellar liposome) or a number of bilayers (multilamellar liposome) of various phospholipids and similar amphipathic lipids concentrically oriented around an aqueous core. Liposomes are spontaneously formed when certain phospholipids are dispersed in excess water. Liposomes have been utilized as drug delivery systems to carry lipophilic compounds within their lipid bilayers.

Loading dose: The one-time dose administered at the beginning of therapy in conjunction with a regular dose regimen, to achieve therapeutic drug concentration faster than the regular dose regimen alone.

Locus: Any genetically defined site.

Log D (see also Distribution coefficient): Logarithm of the distribution coefficient. A rough estimate of log D of a compound at any given pH can be obtained by subtracting one unit from log P for every unit of pH above (for acids) or below (for bases) the pK_a:

$$\boxed{\log D \approx \log P - \Delta|pK_a - pH|}$$

Log $D_{7.4}$: Log D measured at pH 7.4 in an aqueous phase.

Log P (see Partition coefficient): Logarithm of the partition coefficient.

MAD: Maximum absorptive dose.

MAO (see Monoamine oxidase).

Mass balance study: Study performed in humans or animals, usually small laboratory animals such as rats or mice, using a radiolabeled compound to elucidate the pharmacokinetic profiles of the compound. Mass balance studies in small animals are conducted in metabolic cages, which enable the collection of urine and feces samples for estimating the degree of drug recovery during the experiments. Levels of radioactivity in various tissues and organs can be determined by sacrificing animals at various time points. The main purposes of such studies are to investigate the routes and extent of elimination of a drug and its metabolites from the body after administration and to understand tissue distribution profiles of a drug and its metabolites when needed. The information from these studies can be useful for the interpretation of potential organ-specific toxicity. Mass balance studies with radiolabeled compounds in humans involve many regulatory and ethical issues, and there are many different guidelines in various countries. In general, a mass balance study is performed using a limited number of (usually male) volunteers, e.g., three or four, and women at the age of procreation are excluded.

MAT (see Mean absorption time).

MDR (see Multidrug resistance gene product and P-glycoprotein).

Mean absorption time (MAT): The average time it takes for a drug molecule to be absorbed into the systemic circulation from the site of administration, e.g., into the gastrointestinal tract after oral dosing. The MAT of a drug can be determined from the difference between the mean residence time (MRT) values after oral and intravenous bolus administration.

Mean input time (MIT): The average time it takes a drug molecule to reach the systemic circulation from the site of extravascular administration. The MIT of a drug following extravascular administration can be determined as the difference in the mean residence time (MRT) values between extravascular and intravenous bolus administration.

Mean residence time (MRT): The average time after administration that a drug molecule spends in the body before being eliminated.

Mechanism-based inhibitor: A compound inhibiting the activity of metabolizing enzymes, such as cytochrome P450s, by forming covalent bonds with the enzymes as a result of its own metabolism by the enzyme(s). For instance, 1-aminobenzotriazole (ABT) is a mechanism-based inhibitor of various cytochrome P450 isozymes. To be activated, ABT undergoes a P450-catalyzed oxidation to form benzyne, a reactive intermediate, which covalently binds to the prosthetic heme group of cytochrome P450 and thereby causes an irreversible loss of enzyme activity.

Membrane: Cell membranes are basically composed of two different kinds of molecules, i.e., lipids and proteins. There are three types of lipids, i.e., phospholipids, cholesterol, and glycolipids. Phospholipids, which have both hydrophilic and lipophilic groups in their

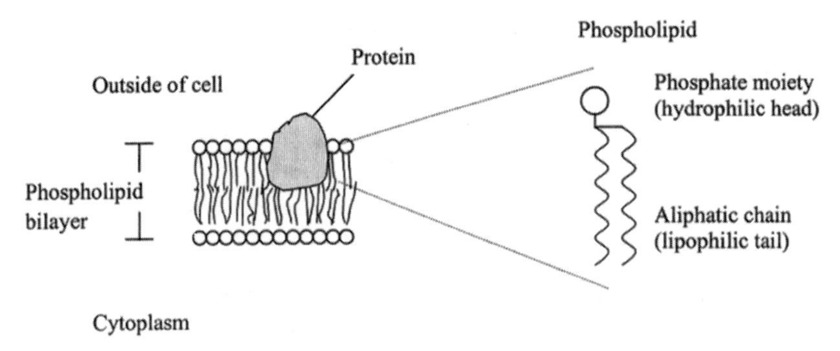

Figure G.7

structures, form a bilayer (ca. 70 Å in thickness) such that their polar groups constitute the outer surfaces of the membrane, whereas nonpolar groups are buried within the interior region of the membrane. Protein molecules may lie at or near the inner or outer membrane surface or penetrate partially or entirely through the membrane (Fig. G.7).

Metabolic ratio (MR, see also Polymorphism): A ratio between the amount of parent drug and its particular metabolite(s) excreted in urine. When MR values are measured for metabolites known to be produced by polymorphic enzyme(s), they can be used for polymorphic phenotype screening for drugs subject to polymorphic metabolism:

$$MR = \frac{\text{Amount of parent drug excreted in urine}}{\text{Amount of metabolite excreted in urine}}$$

Metastable form (see also Polymorphism): Various types of less stable crystalline forms of a chemical than its most stable crystalline form at a given temperature and pressure. The metastable forms of a compound convert in time to the stable crystalline form.

MFO: Mixed function oxidase.

MIC (see Minimum inhibitory concentration).

Michaelis–Menten equation: An equation, shown below, that characterizes certain concentration–dependent biological or pharmacokinetic processes such as, e.g., protein binding, metabolism, and active transport:

$$\frac{dC}{dt} = \frac{V_{max} \cdot C}{K_m + C}$$

where C is the concentration of substrate, K_m is the Michaelis–Menten constant (concentration of substrate at which the rate of the process is one half of V_{max}), and V_{max} is the maximum rate of the process. At substrate concentrations well below K_m, the rate of change in concentration is a function of the concentration [$dC/dt = (V_{max}/K_m) \cdot C$], i.e., the first-order kinetics), whereas at concentrations well above K_m, it becomes constant ($dC/dt = V_{max}$, i.e., the zero-order kinetics).

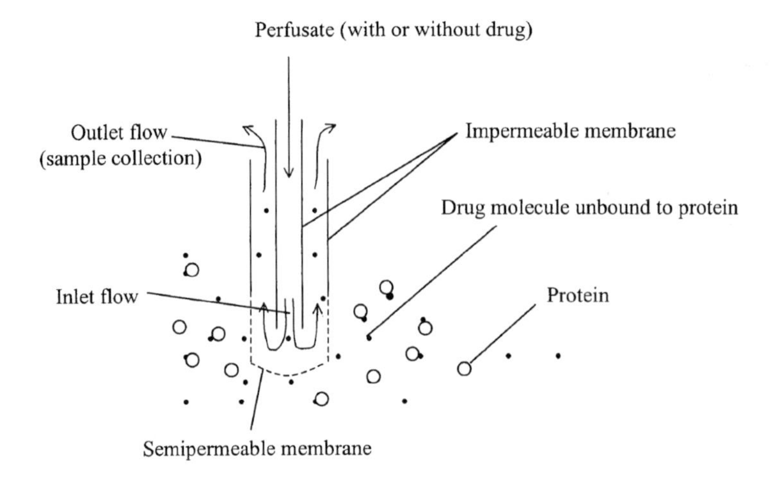

Figure G.8

Microdialysis: A technique for measuring the extracellular concentration of a drug(s) not-bound to proteins by implanting a semipermeable membrane probe into tissues or blood vessels in an animal (freely moving). The typical microdialysis probe consists of an impermeable tube attached to a semipermeable, hollow fiber usually 200–400 μm in diameter with a molecular-weight cutoff of 10,000–30,000 Da, inside of which is placed an impermeable tube for perfusate. The design of the probes can be specialized to match different tissues. The microdialysis probe is bidirectional. That is, it can collect drug-containing samples from the site of implantation by flushing the tube inside of the probe with an isotonic perfusion fluid, but also can deliver the drug by perfusing the tube with the drug solution. Microdialysis can be also used for *in vitro* experiments, such as, e.g., protein binding and drug metabolism in microsomes.

Middle molecules (see also Middle-molecule hypothesis): The specific uremic toxins identified in uremia with a molecular weight approximately between 300 and 12,000. Uremic retention solutes representing middle molecules include parathormone, β_2-microglobulin, some peptides, and glucuronated conjugates among others, of which most are still without proven toxicity.

Middle-molecule hypothesis: A hypothesis that uremic syndrome (a progressive deterioration of physiological and biochemical functions of the kidneys that accompanies the development of renal failure) is the result of the retention of metabolic compounds with a molecular weight approximately between 300 and 12,000 (so-called "middle molecules") that are normally cleared by healthy kidneys.

Minimum inhibitory concentration (MIC): The lowest concentration of an antimicrobial agent that prevents the growth of microorganisms on an agar plate after incubating for a certain period of time, usually 18 to 24 hr.

MIT (see Mean input time).

Molarity (M, see also Mole): The number of moles of solute dissolved in 1 liter of solution. Solutions of 1 M, 1 mM, and 1 μM contain 1 mole, 1 mmole, and 1 μmole of drug in 1 liter

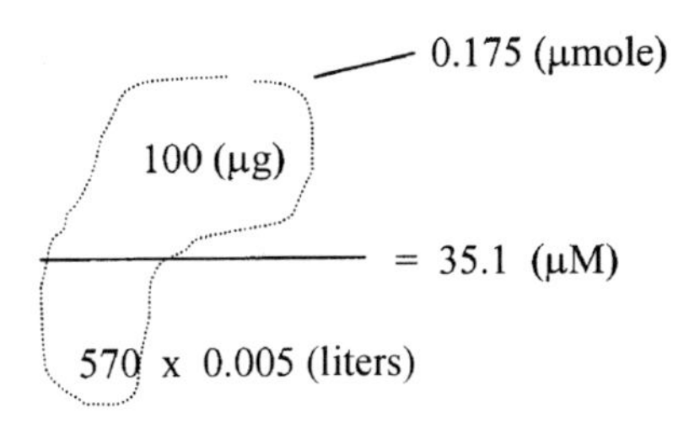

Figure G.9

of solution. Thus, molarity has the units of concentration, i.e., moles per liter:

$$\text{Molarity (M, mM or } \mu\text{M)} = \frac{\text{Number of moles of solute (mol, mmol, } \mu\text{mol)}}{\text{Number of liters of solution (liters)}}$$

For instance, a molar concentration of 100 μg of compound A with a molecular weight of 570 dissolved in 5 ml of water is 35.1 μM (Fig. G.9). If one needs 5 ml of a 10 μM of compound A in 1% DMSO and water, one dissolves 28.5 μg of the compound in 50 μl of DMSO and adds 4.95 ml of water:

$$\frac{10(\mu\text{mole}) \times 570}{1000\text{ ml}} \times 5(\text{ml}) = 28.5(\mu\text{g})$$

Mole: A mole of any substance that contains Avogadro's number (6.022×10^{23}) of atoms or molecules. For instance, 1 mole of water molecules contains 6.022×10^{23} H_2O molecules, or 12.044×10^{23} H atoms and 6.022×10^{23} O atoms. The mass of 1 mole of the substance is equal to its molecular weight in grams. For instance, 1 mole, 1 mmole, and 1 μmol of water are equal to 18 g, 18 mg, and 18 μg of water, respectively:

$$\text{Mole (mole, mmole, or } \mu\text{mole)} = \frac{\text{Amount of substrate (g, mg, or } \mu\text{g)}}{\text{Molecular weight}}$$

Monoamine oxidase: Monoamine oxidase (MAO) is known to be related to the metabolism of exogenous tyramine and the "cheese effect" produced by the ingestion of large amounts of tyramine-containing foods under certain conditions. MAO catalyzes the oxidative deamination of biogenic amines. It is primarily a mitochondrial enzyme, although some of its activity has also been reported in the microsomal fraction.

MR (see Metabolic ratio).

MRM: Multiple reaction monitoring (see also SRM): Selected reaction monitoring applied simultaneously to two substances as in the case of a drug accompanied by an internal standard.

MRP (see Multidrug resistance-associated protein).

MRT (see Mean residence time).

MS: Mass spectrometry.

MTD: Maximum tolerated dose.

Table G.10. Nomenclature of MDR Gene Product

Species	Class I	Class II
Human	MDR1	MDR3
Rat, mouse, hamster	mdr1a, mdr1b	mdr2

Multidrug resistance–associated protein (MRP): A member of the ATP binding cassette (ABC) family of transporters with molecular weight of about 190 kDa, which can confer drug resistance in tumor cells with a broad spectrum of substrates. Multidrug resistance–associated protein (MRP) was first found in some multidrug resistant cell lines, in which no overexpression of either P-glycoprotein or mRNA from the encoding multidrug resistance (MDR1) gene could be detected. MRP is also found in normal cells, and its tissue distribution patterns are similar to those of P-glycoprotein; for instance, MRP can be found, e.g., in liver (canaliculus), erythrocyte membranes, heart, kidneys and intestinal brush border membranes, and lungs. At least two different isoforms of MRP, i.e., MRP1 and MRP2, are present in human and rodent hepatocytes. MRP1 is present in the lateral membrane domains of normal hepatocytes at a very low level, whereas MRP2 is localized exclusively in canalicular membrane of hepatocytes. MRP1 and MRP2 are considered to be the same as the ATP-dependent glutathione (GSH) S-conjugate transporter and the canalicular multispecific organic anion transporter (cMOAT) responsible for active biliary excretion of amphipathic anionic conjugates such as glutathione-, glucuronide-, and sulfate conjugates of various substrates in an ATP-dependent manner, respectively.

Multidrug resistance gene product (see also P-glycoprotein): Multidrug resistance (MDR) gene product (also known as P-glycoprotein or P-gp) functions as an ATP-dependent drug efflux pump in the membrane, which lowers the intracellular concentrations of cytotoxic drugs, and it is one of potential causes of resistance of cancer cells to a wide range of chemotherapeutic agents. MDR gene products are also present in normal cells and their locations are confined to the lumenal domains of cells in specific organs such as the liver, intestine, kidney, and brain. Physiological functions of MDR gene products in normal cells appear to be related to active transport of organic cations. There are two different isoforms in normal human cells, i.e., MDR1 and MDR3, and three *mdr* gene products in rats and mice, i.e., mdr1a, mdr1b, and mdr2. MDR1, mdr1a, and mdr1b confer drug resistance on otherwise drug-sensitive cells, but MDR3 and mdr2 do not. Recent studies have suggested that MDR1 and MDR3 in normal hepatocytes might mediate active biliary excretion of hydrophobic organic (cationic) compounds and phosphatidylcholine across canalicular membrane, respectively (Table G.10).

NADPH (Nicotinamide-adenine-dinucleotide-phosphate): A cofactor required along with oxygen for the oxidation reaction by cytochrome P450 enzymes.

NAT: N-acetyltransferase.

NCE: New chemical entity.

NDA (New drug application, see also Phase III study): After successful phase III clinical trials, the new drug application (NDA) for an investigational new drug (IND) is filed for review by the government regulatory agency.

Neoteny: The retention of formerly juvenile characters by adult descendants produced by retardation of somatic development, i.e., a sort of sustained juvenilization or slow development of animals to adulthood. If related to humans, neoteny is the preservation in adults of shapes and growth rates that characterize juvenile stages of ancestral primates.

N-in-1 dosing (see Cassette dosing).

NOAEL (No adverse effect level): Plasma or blood exposure level of the test compound at which no adverse (toxic) effect is observed during toxicity studies in the test animal species.

Noncompartmental model: A pharmacokinetic approach for estimating pharmacokinetic parameters such as clearance, volume of distribution, mean residence time, and bioavailability from plasma drug concentration *vs.* time profiles without any assumptions of a specific compartmental model for the body. There are various noncompartmental techniques including statistical moment analysis, and a noncompartmental recirculatory model of drug plasma concentration *vs.* time profiles.

Nonlinear condition (see also Nonlinear kinetics): The conditions in which pharmacokinetic processes of a drug during absorption, distribution, metabolism, and excretion cannot be properly described with first-order kinetics. The nonlinearity of the system can be assumed when the dose-normalized plasma drug concentration–time profiles at different dose levels are not superimposible and cannot be properly described by the same first-order kinetics.

Nonlinear kinetics (see also Michaelis–Menten equation and Zero-order kinetics): Any pharmacokinetic process that does not follow first-order kinetics. Nonlinear kinetics of a drug in the body can be recognized when there is no superimposibility of plasma drug concentration *vs.* time profiles at different dose levels. Virtually all drug pharmacokinetic processes can be considered to be nonlinear at high dose (or concentration) levels owing to saturation of the various enzymatic or carrier-mediated processes during absorption, distribution, metabolism, or excretion.

NONMEM (Nonlinear mixed effect modeling): A computer program provided as FORTRAN source code, developed by the NONMEM Project Group, University of California at San Francisco, CA. NONMEM performs nonlinear regression analysis of sparse data sets from individual or population pharmacokinetic studies.

NSAID: Nonsteroidal anti-inflammatory drug.

OROS (see Osmotic pump): Oral osmotic pump drug delivery system.

Osmolarity: The osmolar concentration of solution (osmole per liter of water, osm/liter). One osmole is equal to 1 mole of solute, i.e., 6×10^{23} molecules. For instance, osmolarity of a solution containing 1 mole of glucose in 1 liter of water is 1 osmole/liter (osm/liter) of glucose. If a molecule dissociates into more than one ion, osmolarity of a solution is affected by the number of osmotically active molecules or ions in the solution. For instance, a solution that contains 1 mole of sodium chloride has an osmotic concentration of 2 osm/liter, because one molecule of sodium chloride dissociates into two ions, i.e., sodium and chloride.

Osmosis: The net diffusion of water from a solution of low solute concentration to a solution of higher solute concentration across a semipermeable membrane between the two solutions that allows only water molecules to diffuse through.

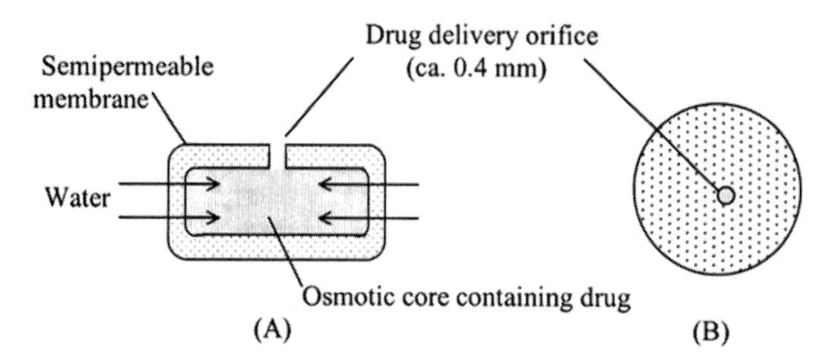

Figure G.11. Cross-sectional (A) and bird's eye view (B) of the oral osmotic pump drug delivery system.

Osmotic pressure: The exact amount of pressure that has to be applied to a solution to prevent osmosis (the net diffusion of water through a semipermeable membrane). The higher the osmotic pressure of a solution is, the higher its solute concentration.

Osmotic pump (OROS, oral osmotic drug delivery system): A solid oral dosage form developed by Alza Co., Palo Alto, CA. An osmotic pump utilizes osmotic pressure built up by the osmotic drug core and water drawn from the gastrointestinal tract after oral administration, via a semipermeable membrane, and thereby releases a drug solution at a constant rate through single or multiple drug delivery orifices (Fig. G.11).

OTC: Over-the-counter drug.

PAH: Polycyclic aromatic hydrocarbons.

PAPS (Phosphoadenosine-5-phosphosulfate): A cofactor required for sulfation by sulfotransferase.

Parenteral administration: Administration of a drug by injection other than into the gastrointestinal tract, such as intraarterial (artery), intraarticular (joint), intracardiac (heart), intradermal or intracutaneous (skin), subcutaneous (beneath the skin), intramuscular (muscle), intraosseous (bone), intraperitoneal (peritoneal cavity), intraspinal (spine), intravenous (vein), or intrasynovial (joint-fluid region) injection.

Partition coefficient (see also Distribution coefficient): Partition coefficient (P, or log P as generally described) is defined as the ratio of concentrations of the un-ionized compound between organic and aqueous phases at equilibrium. The partition coefficient can be viewed as an indicator for intrinsic lipophilicity of the compound in the absence of ionization or dissociation of the compound.

$$P = \frac{[\text{Un-ionized}]_{\text{organic phase}}}{[\text{Un-ionized}]_{\text{aqueous phase}}}$$

PC-NONLIN: A computer program that performs nonlinear regression analyses of pharmacokinetics and pharmacodynamics data.

PCR: Polymerase chain reaction.

PEG: Polyethylene glycol.

Peptidomimetic compounds: Peptidelike compounds in terms of their chemical structures. Compounds with more than one amide moiety bonded in a sequence similar to those of peptides.

Peripheral compartment (see also Deep compartment): The compartment representing organs and tissues in the body into which distribution of a drug in plasma is slower than that into the organs and tissues represented by the central compartment (usually blood and highly perfused organs).

Peroral administration: Administration of drug into the gastrointestinal tract via the mouth.

PD (see Pharmacodynamics).

P-glycoprotein (P-gp, gp170, or *MDR* gene products, see also Multidrug resistance (MDR) gene product): A transmembrane protein with molecular weight of 170 kDa classified as an ATP-dependent primary active transporter belonging to the ATP binding cassette (ABC) transporter superfamily. P-glycoprotein (P-gp) was originally identified as a multi-drug resistance (MDR) gene product, which pumps out numerous anticancer agents such as, e.g., vinblastine and daunomycin from tumor cells causing a decrease in their intracellular concentrations and thus resistance to those drugs. Its wide range of substrate specificity has been recognized as one of mechanisms for multidrug resistance of cancer cells with enhanced expression of P-gp. Its substrates other than anticancer agents include several peptides such as cyclosporin, calcium channel blockers such as verapamil, and various cations. P-gp has been also found in lumenal membranes of normal tissues including liver (canalicular membrane of hepatocytes), intestine (brush border membrane of enterocytes), kidneys, adrenals, and brain. One of physiological functions of P-gp has been identified as phosphatidylcholine transporter in canalicular membrane of hepatocytes based on findings from mdr2 knockout mice experiments.

pH: The negative logarithm of the hydrogen ion concentration of a chemical substance in water as a measure of the acidity or the alkalinity of the compound expressed as a number from 0 to 14: $pH = -\log[H^+]$.

Phagocytosis: The process of intake ("engulfment") of solid particles by a cell.

Pharmacodynamics: The study that examines relationships between drug concentrations at the effect site(s) where target enzymes or receptors are located and the magnitude of the pharmacological efficacy of the drug.

Pharmacogenetics: The study of the hereditary (genetic) basis of the differences in responses to or metabolism of various pharmaceutical agents.

Pharmacokinetics: The study of the behavior of drug molecules in the body after administration in terms of absorption, distribution, metabolism, and excretion (ADME) processes through examination of the concentration profiles of drug in readily accessible body fluids, such as blood or plasma as a function of time.

Phase I study: First-in-man (FIM) trials of an investigational new drug (IND). The trials consist of short-term studies in a small number of healthy male subjects or patients suffering from the target disease to be treated. The primary objectives of phase I clinical trials are to establish a dose–tolerance relationship and evaluate pharmacological properties and efficacy of the drug, if possible. The number of subjects usually ranges from 20 to 80.

Phase II study: Pilot therapeutic studies in patients following a phase I study. The main objectives of phase II trials are to assess the effectiveness and determine the common short-term side effects of an investigational new drug (IND). The studies provide information on clinical pharmacokinetic and pharmacodynamic relationships, often with short-term response parameters, the so-called surrogate endpoints. In general, phase IIa is exploratory (controlled or not) in nature and phase IIb is controlled. The number of patients with the target disease to be treated ranges from 100 to 200, and the controlled studies are usually conducted under double-blind and placebo-treated conditions.

Phase III study: Upon obtaining preliminary evidence of the effectiveness of an investigational new drug (IND) from phase II studies, expanded controlled and uncontrolled clinical trials are initiated. The main purposes of phase III studies are to gather additional information about the effectiveness and side effects (safety) for the overall benefit and risk relationship of the IND and to verify its dosage range. The number of patients with the target disease or more than one disease condition usually ranges from 600 to 800. After successful phase III trials, the new drug application (NDA) for the IND is filed for review by the government regulatory agency.

Phase IV study (postmarketing surveillance): Postmarketing surveillance and/or clinical trials of a drug already on the market. During this stage, new indications, pharmaceutical formulations, methods of administration, dosage regimen, and target population of the drug on the market are continuously surveyed and studied. If required, clinical trials can be performed as trials of new medicinal products with similar objectives as premarketing trials in some countries.

Phenotype: The observable structural and functional properties of an organism produced by the interaction between an organism's genetic potential (genotype) and the environment surrounding it.

Physiologically based pharmacokinetic (PBPK) model (or Physiological pharmacokinetic model): A pharmacokinetic model based on actual animal physiology and anatomy. Unlike conventional compartment models, e.g., one- or two-compartment models for drug disposition in the entire body without a detailed understanding of animal physiology, the physiologically based pharmacokinetic model describes the body or the organs with compartments relevant to their anatomical location and physiological function. In other words, compartments in these physiologically more realistic models represent actual organs or tissues in the body with actual volumes and are connected according to their anatomical locations in the body with appropriate organ blood flows. Drug concentrations in different organs or tissues are experimentally measured at the same time, so that an exact description of the time course of drug concentrations in organs or tissues of interest becomes available. In addition, the parameters estimated in these models correspond to actual physiological and physicochemical measures such as organ blood flow rates and volumes and partition coefficients of drugs between blood and tissues. Thus, any changes in the disposition kinetics of a drug as a result of physiological or pathological alterations in the functions of particular organs or tissues can be estimated and/or predicted. Since the parameters of these models reflect actual physiological and anatomical measurements, animal scale-up based on physiological pharmacokinetics provide a rational basis for parameter extrapolation between different species.

p*I* (see Isoelectric point).

Pinocytosis: The active intake process of fluid by the cell.

Pittsburgh cocktail: Five probe drugs, i.e., caffeine (a probe for CYP1A2), mephenytoin (CYP2C19), debrisoquin (CYP2D6), chlorzoxazone (CYP2E1), and dapsone (CYP3A and N-acetylation), simultaneously administered as a metabolic cocktail to estimate phenotypic activities of cytochrome P450 and N-acetyltransferase enzymes in humans *in vivo* (Frye R. F. *et al.*, Validation of the five-drug "Pittsburgh cocktail" approach for assessment of selective regulation of drug-metabolizing enzymes, *Clin. Pharmacol. Ther.* **62**: 365–376, 1997).

PK (see Pharmacokinetics).

pK_a (see also Henderson–Hasselbalch equation): The negative logarithm of K_a, the equilibrium dissociation constant of acids or bases. The pK_a of an ionizable compound is the same as the pH at which the concentration of an ionized compound is the same as that of un-ionized compound. The smaller the pK_a value, the stronger the acid, whereas the larger the pK_a, the stronger the base.

Placebo effect: Apparent (usually beneficial) therapeutic effect observed in patients, which arises from psychological factors following administration of an inert substance (placebo).

Plasma (see also Serum): The clear supernatant after centrifuging blood. Plasma still contains the coagulating factors.

Plasma protein binding: Plasma protein binding indicates how much of the total amount of drug in plasma is bound to plasma proteins such as albumin or α_1-acid glycoprotein.

Poor metabolizer [PM, see also Extensive metabolizer (EM)]: An individual with deficient metabolic ability of a particular drug in a certain metabolizing enzyme(s) owing to a genetic defect.

P.O. administration: Oral administration.

Polymorphism: (1) A Mendelian or monogenic trait that exists in the population in at least two phenotypes (and presumably at least two genotypes), neither of which shows a frequency of less than 1–2 % of the population. If the frequency is lower than 1–2%, it is called a rare trait. **(2)** More than one crystalline form of the same chemical substance. Some chemicals can exist in several different types of crystalline forms, depending on the conditions for inducing crystallization including, e.g., temperature, solvent, time, and pressure. There is only one form of a pure chemical substance most stable at a given temperature and pressure as compared with other less stable forms. These are called metastable forms and convert in time to the more stable crystalline form. The various polymorphic forms of the same chemical can differ in many physical properties including, e.g., aqueous solubility and melting point. In general, the metastable forms exhibit higher kinetic aqueous solubility and thus higher dissolution rates than the stable crystal form of the same drug. However, the most stable crystalline form in a given storage condition is frequently preferred in a pharmaceutical dosage formulation because of its greater resistance to chemical degradation.

Postmarketing surveillance (see Phase IV study).

Precision of assay (see also Accuracy of assay): Reproducibility of replicate determination within a run (intrarun method precision) and reproducibility between determinations from separate runs (interrun precision).

Presystemic elimination (see First-pass effect).

Primary cell culture: A cell culture started from cells, tissues, or organs taken directly from an organism.

Primary metabolite (see also Secondary metabolite): A metabolite originally produced from the parent compound that has not been further metabolized.

Principles of superposition (see Superposition).

Prodrug: Any compound that undergoes biotransformation and/or chemical degradation *in vivo* to produce the active parent drug. Prodrugs can be divided into two classes, i.e., the carrier-linked prodrug (commonly known as prodrug) and the bioprecursors. The carrier-linked prodrug contains a specific nontoxic moiety (carrier), which is mostly of a lipophilic nature, linked with the active parent drug, in order to alter undesirable physicochemical properties of the parent drug usually related to poor aqueous solubility and/or membrane permeability. A simple chemical or enzymatic hydrolysis can cleave this carrier linkage and release the active drug *in vivo* at the right moment. The carrier is usually linked via an ester or amide bond. The bioprecursor is a compound for which a metabolite *in vivo* is expected to be active. It is different than the carrier-linked prodrug in that it does not imply a simple temporary linkage between the active drug and a carrier moiety, but involves a chemical modification of the active molecule, which becomes subject to *in vivo* metabolism. Typical criteria for designing a carrier-linked prodrug are: (1) the linkage of the carrier-linked prodrug is usually a covalent bond; (2) the linkage is cleaved mainly *in vivo*; (3) the production of the active parent drug must occur with the right kinetics to ensure effective drug levels at the site of action and to minimize metabolism of prodrug itself; and (4) the prodrug and the carrier released *in vivo* are nontoxic.

Product inhibition: Inhibition of metabolism of a parent compound by its own metabolite(s).

Prosthetic group: A nonpeptide portion of certain protein molecules that may be intimately concerned with the specific biological activities of the protein. For instance, *heme* is the prosthetic group of hemoglobin.

Proteresis (see Hysteresis).

Pseudodistribution equilibrium phase [see also Beta (β) phase]: When a semilogarithmic plot of plasma drug concentration *vs.* time [log $C_p(t)$ *vs.* t] after intravenous bolus drug administration exhibits a biexponential decline, the straight terminal segment of the plot is called the pseudodistribution equilibrium phase. This phase is also called a β-, postdistribution, elimination, or terminal phase. During this phase, the ratio of amounts of the drug between the plasma pool and all the tissues in the body becomes constant.

Quantitative structure–activity relationship (QSAR): The relationship between various physicochemical parameters of a series of congeneric compounds and the quantitative potency of a particular biological or pharmacological activity. If the studies on the quantitative structure–activity relationship (QSAR) demonstrate that certain physicochemical properties are important for a particular pharmacological activity, a series of structural modifications that will enhance such properties can lead to compounds with greater potency. One of most widely used QSAR methods in drug design is the Hansch approach (Hansch C. and Leo A., *Substituent Constants for Correlation Analysis in Chemistry and Biology*, John Wiley & Sons, New York, 1979).

Racemate (Racemic mixture, see also Enantiomers): A mixture of equimolar parts of enantiomers. A racemic mixture does not rotate the plane of polarized light. Often, the prefix ($\pm$) is used to specify the racemic nature of the particular sample. For instance, 1:1 mixture of (+) lactic acid and (−) lactic acid is a racemic mixture of lactic acid and can be indicated as ($\pm$)-lactic acid.

Rate constant: Proportionality constant relating the rate of change in the amount of drug (dA/dt) to the amount of drug (A), e.g., $dA/dt = k \cdot A$ for the simplest situation, where k is a first-order rate constant.

Rate-limiting step: The process with the slowest rate constant in sequential kinetic processes, which governs the overall rate of the processes to the final outcome.

Regiospecific reaction: Characteristics of the particular reaction, which yields exclusively or nearly exclusively one of several possible isomeric products.

Relative bioavailability (see also Absolute bioavailability): The proportion of a drug reaching the systemic circulation upon extravascular administration as compared to that of a standard dose of the drug administered via the same route. The relative bioavailability can be assessed for several dosage forms of the same drug without knowing its exposure levels after intravenous administration. For instance, the relative bioavailability of a drug in a dosage form A compared to a dosage form B after oral administration can be obtained as follows:

$$F = \frac{AUC_{0-\infty,A} \cdot D_B}{AUC_{0-\infty,B} \cdot D_A}$$

where $AUC_{0-\infty,A}$ and $AUC_{0-\infty,B}$ are AUC from time zero to infinity after oral administration of a drug in the dosage forms A or B, respectively; D_A and D_B are oral doses of a drug in the dosage forms A or B, respectively; F is the bioavailability (in this case, relative bioavailability).

Renal clearance: Clearance of a drug via the kidneys, which is composed of glomerular filtration, active secretion, and passive reabsorption. Renal clearance can be estimated by the following equation:

$$\text{Renal clearance} = \frac{\text{Amount of unchanged drug excreted in the urine}}{\text{AUC of drug}}$$

Reversible metabolism: Interchangeability between a substrate and its metabolite(s) by a single metabolizing enzyme.

Rule of 5: A general guideline proposed by Lipinski (Lipinski C. A. *et al.*, Experimental and computational approaches to estimate solubility and permeability in drug discovery and development settings, *Adv. Drug Del. Rev.* **23**: 3–25, 1997) to spot potential oral absorption problems of a compound in humans during drug discovery processes, based on its physicochemical properties. According to this hypothesis, poor absorption of a compound *in vivo* owing to its limited aqueous solubility and/or membrane permeability is more likely when any combination of two or more of the following conditions are observed: (1) the molecular weight of the compound is greater than 500; (2) there are more than 5 H-bond donors in the molecule; (3) there are more than 10 H-bond acceptors (the sums of N's and O's) in the molecule; and (4) the calculated log P (ClogP) of the compound is greater than 5 (or $M \log P > 4.15$). These guidelines are not appropriate for peptidomimetic compounds or compounds subject to active transport mechanisms in the intestinal membranes.

Safety margin (see also Therapeutic index): The ratio between the no adverse effect (toxic) exposure level (NOAEL) in animals and the exposure level required to produce a satisfactory pharmacological response in humans.

SAR: Structure-activity relationship (see also Quantitative structure–activity relationship).

S.C. injection: Subcutaneous injection.

Secondary metabolite (see also Primary metabolite): A metabolite produced from the metabolite originally generated from the parent compound.

Selectivity: The intrinsic ability of an assay method to differentiate and quantify the analyte of interest in the presence of other constituents in a sample.

Serum (see also Plasma): The clear supernatant after centrifuging blood without an anticoagulant, allowing the red blood cells to clot. Serum does not contain coagulating factors.

SHAM analysis: Data analysis to obtain information on slope, height, area, and moment (SHAM) from plasma drug concentration *vs.* time profiles after drug administration.

Sham operation: Surgical operation usually done on control animals to measure its effects on pharmacokinetic or pharmacological properties of compounds tested on study animals that have undergone the same type of the operation.

SIM (Selected ion monitoring): A mass spectrometry operating mode in which a single ion specific for the target compound is monitored.

Soft drug: A pharmacologically efficacious compound that undergoes a predictable and controllable metabolic deactivation *in vivo* to nontoxic moieties after playing its therapeutic role.

Solution: A liquid dosage formulation that contains one or more soluble chemical substances usually dissolved in water.

SRM (Selected reaction monitoring): A sequential sample monitoring process for an analyte of interest used in tandem-mass spectrometry (MS/MS). It consists of three sequential steps: (1) isolation of the precursor ion (parent ion) of the analyte in the first quadrupole of the instrument, (2) fragmentation of this ion in a collision chamber with argon atoms, and (3) isolation of an intense and/or characteristic product ion (daughter ion) of the precursor in the third quadrupole of the instrument.

Steady state condition: The condition at which the rate of input (rate of administration) of a drug into the system (e.g., body) equals the rate of output (rate of elimination) of that drug from the system. For instance, the steady state of a drug in the body can be achieved after constant-rate intravenous infusion. When a drug is infused into the systemic circulation (blood) at a constant rate, the plasma drug concentration starts to increase until the rate of infusion of the drug (rate of input) is equal to the rate of elimination (rate of output). At steady state, the rate of change in the net amount of the drug in the body becomes zero.

Stereoisomers (see also Diastereomers and Enantiomers): The particular kind of structural isomers that differ from each other only in the spatial orientation of their atoms. There are two different types of stereoisomers, i.e., diastereomers and enantiomers.

Stereoselective reaction (see also Diastereomers and Enantiomers): Characteristics of the particular reaction that produces one diastereomer preferentially (or a pair of enantiomers) over all other possible diastereomers.

Stereospecific reaction (see also Diastereomers and Enantiomers): Characteristics of the particular reaction in which stereoisomeric starting materials (substrates) yield stereoisomerically different products.

Stoichiometry: The relative quantities of the substances participating in any chemical reaction measured according to their molar proportions in the reaction.

Structure–activity relationship (SAR, see also Quantitative structure–activity relationship): Quantitative or qualitative relationship between various molecular structures and their pharmacological and/or biochemical activity.

Subcutaneous injection: Injection beneath the skin.

Sublingual administration: Administration of a drug under the tongue.

Suicide inhibitor (see Mechanism-based inhibitor).

Superposition (dose-proportionality, see also First-order kinetics): The phenomenon that plasma concentration *vs.* time profiles of a drug at different dose levels are superimposed when normalized to the dose. Superposition of exposure profiles of a drug can be observed when its pharmacokinetic processes follow first-order (linear) kinetics. Superposition can be indicative of the linear pharmacokinetic behavior of a drug in the body.

Suspension: Fine solid particles of a drug suspended in a suitable vehicle. Oral suspensions are usually formulated in an aqueous vehicle, whereas suspensions for other purposes, e.g., intramuscular injection, can be prepared in nonaqueous vehicles such as oil. Suspensions are useful for administering large amounts of a drug, where conventional solid dosage forms such as tablets or capsules are not convenient.

Sustained-release dosage form (see also Controlled-release dosage form): A solid dosage form designed to release a drug substance over an extended period of time at a considerably slower rate *in vivo* compared to a conventional dosage form containing an equivalent dose. The conventional solid dosage forms such as tablets or capsules are designed to release their medication rapidly and completely *in vivo*. The advantage of sustained-release dosage forms is that they require less frequent administration than ordinary dosage forms to maintain therapeutic drug concentrations, and thereby improve patient compliance.

Synergism: An apparent total effect of two different drugs after their coadministration that is greater than the addition of their individual effects (the additive effect) after separate administration.

Tablet: A solid dosage form prepared by compression or molding of solid drug particles with or without other pharmaceutical adjuncts such as diluents, disintegrants, coatings, or colorants.

Tachyphylaxis (see also Tolerance): The loss of response to a drug following its rapid and repeated administration.

TDR (see Therapeutic drug monitoring).

Terminal half-life (see also Half-life): The half-life of a compound during the terminal phase of its plasma concentration *vs.* time profile:

$$t_{1/2} = \frac{-0.693}{\lambda_z}$$

where $t_{1/2}$ is the terminal half-life; and λ_z is the negative slope during the terminal phase of the log–linear plot of the plasma drug concentration *vs.* time. Note that λ_z is the same as k or β in the case of one- or two-compartment models, respectively.

Therapeutic drug monitoring: Monitoring the concentration(s) of a particular drug(s) of interest followed by proper adjustment of the dosage regimen to achieve optimal therapy in the individual patient. The rationale for therapeutic drug monitoring (TDM), sometimes called drug concentration monitoring (DCM), is that there is a significant interindividual variability in pharmacokinetics, which results in a wide range of steady state drug concentrations at any given dose rate. Adequate TDM becomes especially important in connection with drugs with a relatively narrow therapeutic index, i.e., a relatively small ratio of toxic to therapeutic concentrations.

Therapeutic equivalence: Comparable clinical efficacy and safety of a drug in different products or formulations.

Therapeutic index (see also Safety margin): The ratio between the maximum drug concentration in blood or plasma that can be tolerated and the minimum drug concentration to show a satisfactory pharmacological response in humans.

Tissue compartment (see Peripheral compartment).

Tolerance: A decrease in the magnitude of drug response after one or more doses. The damped response can be due to changes in drug concentration at the effect site and/or desensitization of the drug receptors.

Toxicokinetics: Pharmacokinetic principles and techniques applied to concentration *vs.* time data generated at the (high) dose levels that are customary in toxicity studies, in order to determine the rate, extent, and duration of exposure of the test compound in the test animal species.

Tween 80: Polyoxyethylene sorbitan monooleate, an emulsifying agent.

UDPGA (uridine disphosphoglucuronic acid): A cofactor for uridine diphosphate glucuronosyl transferase.

UDPGT: Uridine diphosphate glucuronosyl transferase.

Unstirred water layer (UWL): A stagnant water layer adjacent to the luminal surface of the intestinal membrane, which behaves as a barrier against absorption of drug molecules from the intestinal lumen.

Uremic middle molecules (see Middle molecules).

Volume of distribution: A proportionality constant relating the amount of drug present in the body to drug concentration measured in a reference body fluid such as blood or plasma. The volume of distribution does not necessarily represent identifiable physiological organs/tissues or volumes, but rather a hypothetical volume accounting for the total amount of a drug in the body referred to the body fluid in which the drug concentration is measured. The extent of the volume of distribution is dependent on various physicochemical properties of a compound, such as lipophilicity, as well as physiological factors, including protein binding in plasma and tissue and actual volumes of blood and tissue. Three important quantities in regard to the drug are the volume of distribution of the central compartment (V_c), at steady state (V_{ss}), and at pseudodistribution equilibrium (V_β).

WBA: Whole body autoradiography.

Zero-order kinetics (see also First-order kinetics): A kinetic process in which the rate of change in concentration or the amount of drug with time is constant and independent of both drug concentration and time, i.e.,

$$\text{Rate of change of drug concentration} = k_0$$

where k_0 is the zero-order rate constant.

Appendix

A. Important Pharmacokinetic Equations

1. Plasma Drug Concentration at Time t in a One-Compartment Model under Linear Kinetics

AFTER INTRAVENOUS ADMINISTRATION:

$$\boxed{C_p(t) = C_0 \cdot e^{-t}}$$

C_0: Estimated plasma concentration at time zero (intravenous dose/volume of distribution in the central compartment).

$C_p(t)$: Plasma drug concentration at time t after intravenous bolus injection.

AFTER ORAL ADMINISTRATION

$$\boxed{C_p(t) = \frac{k_a \cdot F \cdot D_{po}}{V \cdot (k_a - k)} \cdot (e^{-k \cdot t} - e^{-k_a \cdot t})}$$

$C_p(t)$: Plasma drug concentration at time t after oral administration.

D_{po}: Oral dose.

F: Oral bioavailability.

k_a, k: Absorption and elimination rate constants, respectively.

V: Apparent volume of drug distribution.

2. Systemic Plasma Clearance

$$\boxed{Cl_s = \frac{D_{iv}}{AUC_{iv,0-\infty}}}$$

Cl_s: Systemic (plasma) clearance.

$AUC_{iv,0-\infty}$: Area under the plasma drug concentration *vs.* time curve from time zero to infinity after intravenous bolus injection.

D_{iv}: Intravenous dose.

3. Systemic Blood Clearance vs. Systemic Plasma Clearance

$$Cl_b \cdot AUC_b = Cl_p \cdot AUC_p$$
$$Cl_b \cdot C_b = Cl_p \cdot C_p$$

AUC_b, AUC_p: Area under the blod or plasma drug concentration *vs.* time curves after intravenous bolus injection, respectively.
C_b, C_p: Blood or plasma drug concentrations, respectively.
Cl_b: Systemic blood clearance.
Cl_p: Systemic plasma clearance.

4. Hepatic Blood Clearance

WELL-STIRRED MODEL:

$$Cl_h = \frac{Q_h \cdot f_{u,b} \cdot Cl_{i,h}}{Q_h + f_{u,b} \cdot Cl_{i,h}}$$

PARALLEL-TUBE MODEL:

$$Cl_h = Q_h \cdot (1 - \exp(-f_{u,b} \cdot Cl_{i,h}/Q_h)$$

Cl_h: Hepatic clearance.
$Cl_{i,h}$: Intrinsic hepatic clearance.
$f_{u,b}$: Ratio between unbound and total drug concentrations in blood.
Q_h: Hepatic blood flow rate.

5. Renal Plasma Clearance

$$Cl_r = \frac{A_{e,0-\infty}}{AUC_{0-\infty}}$$

$A_{e,0-\infty}$: Cumulative amount of drug excreted unchanged in urine from time zero to infinity.
$AUC_{0-\infty}$: Area under the plasma drug concentration–time curve from time zero to infinity regardless of the route of administration.
Cl_r: Renal clearance.

6. Volume of Distribution at Steady State

$$V_{ss} = \underbrace{\frac{D_{iv}}{AUC_{iv,0-\infty}}}_{Cl_s} \cdot \underbrace{\frac{AUMC_{iv,0-\infty}}{AUC_{iv,0-\infty}}}_{MRT}$$

$AUC_{iv,0-\infty}$: Area under the plasma drug concentration *vs.* time curve from time zero to infinity after intravenous bolus injection.
$AUMC_{iv,0-\infty}$: Area under the first-moment curve of plasma drug concentration *vs.* time curve from time zero to infinity after intravenous bolus injection.
Cl_s: Systemic plasma clearance.
MRT: Mean residence time.
V_{ss}: Volume of distribution at steady state.

or

$$V_{ss} = V_P + \frac{f_u}{f_{u,t}} \cdot V_t$$

V_P: Actual physiological volume of plasma.
V_t: Actual physiological volume of extravascular space (interstitial fluid and tissue) outside plasma, which drug molecules distribute into.
f_u: Ratio between unbound and total drug concentrations in plasma.
$f_{u,t}$: Averaged ratio between unbound and total drug concentrations in extravascular space.

7. Terminal Half-Life

$$t_{1/2} = \frac{-0.693}{\lambda_z}$$

λ_z: The negative slope during the terminal phase of the log–linear plot of plasma drug concentration *vs.* time. λ_z is the same as k or β, exponential coefficients of mono- or biexponential differential equations describing for plasma concentration–time profiles of drug, respectively.
$t_{1/2}$: Terminal half-life.

or

$$t_{1/2} = \frac{-0.693 \cdot V_{ss}}{Cl_s}$$

which is true only when the plasma drug concentration *vs.* time curve after intravenous injection can be properly described by a one-compartment model. Otherwise, $t_{1/2}$ is a function of V_β, the volume of distribution at the terminal phase, not V_{ss}, the volume of distribution at steady state ($V_\beta > V_{ss}$).

8. Oral Bioavailability

$$F = \frac{AUC_{po,0-\infty} \cdot D_{iv}}{AUC_{iv,0-\infty} \cdot D_{po}}$$

$AUC_{iv,0-\infty}$: AUC from time zero to infinity after intravenous bolus injection.
$AUC_{po,0-\infty}$: AUC from time zero to infinity after oral administration.
D_{iv}: Intravenous dose.
D_{po}: Oral dose.
F: Oral bioavailability.

9. Mean Residence Time

$$MRT = \frac{AUMC_{0-\infty}}{AUC_{0-\infty}}$$

$AUC_{0-\infty}$: AUC from time zero to infinity regardless of the route of administration.
$AUMC_{0-\infty}$: AUMC from time zero to infinity regardless of the route of administration.
MRT: Mean residence time.

10. Mean Absorption Time

$$MAT = MRT_{po} - MRT_{iv}$$

MAT: Mean absorption time.
MRT_{iv}, MRT_{po}: MRT after intravenous bolus injection or oral administration, respectively.

11. Plasma Drug Concentration at Steady State after Continuous Intravenous Infusion

$$C_{p,ss} = \frac{k_0}{Cl_s}$$

$C_{p,ss}$: Plasma drug concentration at steady state after continuous intravenous infusion.
Cl_s: Systemic plasma clearance.
k_0: Infusion rate.

12. Metabolite Kinetics

$$AUC_{0-\infty} \cdot f_m \cdot Cl_s = AUC_{0-\infty,m} \cdot Cl_m$$

$AUC_{0-\infty}$, $AUC_{0-\infty,m}$: AUC from time zero to infinity of the drug and its metabolite, respectively regardless of the route of administration.
Cl_s, Cl_m: Systemic plasma clearance of the drug and its metabolite, respectively.
f_m: Fraction of the dose of the drug transformed to the metabolite.

13. Relationship between Blood and Plasma Concentrations

$$C_b = Hct \cdot C_{rbc} + (1 - Hct) \cdot C_p$$

C_b: Drug concentration in blood.
C_{rbc}: Drug concentration in red blood cells.
C_p: Drug concentration in plasma.
Hct: Hematocrit.

14. Amount of a Drug Absorbed into the Portal Vein after Oral Administration

MASS BALANCE METHOD:

$$A_a = Q_{pv} \cdot (AUC_{po,pv} - AUC_{po,sys})$$

A_a: Amount of the drug absorbed into the portal vein after oral administration.
$AUC_{po,pv}$, $AUC_{po,sys}$: AUC of the drug from time zero to infinity in the portal vein and systemic blood (or plasma when blood concentrations are the same to plasma concentrations) after oral administration, respectively.
Q_{pv}: Portal vein blood flow rate.

CLEARANCE METHOD:

$$A_a = Cl_b \cdot AUC_{po,pv}$$

Cl_b: Systemic blood (or plasma when blood concentrations are the same as plasma concentrations) clearance.
$AUC_{po,pv}$: AUC of drug in portal vein blood (or plasma when blood concentrations are the same as plasma concentrations) after oral administration.

B. Typical Pharmacokinetic Issues and Their Potential Causes

Route of administration	Issues	Potential causes
Oral administration	Low bioavailability	Limited absorption Poor aqueous solubility (dissolution rate-limited) Poor membrane permeability (permeation rate-limited) Efflux by P-glycoprotein or multidrug resistance-associated protein in the intestine Microfloral metabolism in the intestine, e.g., reduction of azo compounds Extensive first-pass effect Presystemic intestinal metabolism (CYP3A4, CYP2C9, UDPGT, etc) Presystemic hepatic clearance (metabolism and/or biliary excretion)
	Multiple peaks in exposure profile	Enterohepatic circulation Enterohepatic circulation of parent drug Biliary excretion of metabolites followed by subsequent conversion to the parent drug in intestinal lumen (e.g., biliary excretion of glucuronide conjugates of drug and subsequent deconjugation in gut lumen) Variability in pH in the stomach Delayed gastric emptying
	Flip-flop kinetics (a longer terminal half-life after P.O. than I.V. dosing)	Slow absorption of drug The rate of absorption of drug from the intestine is slower than the rate of elimination from body
	Less than dose-proportional increase in exposure as dose increases	Nonlinear absorption and/or clearance Limited aqueous solubility Saturation of active transporters in enterocytes during absorption Saturation of protein binding
	More than dose-proportional increase in exposure as dose increases	Nonlinear absorption and/or clearance Saturation of efflux mechanism(s) in enterocytes during absorption Saturation of clearance mechanisms
	Decreasing exposure after multiple doses	Induction Autoinduced metabolism Induction of P-glycoprotein for biliary and/or intestinal excretion
Intravenous administration	Biexponential declining of exposure profile	Multicompartmental distribution Nonlinear protein binding Saturated protein binding at high concentrations during the initial phase Nonlinear metabolism Product inhibition on metabolism of parent drug during the later phase

Route of administration	Issues	Potential causes
	Sustained or elevated exposure at the beginning and then declining	Nonlinear clearance Substrate inhibition of metabolism during the initial phase Precipitation of drug at the injection site and subsequent dissolution
	Short half-life	Rapid clearance Extensive metabolism, biliary and/or renal elimination Low protein binding Small volume of distribution Confinement of drug in plasma More extensive protein binding in plasma than in tissues
	Longer half-lives at higher dose levels	Nonlinear clearance Limitation of assay sensitivity at low concentrations (no dose dependent half-life changes with better assay sensitivity) Product inhibition

C. References for Laboratory Animal Experiments

Animal Physiology

Altman P. L. and Dittmer D. S., *Respiration and Circulation*, Federation of American Society for Experimental Biology, Bethesda, 1970.

Dressman J. B., Comparison of canine and human gastrointestinal physiology, *Pharm. Res.* **3**: 123–131, 1986.

Frank D. W., Physiological data of laboratory animals, in E. C. Melby Jr. and N. H. Altman (eds.), *Handbook of Laboratory Animal Science, Vol II*, CRC Press, Cleveland, 1974, pp. 23–64.

Havenaar *et al.*, Biology and husbandry of laboratory animals, in L. F. M. van Zutphen, V. Baumans and A. C. Beynen (eds.), *Principles of Laboratory Animal Science: A Contribution to the Humane Use and Care of Animals and to the Quality of Experimental Results*, Elsevier, New York, 1993, pp. 17–74.

Kararli T. T., Comparison of the gastrointestinal anatomy, physiology, and biochemistry of humans and commonly used laboratory animals, *Biopharm. Drug Dispos.* **16**: 351–380, 1995.

Lindstedt S. L. and Calder III W. A., Body size, physiological time and longetivity of homeothermic animals, *Q. Rev. Biol.* **56**: 1–16, 1981.

Animal Handling

Andrews E. J. *et al.*, Report of the AVMA panel on Euthanasia, *JAVMA* **202**: 229–249, 1993.

Coates M. E., Feeding and watering, in A. A. Tuffery (ed.), *Laboratory Animals: An Introduction for New Experiments, 2nd Ed.*, John Wiley & Sons, New York, 1995, pp. 107–128.

Gregory J. A., Principles of animal husbandry, in A. A. Tuffery (ed.), *Laboratory Animals: An Introduction for New Experiments, 2nd Ed.*, John Wiley & Sons, New York, 1995, pp. 87–106.

Laboratory Animal, Reference Series [*The Laboratory Mouse* (T. L. Cunliffe-Beamer, 1998), *The Laboratory Guinea Pig* (L. Terril, 1998), *The Laboratory Hamster and Gerbil* (K. Field, 1998), *The Laboratory Cat* (B. J. Martin, 1998), *The Laboratory Rat* (P. Sharp and M. LaRegina, 1998), and *The Laboratory Rabbit* (M. A. Suckow and F. A. Douglas, 1997)], CRC Press, Boca Raton, FL.

Mann M. D. *et al.*, Appropriate animal numbers in biomedical research in light of animal welfare considerations, *Lab. Animal Sci.* **41**: 6–14, 1991.

Morton D. B. *et al.*, Removal of blood from laboratory mammals and birds, First report of the BVA/FRAME/RSPCA/UFAW joint working group in refinement, *Lab. Animals* **27**: 1–22, 1993.

Morton D. B. and Griffiths P. H. M., Guidelines on the recognition of pain, distress and discomfort in experimental animals and an hypothesis for assessment, *Vet. Record* **16**: 431–436, 1985.

Runciman W. B. *et al.*, A sheep preparation for studying interactions between blood flow and drug disposition. I: physiological profile, *Br. J. Anaesth.* **56**: 1015–1028, 1984.

Animal Disease Models

Cawthorne M. A. *et al.*, Adjuvant induced arthritis and drug-metabolizing enzymes, *Biochem. Pharmacol.* **25**: 2683–2688, 1976.

Cypess R. H. and Hurvitz A. I., Animal models, in E. C. Melby Jr. and N. H. Altman (eds.), *Handbook of Laboratory Animal Science, Vol. II*, CRC Press, Cleveland, 1974, pp. 205–228.

Gurley B. J. *et al.*, Extrahepatic ischemia-reperfusion injury reduces hepatic oxidative drug metabolism as determined by serial antipyrine clearance, *Pharm. Res.* **14**: 67–72, 1997.

Ramabadran K. and Bansinath M., A critical analysis of the experimental evaluation of nociceptive reactions in animals, *Pharm. Res.* **3**: 263–270, 1986.

Sofia R. D., Alteration of hepatic microsomal enzyme systems and the lethal action of non-steroidal anti-arthritic drugs in acute and chronic models of inflammation, *Agents Actions* **7**: 289–297, 1977.

Transgenic Animal Models

Cameron E. R. *et al.*, Transgenic science, *Br. Vet. J.* **150**: 9–24, 1994.

Nebert D. W. and Duffy J. J., How knockout mouse lines will be used to study the role of drug-metabolizing enzymes and their receptors during reproduction and development, and in environmental toxicity, cancer and oxidative stress, *Biochem. Pharmacol.* **53**: 249–254, 1997.

Animal Surgery and Experiments

General Surgery, Sample Collection, and Anesthesia

Beyner A. C. *et al.*, Design of animal experiments, in L. F. M. van Zutphen, V. Baumans and A. C. Beynen (eds.), *Principles of Laboratory Animal Science: A Contribution to the Humane Use and Care of Animals and to the Quality of Experimental Results*, Elsevier, New York, 1993, pp. 209–240.

Castaing D. *et al.*, *Hepatic and Portal Surgery in the Rat*, Masson, Paris, 1980.

Chaffee V. W., Surgery of laboratory animals, in E. C. Melby Jr. and N. H. Altman (eds.), *Handbook of Laboratory Animal Science*, Vol. I, CRC Press, Cleveland, 1974, pp. 231–274.

Cocchetto D. M. and Bjornsson T. D., Methods for vascular access and collection of body fluids from the laboratory rat, *J. Pharm. Sci.* **72**: 465–492, 1983.

Heavner J. E., Anesthesia, analgesia and restraint, in W. I. Gay (ed.), *Methods of Animal Experimentation*, Academic Press, New York, pp. 1–36, 1986.

McGuill M. W. and Rowan A. N., Biological effects of blood loss: implications for sampling volumes and techniques, *ILAR News* **31**: 5–20, 1989.

Scobie-Trumper P., Animal handling and manipulations, in A. A. Tuffery (ed.), *Laboratory Animals; An Introduction for Experiments, 2nd Ed.*, John Wiley & Sons, New York, 1995, pp. 233–254.

Sharp P. E. and LaRegina M. C., *Veterinary Care*, pp. 105–107, *Experimental Methodology*, pp. 129–159, *The Laboratory Rat*, CRC Press, New York, 1998.

Steinbruchel D. A., Microsurgical procedures in experimental research, in P. Svendsen and J. Hau (eds.), *Handbook of Laboratory Animal Science, Vol. I: Selection and Handling of Animals in Biomedical Research*, CRC Press, London, 1974, pp. 371–382.

Tuffery A. A., *Laboratory Animals: An Introduction for Experiments, 2nd Ed.*, John Wiley & Sons, New York, 1995.

Wilsson-Rahmberg M. *et al.*, Method for long-term cerebrospinal fluid collection in the conscious dog, *J. Inves. Sur.* **11**: 207–214, 1998.

Metabolism/Liver Perfusion/Portal Vein Cannulation/Liver Injury

Effeney D. J. *et al.*, A technique to study hepatic and intestinal drug metabolism separately in the dog, *J. Pharmacol. Exp. Ther.* **221**: 507–511, 1982.

Gerkens J. F. *et al.*, *Hepatic and extrahepatic glucuronidation of lorazepam in the dog*, *Hepatology* **1**: 329–335, 1981.

Gores G. J. *et al.*, The isolated perfused rat liver: conceptual and practical considerations, *Hepatology* **6**: 511–517, 1986.

Iwasaki T. *et al.*, Regional pharmacokinetics of doxorubicin following hepatic arterial and portal venous administration: evaluation with hepatic venous isolation and charcoal hemoperfusion, *Cancer Res.* **58**: 3339–3343, 1998.

Maza A. M. *et al.*, Influence of partial hepatectomy in rats on the activity of hepatic microsomal enzymatic systems, *Eur. J. Drug Metab. Pharmacokinet.* **22**: 15–23, 1997.

Plaa G. L., A four-decade adventure in experimental liver injury, *Drug Metab. Rev.* **29**: 1–37, 1997.

Sahin S. and Rowland M., Development of an optimal method for the dual perfusion of the isolated rat liver, *J. Pharmacol. Toxicol. Methods* **39**: 35–43, 1998.

Sloop C. H. and Krause B. R., Portal and aortic blood sampling technique in unrestrained rats, *Physiol. Behav.* **26**: 529–533, 1981.

Urban E. and Zingery A. A., A simple method of cannulation of the portal vein and obtaining multiple blood samples in the rat, *Experienta* **37**: 1036–1037, 1981.

Absorption/Enterohepatic Circulation/Intestinal Perfusion

Fujieda Y. *et al.*, Local absorption kinetics of levofloxacin from intestinal tract into portal vein in conscious rat using portal-vein concentration difference, *Pharm. Res.* **13**: 1201–1204, 1996.

Kuipers F. *et al.*, Enterohepatic circulation in the rat, *Gastroenterology* **88**: 403–411, 1985.

Pang K. S. *et al.*, Disposition of enalapril in the perfused rat intestine-liver preparation: absorption, metabolism, and first-pass effect, *J. Pharmacol. Exp. Ther.* **233**: 788–795, 1985.

Tabata K. *et al.*, Evaluation of intestinal absorption into the portal system in enterohepatic circulation by measuring the difference in portal-venous blood concentrations of diclofenac, *Pharm. Res.* **12**: 880–883, 1995.

Tsutsumi H. *et al.*, Method for collecting bile with a T-cannula in unrestrained conscious Beagles, *Exp. Anim.* **45**: 261–263, 1996.

Windmueller H. G. and Spaeth A. E., Vascular autoperfusion of rat small intestine *in situ Methods Enzymol.* **77**: 120–129, 1981.

Bioavailability

Dressman J. B. and Yamada K., Animal models for oral drug absorption, in P. G. Welling, F. L. S. Tse, and S. V. Dighe (eds.), *Pharmaceutical Bioequivalence, Vol. 48*, Dekker, New York, 1991, pp. 235–266.

Humphreys W. G. *et al.*, Continuous blood withdrawal as a rapid screening method for determining clearance and oral bioavailability in rats, *Pharm. Res.* **15**: 1257–1261, 1998.

Krishnan T. R. *et al.*, Use of the domestic pig as a model for oral bioavailability and pharmacokinetic studies, *Biopharm. Drug Dispos.* **15**: 341–346, 1994.

Lukas G. *et al.*, The route of absorption of intraperitoneally administered compounds, *J. Pharmacol. Exp. Ther.* **178**: 562–566, 1971.

Microdialysis

Deguchi Y. *et al.*, Muscle microdialysis as a model study to relate the drug concentration in tissue interstitial fluid and dialysate, *J. Pharmacobio-Dyn.* **14**: 483–492, 1996.

Evrard P. A. *et al.*, Simultaneous microdialysis in brain and blood of the mouse: extracellular and intracellular brain colchicine disposition, *Brain Res.* **786**: 122–127, 1998.
Telting-Diaz M. *et al.*, Intravenous microdialysis sampling in awake, freely-moving rats, *Anal. Chem.* **64**: 806–810, 1992.
Terasaki T. *et al.*, Determination of *in vivo* steady-state unbound drug concentration in the brain interstitial fluid by microdialysis, *Int. J. Pharm.* **81**: 143–152, 1992.

Placenta Perfusion

Poranen A. K. *et al.*, Vasoactive effects and placental transfer of nifedipine, celiprolol, and magnesium sulfate in the placenta perfused *in vitro*, *Hyper. Preg.* **17**: 93–102, 1998.
Schneider H. *et al.*, Transfer across the perfused human placenta of antipyrine, sodium and leucine, *Am. J. Obstet. Gynecol.* **114**: 822–828, 1972.

Miscellaneous

Beyssac E., The unusual routes of administration, *Eur. J. Drug Metab. Pharmacokinet.* **21**: 181–187, 1996.
Cocchetto D. M. and Wargin W. A., A bibliography for selected pharmacokinetic topics, *Drug Intel. Clin. Pharm.* **14**: 769–776, 1980.
Ramabadran K. and Bansinath M., A critical analysis of the experimental evaluation of nociceptive reactions in animals, *Pharm. Res.* **3**: 263–270, 1986.

D. Abbreviations

α: exponential coefficient of a biexponential differential equation
A_a: total amount of a drug absorbed into the portal vein after oral administration
$A_c(t)$: amount of the drug in the central compartment at time t
A_e: amount of the drug excreted unchanged in the urine
$A_{e,0-\infty}$: cumulative amount of a drug excreted unchanged in the urine from time zero to infinity
$A_m(t)$: amount of the metabolite produced from the drug in the body at time t after intravenous administration
A(t): amount of a drug in the body at time t
ADME: absorption, distribution, metabolism, and excretion
AUC: area under the plasma drug concentration *vs.* time curve
AUC_m: area under the plasma metabolite concentration *vs.* time curve
AUC_{ia}: area under the plasma drug concentration *vs.* time curve after intraarterial injection
AUC_{ip}: area under the plasma drug concentration *vs.* time curve after intraportal vein (or intraperitoneal) injection
AUC_{iv}: area under the plasma drug concentration *vs.* time curve after intravenous injection
AUC_{po}: area under the plasma drug concentration *vs.* time curve after oral administration
$AUC_{po,pv}$: AUC of a drug in portal vein blood (or plasma) after oral administration
$AUC_{po,vc}$: AUC of a drug in vena cava blood (or plasma) after oral administration
AUMC: area under the first-moment curve of plasma drug concentration *vs.* time curve
β: exponential coefficient of a biexponential differential equation

C: concentration

$C_{avg,ss}$: average drug concentration in plasma during a dosing interval at steady state after multiple dosing of a fixed drug dose at the same dosing interval

$C_b(t)$: drug concentration in blood at time t

$C_e(t)$: drug concentration at the effect site at time t

$C_{in,ss}$: drug concentration in blood entering the eliminating organ at steady state

C_{int}: drug concentration in the gastrointestinal fluid

$C_{l,u}$: unbound drug concentration within hepatocytes or available for metabolizing enzyme(s) and/or biliary excretion

$C_m(t)$: metabolite concentration in plasma at time t

C_{max}: the highest drug concentration in plasma after extravascular administration

$C_{out,ss}$: drug concentration in blood leaving the eliminating organ at steady state

$C_p(0)$: imaginary drug concentration in plasma at time zero after intravenous injection of a drug, estimated by extrapolation of the plasma drug concentration–time curve to time zero

$C_p(t)$: drug concentration in plasma at time t

$C_{p,ss}$: plasma drug concentration at steady state after intravenous infusion

C_{rbc}: drug concentration in red blood cells

C_{ss}: drug concentration at steady state

$C_t(t)$: average drug concentration in the extravascular space, into which the drug distributes at time t

$C_T(t)$: average drug concentration in the peripheral (tissue) compartment at time t

$C_u(t)$: concentration of drug not bound to blood components at time t

Cl_b: systemic blood clearance

Cl_{bl}: biliary clearance

Cl_d: distributional clearance

Cl_g: intestinal clearance

Cl_h: hepatic (blood) clearance

$Cl_{i,h}$: intrinsic hepatic clearance

Cl_m: metabolic drug clearance to produce a particular metabolite

$Cl_{(m)}$: systemic metabolite clearance

Cl_{nr}: nonrenal clearance

Cl_{other}: clearance other than via the metabolic pathway of a drug to a particular metabolite

Cl_p: plasma clearance

Cl_r: renal clearance

Cl_s: systemic (plasma) clearance

Cl_u: drug clearance based on unbound drug concentration

CYP: cytochrome P450

D_{ia}: intraarterial dose

D_{ip}: intraportal vein (or intraperitoneal) dose

D_{iv}: intravenous dose

D_N: dispersion number

D_{po}: oral dose

E: extraction ratio or drug effect

E_g: intestinal extraction ratio
E_h: hepatic extraction ratio
E(t): pharmacological effect of drug at time t
EC_{50}: concentration of drug showing 50% of its maximum effect
E_{max}: maximum effect of drug
F: (oral) bioavailability
F_a: fraction of the dose absorbed into gastrointestinal epithelial cells (enterocytes) from the intestinal lumen after oral administration of a drug
f_e: fraction of the dose excreted unchanged in the urine
F_g: fraction of the amount of a drug absorbed into enterocytes after oral administration of a drug that escapes the presystemic intestinal elimination
F_h: fraction of the amount of a drug entering the liver that escapes elimination by the liver on a single pass through the organ, or the fraction of a drug entering the liver that escapes the presystemic hepatic elimination after oral administration
F_l: fraction of the amount of a drug entering the lung that escapes elimination by the lung on a single pass through the organ, or the fraction of a drug entering the lung that escapes the presystemic pulmonary elimination after oral administration
f_m: fraction of the dose metabolized
F_r: fraction of the amount of a drug reabsorbed from the renal distal tubule after being filtered and secreted in the glomerulus and the proximal tubule
F_s: fraction of a dose reaching the systemic circulation as unchanged drug after oral administration (considering first-pass effect by the lung as well)
f_u: ratio between the unbound and total drug concentrations in plasma
$f_{u,b}$: ratio between the unbound and total drug concentrations in blood
$f_{u,t}$: average ratio between unbound and total drug concentrations in tissues (extravascular space)
GFR: glomerular filtration rate
Hct: hematocrit
IC_{50}: concentration of drug showing 50% of its maximum inhibitory effect
k: first-order rate constant
k_{10}: first-order elimination rate constant from the central compartment
k_{12}, k_{21}: first-order distribution rate constants from the central to the peripheral compartments or from the peripheral to the central compartments, respectively
k_a: first-order absorption rate constant
K_m: Michaelis–Menten constant or the apparent Michaelis–Menten constant for metabolizing enzyme(s) and/or biliary excretion
k_m: first-order rate constant associated with the formation of metabolites from the parent drug
$k_{(m)}$: first-order rate constant associated with the elimination of metabolites
k_0: drug infusion rate or zero-order rate constant
MAT: mean absorption time of a drug after oral administration
MIT: mean input time of a drug

MRT: mean residence time of a drug
MRT_{abs}: MRT for the absorption of drug molecules dosed in solution into the systemic circulation
MRT_{disint}: MRT for the disintegration of the orally dosed solid dosage form of a drug to a suspension
MRT_{diss}: MRT for the dissolution of the orally dosed solid drug particles to solution
MRT_{iv}: MRT after intravenous injection
MRT_{po}: MRT after oral administration
P450: cytochrome P450
P_{app}: apparent membrane permeability
P_{int}: intestinal membrane permeability
Q: blood flow rate
Q_h: hepatic blood flow rate
Q_{pv}: portal vein blood flow rate
Q_r: renal blood flow rate
R: accumulation factor
S_{int}: effective surface area of intestinal membranes available for drug absorption
$t_{1/2}$: half-life of a drug during the terminal phase of plasma drug concentration–time profile
t_{last}: the last time point when a quantifiable drug concentration can be measured
t_{max}: time at which C_{max} is observed following extravascular administration of drug
V: apparent volume of distribution of a drug
V_β: apparent volume of distribution at the β phase (or terminal phase) based on drug concentration in plasma
V_c: apparent volume of the central compartment based on the drug concentration in plasma
$V_{extrapolated}$: initial dilution volume of a drug

V_m: apparent volume of distribution of metabolite
V_{max}: maximum rate of enzymatic reaction or the apparent maximum rate of metabolizing enzyme(s) and/or biliary excretion
V_P: actual physiological volume of plasma
V_{ss}: apparent volume of distribution at steady state based on the drug concentration in plasma
$V(t)$: apparent volume of distribution of a drug at time t
V_t: actual physiological volume of extravascular space (blood cells, interstitial fluids, and tissues) outside the plasma into which the drug distributes
V_T: apparent volume of the peripheral (tissue) compartment of drug
λ_z: the negative slope of the terminal phase of a plasma drug concentration–time profile on a semilogarithmic scale.

INDEX